Sustainable Development Goals Series

The **Sustainable Development Goals Series** is Springer Nature's inaugural cross-imprint book series that addresses and supports the United Nations' seventeen Sustainable Development Goals. The series fosters comprehensive research focused on these global targets and endeavours to address some of society's greatest grand challenges. The SDGs are inherently multidisciplinary, and they bring people working across different fields together and working towards a common goal. In this spirit, the Sustainable Development Goals series is the first at Springer Nature to publish books under both the Springer and Palgrave Macmillan imprints, bringing the strengths of our imprints together.

The Sustainable Development Goals Series is organized into eighteen subseries: one subseries based around each of the seventeen respective Sustainable Development Goals, and an eighteenth subseries, "Connecting the Goals," which serves as a home for volumes addressing multiple goals or studying the SDGs as a whole. Each subseries is guided by an expert Subseries Advisor with years or decades of experience studying and addressing core components of their respective Goal.

The SDG Series has a remit as broad as the SDGs themselves, and contributions are welcome from scientists, academics, policymakers, and researchers working in fields related to any of the seventeen goals. If you are interested in contributing a monograph or curated volume to the series, please contact the Publishers: Zachary Romano [Springer; zachary.romano@springer.com] and Rachael Ballard [Palgrave Macmillan; rachael.ballard@palgrave.com].

Patricia Solís • Marcela Zeballos
Editors

Open Mapping towards Sustainable Development Goals

Voices of YouthMappers on Community Engaged Scholarship

Springer

Editors
Patricia Solís
School of Geographical Sciences
Arizona State University
Tempe, AZ, USA

Marcela Zeballos
Center for Geospatial Technology
College of Liberal Arts and Sciences
Texas Tech University
Lubbock, TX, USA

Registered Trademark, YouthMappers®, Reg. No. 5.203.611, U.S. Patent and Trademark Office, Class 41, Ser. No. 87-165,163 to Solis, Patricia.

ISSN 2523-3084   ISSN 2523-3092   (electronic)
Sustainable Development Goals Series
ISBN 978-3-031-05184-5   ISBN 978-3-031-05182-1   (eBook)
https://doi.org/10.1007/978-3-031-05182-1

This Springer imprint is published by the registered company Springer Nature Switzerland AG
The registered company address is: Gewerbestrasse 11, 6330 Cham, Switzerland

Preface

If you have ever used a smartphone to get directions to where you are going, or "flown" around the world using Google Earth while sitting at home, you have benefitted from geospatial data and technology. It is hard to imagine our lives today without these tools. But it hasn't even been fifteen years since we became dependent on them.

The capability that YouthMappers have today is as hopeful to me as exploring space and going to the moon was for my rocket scientist father. As his young daughter, I was more fascinated with the scenes from space coming back to earth. The images of our planet from Landsat and the lower-orbit Space Shuttle are emblazoned in my mind forever. It made me wonder, were young students like me in other countries seeing these spectacular views? Could we use them to help grown-up leaders in the nations of the world find common ground and motivation to get along better?

By the time I got to university, I could only dream of the kind of technology that today allows students anywhere in the world view the same satellite imagery of a location on the ground, and make a map from it, *at the same time*. Because they know their homes and the day-to-day details about the places where they live, they can also add to the map by tagging attributes during fieldwork, giving their local knowledge, too, to the effort. Since YouthMappers are now located all over the world, we have an incredible opportunity to bridge the global and the local, and even connect to each other through the map. YouthMappers are a living example of this dream. They are using geospatial technology that evolved from prior space age investments to map communities today. Historically invisible communities can now be included, giving them a voice they did not previously have. Learning to map and analyze data they collect makes students aware of the issues; being part of the YouthMappers community motivates them to get involved and take action.

But these kinds of dreams don't just happen. It takes purpose to manifest them in our world and on our earth. In the case of YouthMappers, it would take shape through a career long collaboration, meeting a once in a lifetime opportunity.

When I first met Dr. Patricia Solis nearly twenty years ago, she was working at the American Association of Geographers (AAG). I was running a geospatial program at the US Agency for International Development (USAID) in a joint effort with NASA, our nation's space agency. It was clear that we were both passionate about geography and the power of geospatial technology to help address pressing challenges such as climate change, food security, poverty, unprecedented urbanization, and an ever-growing list of environmental threats to the planet.

By 2014, Patricia was working at a university, and I was leading the new GeoCenter at USAID. Two forces were at play then. A geospatial revolution was bringing increased access to remote sensing data and mapping technology, and a "digital" generation of young people had entered the university system. In addition to our shared belief in the transformative impact of geography to illuminate problem areas, Patricia and I both believed in the power of young people to help find the solutions to society's challenges.

With a vision to create a global network of students who would generate and share new geospatial data of unmapped places in the world, we formed a partnership. Three founding universities, each with unique capabilities, established a consortium: Texas Tech University (TTU), George Washington University (GWU), and West Virginia University (WVA). Arizona State University (ASU) later joined to help manage the program that has come to be known as YouthMappers.

The YouthMappers program now boasts participation from nearly 300 universities in more than 60 countries. Since the beginning of the program, student-mappers have contributed more than 12 million data edits to OpenStreetMap (OSM), a web-based platform that is available to anyone with access to the internet. The data on the platform makes a digital map of the world that is by the people, of the people, and for the people. And it is used by USAID and our partners to address malaria, HIV/AIDS, food security, water management, disasters, and more.

But the value of the YouthMappers program goes well beyond generating new data for international development and humanitarian purposes. YouthMappers are trained in digital mapping skills that lead to jobs in technology. They learn about social, economic, and environmental issues at local and international levels. They report feeling empowered to develop solutions and take on leadership roles in their communities. And they gain a means and a purpose to connect with their peers on mutual projects, via a virtual global community of international mappers who are ready to contribute their skills when needed. And they are very much needed now more than ever, especially to advance the Sustainable Development Goals, as we collectively envision them. They are needed now more than ever, as the situations we face are making meeting those goals even more challenging.

For example, when the global COVID-19 pandemic hit, YouthMappers jumped into action. With their local knowledge, they helped map more than 900,000 hospitals, clinics, pharmacies, and doctor's offices, in countries all over the world. In Uganda, YouthMappers used satellite imagery to map border crossing areas that had become hotspots for the Coronavirus. The timely information they created supported efforts by the Ministry of Health, Red Cross, and others responding to the crisis. Many other examples are narrated in the pages that follow.

Through their creativity, innovation, and passion, the youth of today are stepping up to make these dreams happen. I invite you to read the stories of these YouthMappers and their impressive work in the chapters of this book.

As I have said since the launch of this program, we are not just creating a new set of maps for USAID, but a new generation of mappers for the world. Turn the page to explore the efforts of a generation that is changing the world, one map at a time.

Carrie Stokes
United States Agency for International Development
Chief Geographer and GeoCenter Director
Washington, DC, USA

Contents

Preface v

Carrie Stokes

1 Introduction 1

Patricia Solís and Marcela Zeballos

1 The Emergence of YouthMappers 1

2 Where YouthMappers Connect to the SDGs 2

3 Who This Book Is For 5

4 The Framework 5

5 In Their Own Words................................. 7

References... 8

Part I   Mapping for the Goals on Poverty, Hunger, Health, Education, Gender, Water, and Energy

2 Open Data Addressing Challenges Associated with Informal Settlements in the Global South 13

SDGs 1 and 11

Ernest Ruzindana, Federica Gaspari, Erneste Ntakobangize, Chiara Ponti, Carlo Andrea Biraghi, Candan Eylül Kilsedar, Massimo Tadi, Zacharia Muindi, Peter Agenga, and Laura Mugeha

1 Introduction 14

2 Urbanization in the Global South...................... 14

3 Rwanda YouthMappers Chapter Experience 15

3.1 Activities in Kigali 15

3.2 Methodology................................. 16

3.3 Project Results 17

4 Italian YouthMappers Chapter Experience............... 18

4.1 Collaborative Activities 19

4.2 Results, Impressions, Possible Future Activities 21

5 Kenyan YouthMappers Chapter Experience............... 23

5.1 Partnership with Map Kibera Trust 23

5.2 Joint Initiatives Between Map Kibera and YouthMappers 23

     5.3   Methodology 24

     5.4   Results 25

  6   Conclusions 26

References 27

**3   Leveraging Spatial Technology for Agricultural
Intensification to Address Hunger in Ghana** 29
     SDGs 2 and 1
     Prince Kwame Odame and Ebenezer Nana Kwaku Boateng
  1   Intersections of Food, Agriculture, Hunger, and Poverty 29
  2   YouthMappers and the International Institute of Tropical
     Agriculture 31
  3   First Engagements and a Research Method Design 32
  4   Challenges from the Field 34
  5   Findings from Our Work 37
  6   Assessing Results Relative to Other Interventions 38
  7   SDGs and Lessons Learnt 41
References 44

**4   Rural Household Food Insecurity and Child Malnutrition
in Northern Ghana** 47
     SDGs 2 and 12
     Kwaku Antwi, Conrad Lyford, and Patricia Solís
  1   What We Need to Know About Hunger 47
  2   Seeking Answers with YouthMappers 48
     2.1   Locating Food Insecure and Malnourished Regions 48
     2.2   Laying Baseline Data in Poorly Mapped Food
           Insecure Regions 48
  3   Building the Case to Measure and Map Hunger 49
     3.1   Tracking Hunger Through the SDGs 49
     3.2   Current Indicators of Causes of Food Insecurity
           and Malnutrition 50
     3.3   Going Beyond Goals and Indicators Through
           the Map 51
  4   Assessing Household Food Security Status 52
     4.1   Determining Malnutrition Status 53
     4.2   Data Analysis 53
     4.3   Findings 53
  5   Conclusion 55
References 56

**5   Where Is the Closest Health Clinic? YouthMappers Map
Their Communities Before and During the COVID-19
Pandemic** 57
     SDGs 3 and 4
     Adele Birkenes, Siennah Yang, Benjamin Bachman,
     Stephanie Ingraldi, and Ibrahima Sory Diallo
  1   Amid a Global Health Crisis 57
  2   Community Resources in Dutchess County, New York 58
     2.1   Poughkeepsie and the Hudson Valley Mappers 58

2.2 Immediate GIS Response........................... 59
2.3 Longer-Term GIS Support to the Community 61
2.4 Local Perspectives on Challenges and Opportunities.... 62
2.5 Challenges and Opportunities for Open Mapping
from a Youth Perspective 63
3 Health Facilities in Saint-Louis, Senegal.................. 65
3.1 Gaston Berger University YouthMappers............. 65
3.2 Mobilizing to Collect Health Asset Data
in Saint-Louis.................................. 65
3.3 Validating Data to Identify User Stories.............. 66
4 Social Impact of Data in a Pandemic for the SDGs.......... 66
References... 68

6 **Cross-Continental YouthMappers Action to Fight
Schistosomiasis Transmission in Senegal** 69
SDGs 3 and 6
Michael Montani, Fabio Cattaneo, Amadou Lamine Tourè,
Ibrahima Sory Diallo, Lorenzo Mari, and Renato Casagrandi
1 The Complex Transmission Cycle of Schistosomiasis 69
2 Data Collection to Fight Transmission.................... 72
2.1 Mapping the Habitat of Snails 72
2.2 Mapping the Habitat of Humans 76
3 Results of a Cross-Continental Effort 76
3.1 Insights from Drone Mapping 76
3.2 Insights from OSM Mapping...................... 77
4 Conclusions for Joint Youth Action on SDGs 81
References... 83

7 **Understanding YouthMappers' Contributions to Building
Resilient Communities in Asia**............................ 85
SDGs 3 and 13
Feye Andal, Manjurul Islam, Ataur Rahman Shaheen,
and Jennings Anderson
1 Footprints of Youth and Open Data in Asia 85
2 Generation of Open Spatial Data in Asia.................. 86
3 Resilience and the Sustainable Development Goals 87
4 Chapter Contributions in Response to COVID-19........... 87
4.1 Mapping Health Facilities 87
4.2 Mapping Essential Services 88
4.3 Mapping Residential Households.................... 89
5 Chapter Contributions in Response to Climate Change....... 89
6 Digital Innovation During a Pandemic.................... 91
7 Future Outlook, Collaboration, and Engagement 91
References... 91

8 Activating Education for Sustainable Development Goals Through YouthMappers 93
SDGs 4 and 5
Maliha Binte Mohiuddin and Michael Jabot
1   What Does Education for SDGs Mean? Setting the Stage 93
2   YouthMappers Among a Generation of "Solutionaries" 94
    2.1   The Spirit of Solutionaries at the University of Dhaka .. 94
    2.2   The Spirit of Solutionaries at the State University of New York at Fredonia. 95
3   Recasting Science Education for Human Solutionaries. 96
4   Tying It All Together 98
References. 99

9 Seeing the World Through Maps: An Inclusive and Youth-Oriented Approach 101
SDGs 5 and 10
Shraddha Sharma, Courtney Clark, Sandhya Dhakal, and Saugat Nepal
1   Maps and Their Makers Often Exclude Women's Perspectives 101
2   Missing Data → Missing Representation → Missing Societal Development 102
3   Bridging the Gender Gap Through Everywhere She Maps 103
    3.1   Professional Development and Internship Matching 103
    3.2   Mapping Campaigns for Gender Equality 104
    3.3   Women in Technology Leadership Fellowship 104
    3.4   Regional Ambassadors 105
4   Breaking Stereotypes: Mapping Modi Gaupalika 106
    4.1   Impact on the Community 109
    5   Achievements 109
6   Women in the Spotlight: Hearing from the Ones Leading the Chapters 110
7   Engaging Future Generations of Women Mappers 111
References. 112

10 Youth Engagement and the Water–Energy–Land Nexus in Costa Rica 113
SDGs 6 and 7
Jasson Mora-Mussio
1   Linking SDGs with a Nexus Framework 113
    1.1   Water, Energy, and Land 114
    1.2   The Birris and Paez Basins. 114
    1.3   Land-Use Distribution and Resource Management Context. 114
    1.4   A Stakeholder Governance System Re-imagined by Youth 115
    1.5   The Role of Local Water Associations 116

2   Methodology 116
    2.1   Local Governance. 116
    2.2   Open-Source and Geodata Applications. 117
3   Putting the Methodology in Place 118
    3.1   Geophysical and Spatial Modeling
        Interpretation 118
    3.2   Nexus Application 121
4   Discussion of Strengths and Challenges 121
References. 122

11   Power Grid Mapping in West Africa. 125
SDGs 7 and 9
Tommy G. D. Charles
1   The Status of Power Access in West Africa 125
2   Geospatial Solutions for Power Access 126
    2.1   Trajectory of YouthMappers in Sierra Leone 126
    2.2   Disruption of a Pandemic Drives Innovation 127
    2.3   Stitching Together Open Tools to Innovate
        Fieldwork 127
3   Local Fieldwork Implementation with YouthMappers. 127
    3.1   Capacity Assessment and First-Pass Remote
        Mapping 128
    3.2   Field Mapping and Ground Truthing 128
    3.3   Setting Up and Deploying Mapping Teams
        per Location 128
    3.4   Authorization and Identification. 129
    3.5   Field Mapping and Monitoring 131
    3.6   Second Remote Mapping 132
    3.7   Summary of Output Metrics and Mapping Outcomes . . . 132
4   Beyond the Map: Powerful Mapping for the SDGs 134
    4.1   Affordable and Clean Energy 135
    4.2   Build Resilient Infrastructure and Foster Innovation 136
5   Challenges and Recommendations 138
References. 140

12   Mapping Access to Electricity in Urban and Rural Nigeria 141
SDGs 7 and 11
Emmanuel Jolaiya, Mercy Akintola, and Opeyemi Nafiu
12.1   Interrupted, Unstable Electricity Access Across
    Nigeria 141
12.2   Geospatial Solutions for Electricity Provision
    and Management 142
    12.2.1   Trajectory of YouthMappers in Nigeria 142
    12.2.2   Disruption of Power Sparks Action
          Toward SDGs. 143
    12.2.3   Geospatial Knowledge Needed to Understand
          Electricity Access. 144

12.3   Geospatial Data Implementation Projects
        with YouthMappers 144
        12.3.1   Remote Mapping Results. 144
        12.3.2   Field Mapping Results. 145
        12.3.3   Mapping Challenges 146
12.4   Beyond Making a Map: Mappers for the SDGs. 148
References. 148

**Part II   Youth Action on Work, Leadership, Innovation,
            Inequality, Cities, Production and Land**

**13   Stories from Students Building Sustainability Through
      Transfer of Leadership** 153
SDGs 8 and 9
Saurav Gautam, K. C. Aman, Rabin Ojha, and Gaurav Parajuli
1   Where the Cycle Began 153
2   YouthMappers in Nepal 154
     2.1   Designing Student-Centered Activities 154
     2.2   Adaptability During the COVID-19 Global Pandemic .. 155
     2.3   Valuable Collaborations Happen Locally. 156
3   Prioritizing Chapter Sustainability. 157
4   Future Outlook for Us and for the SDGs 158
References. 159

**14   Drones for Good: Mapping Out the SDGs Using
      Innovative Technology in Malawi** 161
SDGs 9 and 12
Ndapile Mkuwu, Alexander D. C. Mtambo, and Zola
Manyungwa
1   Evolution of an Innovation. 161
2   Drones to Promote Inclusion, Empower Youth
     and Foster Innovation. 162
     2.1   Youth and Drones. 162
     2.2   An Innovation Region 163
3   Drones to Promote Good Health and Well-Being 163
     3.1   Dzaleka Mapping Project. 164
     3.2   UAV Imagery Giving an Essential Perspective 165
     3.3   Collaborations, Youth and Community Involvement. ... 165
4   Drones for Innovation in Health Prevention Efforts 166
     4.1   Novel Technology to Control Malaria 166
     4.2   Mosquito Breeding Sites in Kasungu. 166
     4.3   Pre-flight Preparation 167
     4.4   Data Collection. 168
     4.5   Working Through Challenges. 169
5   Reflecting on the Potential of Drones for SDGs
     in the Hands of Youth. 169
References. 170

15 Assessing YouthMappers Contributions to the Generation of Open Geospatial Data in Africa 171
SDGs 10 and 8
Ebenezer N. K. Boateng, Zola Manyungwa,
and Jennings Anderson
1 The Landscape of Youth and Open Data in Africa. 171
2 Generation of Open Spatial Data in Africa 173
3 Implications for the Sustainable Development Goals 174
    3.1 On SDG 10: "We Don't Just Build Maps"
        That Bridge the Digital Divide. 174
    3.2 On SDG 8: "We Also Build Mappers"
        with Geospatial Skills. 177
4 Challenges and Opportunities 177
5 Imagining Africa's Tomorrow, Today 178
References. 179

16 Mapping Invisible and Inaccessible Areas of Brazilian Cities to Reduce Inequalities 181
SDGs 10 and 11
Elias Nasr Naim Elias, Everton Bortolini,
Jaqueline Alves Pisetta, Kauê de Moraes Vestena,
Maurielle Felix da Silva, Nathan Damas,
and Silvana Philippi Camboim
1 A Scenario of Digital Inequality 181
2 YouthMappers at the Federal University of Paraná 182
3 Systematized Collaborative Mapping 183
4 Mapping Favelas 184
    4.1 Ensuring Quality in Data Collection 184
    4.2 Open Tools and Best Practices 185
5 Accessibility in Brazil 186
    5.1 Contributing Accessibility Data to OSM 187
    5.2 Contributing Concepts to the OSM Data Model 187
6 Final Considerations. 187
References. 188

17 Visualizing YouthMappers' Contributions to Environmental Resilience in Latin America 189
SDGs 10 and 15
Nayreth Walachosky, Cristina Gómez, Karen Martínez,
Marianne Amaya, Maritza Rodríguez, Mariela Centeno,
and Jennings Anderson
1 A Picture of Youth and Open Data in Latin America. 189
2 Generation of Open Spatial Data for Resilience 190
3 Cases for Environmental Resilience in Latin America 192
    3.1 Mapping a Hidden Paradise During a Pandemic 193
    3.2 Kuna Nega, a Community Living Amidst Pollution 193
    3.3 Mapping 'Where Life Is Born'. 195
    3.4 Ecological Zoning for Conservation
        and Sustainability. 198

3.5   Mapping Trees and Palms Around Old Panama
Historic Monument 203
3.6   Smart Campus as a Model of Resilient Inclusive
Innovation...................................... 203
4   Towards an Equitable Resilient Future 205
References... 206

**Part III   Marking a Path to Goals on Sustainable Communities,
Consumption, Climate, Oceans, Land, and Justice**

18   **Youth Engagement and Participation in Mitigating
Perennial Flooding in Kampala, Uganda Using
Open Geospatial Data**................................. 209
SDGs 11 and 6
Ingrid M. Kintu and Henry N. N. Bulley
1   Rural to Urban Migration in Africa 209
2   Urban Flood Mapping of Informal Settlements............. 210
2.1   Bwaise and Kalerwe Communities 210
2.2   Pre-fieldwork Preparations....................... 211
2.3   Data Collection................................. 212
2.4   Community Mapping Results...................... 213
3   Lessons from Fieldwork in Informal Settlements 214
3.1   Flood Risk from Waste Disposal in Kampala District . . . 216
3.2   Mitigating Perennial Flooding Using Open-Source
Geospatial Data 216
3.3   Mapping with Local Residents of Informal
Settlements.................................... 217
4   Toward Informing SDG 11 Through Flood Mapping 217
References... 219

19   **Sustainable Mobility Through Knowledge Exchange
and Collaborative Mapping of Cycling Infrastructure:
SIGenBici in Medellín, Colombia** 221
SDGs 11 and 9
Natália da Silveira Arruda, Hernán Darío González Zapata,
and Ana Maria Navia Hermida
1   Sustainable Transportation and Open Mapping for SDGs..... 221
2   Context of Cycling in Medellín 222
3   Open Mapping Methods and Activities 223
3.1   Remote Data Collection.......................... 223
3.2   Socio-spatial Survey of Cyclists................... 224
3.3   Thematic Mapping and Data Analysis............... 225
3.4   Essential Participation of the Community
as Researcher 225
4   Main Findings of Cycling Infrastructure 226
5   Reflections About Cycling and Participatory Planning 228
References... 228

20 Wastesites.io: Mapping Solid Waste to Meet Sustainable Development Goals 231
SDGs 12 and 3
Chad Blevins, Elijah Karanja, Sharon Omojah,
Chomba Chishala, and Temidayo Isaiah Oniosun
1 The State of the Trash Problem 231
   1.1 Missing Location Data for Waste Site Management 232
   1.2 The Open Spatial Data Opportunity 233
2 Case Studies of Mapping Solid Waste 233
   2.1 YouthMappers in Nairobi, Kenya 233
   2.2 YouthMappers in Lusaka, Zambia 234
   2.3 YouthMappers in Akure, Nigeria 235
3 Wastesites.io: A Solution for the Global Goals 237
References 238

21 Mapping for Resilience: Extreme Heat Deaths and Mobile Homes in Arizona 241
SDGs 13 and 3
Elisha Charley, Katsiaryna Varfalameyeva,
Abdulrahman Alsanad, and Patricia Solís
1 Local Impacts from a Global Problem 241
2 YouthMappers Making Vulnerability Visible on the Map 242
   2.1 Extreme Heat Deaths 243
   2.2 Mapping the Pattern Beyond Indicators 243
   2.3 Gathering Data on Unseen Climate Vulnerable
       Locations 243
   2.4 Labeling Attributes – The Challenge of a Novel
       Housing Type 244
   2.5 Results of Data Collection Campaigns 245
3 Community Implementation with YouthMappers 246
   3.1 Global Partners, Local Stakeholders, and Mobile
       Home Residents 247
   3.2 Toward Heat Resilient Solutions 247
   3.3 SDGs as Link from Global to Local Climate
       Action for Health 248
References 248

22 Mapping for Women's Evacuation Plans During Climate-Induced Disasters 251
SDGs 13 and 5
Airin Akter and Mobashsira Tasnim
1 Youth and Women Affected by Climate-Induced Disasters 251
2 Present Scenario of Climate Change in Bangladesh 252
   2.1 Impact of Climate Change-Induced Disasters
       on the Coastal Area of Bangladesh 252
   2.2 Gendered Impacts of Climate Change 253
   2.3 Role of Gender in Emergency Response
       and Evacuation Planning During Cyclones 253

3   Voices of Women and Men Facing Climate-Induced
Disasters. 254
4   Integration of Open Data and Youth Participation
in Emergency Response Management Through
YouthMappers 254
    4.1   YouthMappers and OSM 256
    4.2   Climate Change Education and Emergency
          Response Awareness. 257
    4.3   Youth Advocating Climate Action That Inspires Us 257
5   Conclusion 259
References. 259

23 **Sustainable Development in Asia Pacific and the Role
    of Mapping for Women**. 261
    SDGs 5 and 14
    Celina Agaton
    1   Women's Sustainable Development in the Asia
        Pacific Region 261
    2   The Gender Gap in Mobility 262
    3   Key Findings and Analysis. 263
        3.1   Transportation Access and Mobility for Household
              Essentials 263
        3.2   Transportation Access and Mobility for Healthcare
              Services 264
        3.3   Transportation Access and Mobility for Livelihood
              and Microenterprise Purposes 266
        3.4   Transportation Access and Mobility for Education 268
        3.5   Safety and Security. 270
        3.6   Well-Being 272
    4   Lessons Learned. 272
        4.1   Transportation Access and Mobility for Household
              Essentials 273
        4.2   Transportation Access and Mobility for Health
              Care Services 273
        4.3   Transportation Access and Mobility for Education 273
        4.4   Transportation Access and Mobility for Livelihoods. . . . 273
        4.5   Safety and Security Issues 274
        4.6   Gender-Based Violence Issues 275
    References. 275

24 **Sustainable Coastal Communities in the Anthropocene:
    Lessons from Crowd-Mapping Projects in Colombia** 277
    SDGs 14 and 13
    Yéssica De los ríos-Olarte, Maria Fernanda Peña-Valencia,
    Natalia da Silveira Arruda, and Juan Felipe Blanco-Libreros
    1   Coastal Communities and SDGs 277
    2   Anthropocene, Socioecological Systems, and Connectivity . . . 278
    3   YouthMappers in the Anthropocene. 279
        3.1   The Coastmap Project 279

3.2 A Participatory Approach to Mapping Coastal
Communities 279
3.3 Phased Methodological Processes with Community
Participation 280
4 Anthromes 282
5 Rethinking SDGs 14 and 13 for the Coastal Anthropocene. ... 283
References... 285

25 **Collaborative Cartography Making Riparian Communities
Visible in Tefé, Amazonas, Brazil**............................ 287
SDGs 15 and 16
Ana Luisa Teixeira, Silvia Elena Ventorini,
Évelyn Márcia Pôssa, Francisco Davy Braz Rabelo,
Leonardo Cristian Rocha, Mucio do Amaral Figueiredo,
and Paula dos Santos Silva
1 Introduction 287
2 Invisible Communities............................... 288
3 Increased Visibility Through Collaborative Mapping 289
3.1 Methodology 289
3.2 Discussion 290
4 Riparian Communities Seen on the Map for SDGs 291
References... 292

26 **Open Mapping with Official Cartographies in the Americas** ... 295
SDGs 16 and 4
Vivian Arriaga, Adele Birkenes, Daniel Council, Mason Jones,
Enith K. Lay Soler, John Sawyer McCarley, Emily Wulf,
Calvin Zhang, Jean Parcher Wintemute, Nancy Aguirre,
and Patricia Solís
1 The Case for Open Spatial Data in Public Institutions 295
1.1 How Do Governmental Institutions Benefit?.......... 296
1.2 Where and Why the Americas? 296
1.3 Why and How to Engage YouthMappers? 297
2 A Summary of Six Case Studies 297
2.1 Belize and Jamaica............................ 298
2.2 Colombia 299
2.3 Costa Rica 299
2.4 Dominican Republic............................ 300
2.5 Mexico 301
2.6 Panama..................................... 302
3 Challenges and Recommendations 303
References... 304

27 **Cities of the Future Need to Be Both Smart and Just:
How We Think Open Mapping Can Help**.................. 305
SDGs 16 and 11
Stellamaris Nakacwa and Bert Manieson
1 The Scope of the Challenge........................... 305
2 OSM for Urban Governance 306
3 Role of Youth Towards Achieving Positive Urban
Governance... 307

4   YouthMappers Smart, Just Activities in Uganda 307
     4.1   Flood-Risk Mapping in Ggaba 307
     4.2   Project Model 308
     4.3   Remote Mapping 309
     4.4   Data Collection 309
     4.5   Drone Imagery and Validation 310
     4.6   Results and Achievements 310
5   YouthMappers Smart, Just Activities in Ghana 310
     5.1   Accra City Mapping (Open Cities Project) 311
     5.2   OSM Approach 311
     5.3   Results and Achievements 311
6   Smart and Just Cities of the Future 312
References 313

Part IV   Supporting YouthMappers to Advance the SDGs
           Through Institutions and Partnerships

28   Mentoring Experiences in YouthMappers Chapters 317
     SDGs 17 and 4
     Anthony Gidudu, María Adames de Newbill, Jonathon Little,
     Maria Antonia Brovelli, and Serena Coetzee
     1   The Role of Mentorship for a Student-Led Framework 318
     2   Experiences from the PoliMappers, Italy: Advancing
          Curriculum and Interdisciplinary Innovation
          to Benefit Students 319
     3   Experiences from the YouthMappers UP, Panama:
          Local Sector and Regional Partnerships Supporting
          Student Success 320
     4   Experiences from the YouthMappers Chapter in Pretoria,
          South Africa: Generating a Social Circle of Peers
          and Pipelines 321
     5   Experiences from a YouthMappers Chapter in a Two-Year
          Institution in the USA: Turnover Mentoring
          and Alumni Leverage 321
     6   Experiences from YouthMappers in Uganda: A Multifaceted,
          Multinational, and Multidisciplinary Mentor Approach 322
     7   Conclusion 323
     References 324

29   The Ecosystem Where YouthMappers Live and Thrive 325
     SDGs 17 and 8
     Dara Carney-Nedelman and Courtney Clark
     1   The Structure of a Partnership Ecosystem 325
          1.1   University Chapter Network and Regional
                 Ambassador System 325
          1.2   Student, Alumni, and Staff Perspectives 327
          1.3   Connective Tissue for YouthMappers 328

2   Components of the Ecosystem: Sponsors and Partners 329
    2.1   The American Geographical Society 329
    2.2   Mapillary 330
    2.3   TeachOSM 331
    2.4   Locana 332
    2.5   Mapbox. 332
3   Partners Ecosystem for the Implementation of SDGs
    17 and 8 333
References. 334

**30   A Free and Open Map of the Entire World: Opportunities
for YouthMappers Within the Unusual Partnership
Model of OpenStreetMap** 335
SDGs 17 and 9
Mikel Maron and Heather Leson
1   The Unusual Model of OpenStreetMap 335
2   The OSM and YouthMappers Journey 336
    2.1   The Origins of OpenStreetMap 336
    2.2   Expansion of OSM User-Creator Communities. 337
    2.3   A Stage Set for the Emergence of YouthMappers 337
3   Present Innovative Patterns of Collaboration 337
    3.1   The Innovative Model from Individuals to Groups 338
    3.2   Disruption as a Component of Innovation 338
    3.3   Efforts Off the Map 339
    3.4   Communication Channels for a Multicultural
          Movement. 340
4   Recommendations for New Innovations Within
    the OSM + YouthMappers Communities 341
    4.1   Stronger Connections to Civic and Open
          Communities 341
    4.2   OSM Contributions to Locally Defined Priorities
          of the SDGs 341
    4.3   Deeper Links to Corporate Social Impact 342
5   Looking Ahead. 342
References. 342

**31   Youth and Humanitarian Action: Open Mapping
Partnerships for Disaster Response and the SDGs** 345
SDGs 17 and 11
Tyler Radford, Geoffrey Kateregga, Harry Machmud,
Carly Redhead, and Immaculata Mwanja
1   The Need for Digital Mapping in Humanitarian Response 345
2   Early Precedents for Student Engagement in "Crisis"
    Mapping. 346
3   HOT Partnerships Engaging YouthMappers Students 346
    3.1   Data to Inform Both Humanitarian Action
          and the SDGs 347
    3.2   Relationships to Enhance Citizen Generated
          Data in Vulnerable Places. 347

4   Cases of HOT Partnerships Engaging YouthMappers
    Students 348
    4.1   Indonesia: Supporting Disaster Risk Management
          and Contingency Planning 348
    4.2   Tanzania: Flood Resilience Through Open Map Data . . . 351
    4.3   Malaria Elimination Mapping Campaign. 352
    4.4   Peru: Collaboration Among Government, Academia,
          and the Regional Group of Earth Observations
          (AmeriGEO). 354
5   Looking Forward: Youth Leading on Sustainable
    Action in Their Communities Towards the SDGs 354
References. 355

Part V   The Paths Ahead

32 Generation 2030: The Strategic Imperative of Youth
    Civic and Political Engagement. 359
    SDGs 17 and 16
    Michael McCabe and Steven Gale
    1   The State of Youth Engagement. 359
    2   Six Troubling Trends in Youth Engagement 360
        2.1   Lost Confidence in Democracy 360
        2.2   Stepping Back from Political Engagement 360
        2.3   Influences Shifting on Trust and Opinions. 360
        2.4   Absent from Decision-Making. 360
        2.5   COVID-19 Makes Matters Worse 360
        2.6   Voiceless on Key Issues Like Climate Change 361
    3   Building a Meaningful Compact with Young People. 361
    References. 362

33 Reflecting on the YouthMappers Movement 365
    Jennings Anderson, Chad Blevins, Nuala Cowan,
    Dara Carney-Nedelman, Courtney Clark, Michael Crino,
    Ryan Engstrom, Richard Hinton, Michael Mann,
    Brent McCusker, Rory Nealon, Patricia Solís,
    and Marcela Zeballos
    1   Marveling at the Movement as a Digital Public Good. 366
        1.1   What Has Been the Main Contribution
              of YouthMappers to the Potential for Reaching SDGs?. . 366
        1.2   How Does YouthMappers Strengthen Youth
              Links to the UN SDGs? 366
        1.3   In What Ways Do You Believe Leadership Assumed
              by YouthMappers Has Been Able to Contribute,
              Galvanize, or Mobilize Their Communities?. 367
    2   Reflecting on Innovation, a Spirit of Overcoming,
        and Action Towards the SDGs 367
        2.1   What Are Some of the Most Unique Methodologies
              and Technologies Used by YouthMappers That
              Relate to the UN SDGs? And What Is Unique
              About Them? 367

2.2   What New Innovations or Methods Do YouthMappers
       Have with the Potential to Advance? 367
2.3   What Conceptual Ideas Do YouthMappers
       Have the Potential to Advance? 368
3   Identifying Where We Still Have a Lot of Work to Do 368
4   Where Are YouthMappers Going Next 369
5   How Are We to Support This Journey from Here? 369

Index 371

Introduction

Patricia Solís and Marcela Zeballos

Abstract

In an era of global challenges – from climate change to economic unrest to social disruption to pandemics – the need to hear from voices of the next generation of leaders is clear. The time to listen to them is now. The purpose of this book is to assemble, organize, and amplify the knowledge and experiences of some of the world's young people who are working locally and collectively to use scientific results, geospatial technologies, and multi-national collaboration to address some of the most pressing issues facing their local communities and global society. From every region of the world, students have emerged as leaders in the YouthMappers movement, to study such problems by creating and using open data that has a spatial component. The issues they are addressing with these common tools and methods range across the entire scope of topics known as the Sustainable Development Goals, articulated globally through the United Nations. Not only do YouthMappers create new knowledge and bring unique perspectives and experiences, but they are also proposing and taking action based upon what they see and what they know from the map and from each other.

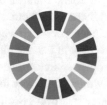

Keywords

Youth · Sustainable Development Goals · OpenStreetMap · Community-engaged scholarship

1 The Emergence of YouthMappers

More than 1 billion people in the world are missing from the maps that are foundational for basic needs and services and simply navigating our world. YouthMappers are university students who create and use open spatial data – mapping – for authentic community development and humanitarian purposes, putting their own communities and nations on the map. This group of knowledgeable students have emerged to fuel a global movement to use digital technologies –

P. Solís (✉)
Knowledge Exchange for Resilience, Global Futures Laboratory, and School of Geographical Sciences and Urban Planning, Arizona State University, Tempe, AZ, USA
e-mail: patricia.solis@asu.edu

M. Zeballos
YouthMappers, Texas Tech University, Lubbock, TX, USA
e-mail: marcela.zeballos@ttu.edu

P. Solís, M. Zeballos (eds.), *Open Mapping towards Sustainable Development Goals*, Sustainable Development Goals Series, https://doi.org/10.1007/978-3-031-05182-1_1

both geospatial platforms and communications technologies – to answer the call for leadership in this next generation of challenges.

The formal narrative of the establishment of YouthMappers as a network of young people, as a consortium of university student groups and their mentors, and as an entity that supports the capacity, activity, and engagement of members who contribute to open data and open thinking is, for the most part, already well documented in various published articles, books, and media, most of which have been co-authored between students and mentors (see especially Solís et al. 2018; Carney-Nedelman 2020; YouthMappers 2016). In particular, the literature about YouthMappers experiences includes peer-reviewed publications reflecting on how they fit into the global geospatial open data and open source movement (Brovelli et al. 2020); how they effectively use open source tools for their career pathways (Solís et al. 2020a); how YouthMappers approaches build capacity and experience in their local peer groups (Coetzee et al. 2018); how this matters for teaching and learning (Price et al. 2019; Larsen et al. 2021; Rees et al. 2020); how what they do matters for global citizenship (Solís and DeLucia 2019); and how their presence impacts what appears on the map where the needs are greatest (Herfort et al. 2021).

This book thus adds a critical yet heretofore missing component of that documented history of the trajectory of this movement. It is envisioned as a collection that conveys a sense of the pre-existing, underlying, robust knowledge and enthusiasm that have been the precondition for the emergence of YouthMappers. Sometimes this stemmed from studies or academic experiences, but often from the overflow of youth energy and ideas that our traditional higher education institutions cannot fully contain or enable (Solís et al. 2020b). While the YouthMappers blog is written by and for students and offers an enriching communications space for immediate sharing and peer learning (see especially Hite et al. 2018; Mugeha 2020; Chishala and Suleiman 2020; Arruda 2021), we are overdue to recruit lengthier, deeper, organized contributions reflecting on

YouthMappers experiences. As such, this book aims to document and share insights about this movement's emergence from the first-person voices of the very students themselves who are among those at the forefront of creating our new people's map of the world.

Using the OpenStreetMap platform as the starting point, a foundational sharing mechanism for creating data together, the authors of the chapters in this book impart the way they are learning about themselves, about each other, and about the world, developing technology skills, and simultaneously teaching the rest of the world about the potential contributions of a highly connected generation of emerging world leaders. The book is timely in that it captures a pivotal moment in the trajectory of the YouthMappers movement's ability to share evolving expertise and one that coincides with a pivotal moment in the geopolitical history of planet earth that needs to hear from them – and needs to listen to what they are saying. Certainly, this book does not portend to be comprehensive of all of voices and perspectives, but it does offer a rich glimpse into the minds and hearts of the youth behind, within, and at the front of the movement.

2   Where YouthMappers Connect to the SDGs

A common thread visible in the research and audible in the stories of many YouthMappers is a desire to make a better world, often in the face of struggle because of resource- and opportunity-poor environments, which they overcome with resourcefulness, spirit, and action. While their specific objectives for any particular mapping activity, whether locally or remotely, may vary depending on the context for their work, YouthMappers activities tend to relate very closely to the United Nations Sustainable Development Goals (SDGs) (Solís et al. 2018, 2020b; Chishala and Suleiman 2020). The SDGs are a comprehensive collection of 17 interlinked global goals designed to be a blueprint to achieve a better and more sustainable future for all, put in place by the United Nations General

Assembly in 2015. While the goals are aspirational, each promotes concrete sets of development objectives, written so that countries may work to accomplish them by the year 2030. This time dimension is matched by a spatial dimension – all of them literally need to "take place" somewhere. It is serendipitous – and advantageous – that the official launch of YouthMappers occurred also in 2015, in November of that same year.

While the SDGs are not legally binding, nations are expected to mobilize action to ensure leadership and solutions under all of the themes represented in the 17 goals. The YouthMappers approach to create open geospatial data is a ubiquitously relevant action that can lead to evidence to track progress across any goal, yield observations of patterns that reveal possible solutions, and involve activities that build leadership capacity, all together, openly. The stories told in this book, in their own words, provide evidence that not only is this ambition ever-present across the range of places where YouthMappers are growing, but it is also a driving motivational force for the choices of mapping projects that youth make under their own volition.

In the years since 2015, when YouthMappers began actively mapping under that name, the growth of the movement has been steady. Within the first 100 weeks, there were an astounding 100 chapters that have joined the network (YouthMappers 2018; USAID 2018). Figures 1.1 and 1.2, respectively, show the growth of activity, represented in terms of edits contributed to OpenStreetMap (by changeset hashtag) by country and accumulated over time. At the time of submission for publication, the network is celebrating its 300th chapter in more than 60 countries.

The viral growth of the network, we assess, is partly attributable to the flexible inclusive design of YouthMappers as a youth-led, faculty-mentored, university-based, student-centered, purpose-driven, no-cost, social chapter model. But we also believe that this growth is due in no small part to the idea that YouthMappers tapped into, crystalized, and catalyzed a new space to accommodate the energy and knowledge of young people, committed to make a difference using geospatial technologies for sustainable development in tangible local ways, speaking to larger global goals. This is a purpose-driven, identity-based community.

This book seeks to guide the reader to understand what underlies this kind of movement. Most publications that cover the topic of sustainability and/or youth development and global issues are written by non-youth authors. Moreover, most are written by non-majoritarian, entrenched academic scholars. Each chapter puts forward the voices of students and recent graduates in countries where YouthMappers works, all over the world. Many of them hail from countries where expertise in geospatial technologies for the SDGs is nascent and needed. They cover topics that range from water, agriculture, and food to waste, education, and gender from their own eyes in working with data, mapping, and humanitarian and development action.

The idea for the book itself was generated from collaborative discussions with and among the organizers of network resources and the youth authors, who have decided to address these themes at various scales of perspective: from individual/local city level to provincial and national level to multi-national regional to global scopes. Often these narratives also cross borders and engage as the youth themselves are engaged, in an intricate multi-national network of ideas and activities, with co-authors spanning these borders. They reflect on some of the lessons learned, unresolved challenges, and insights they have encountered in their efforts – which may be often successful and sometimes not fully realized. They, in fact, were the voices to have elected to map their ideas to the 17 SDGs, as a concerted, purposeful way to connect the youth agenda to the global agenda and speak to leadership audiences. As editors, we have sought to facilitate this vision in the most faithful way possible.

In the end, we believe the book fills a missing niche in the literature on SDGs, science and technology (especially geospatial), and literature written by young voices who are knowledgeable and experienced yet bring a fresh leadership perspective in their own authentic voice. The resources that are available range from very broad audiences of the very young to scholarly texts that

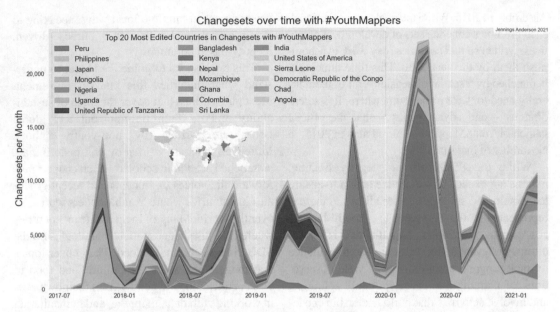

Fig. 1.1 Number of changesets per month show activity from 2017 through 2021 with the hashtag #YouthMappers for the top 20 most edited countries in OpenStreetMap. (Credit: J. Anderson, 2022)

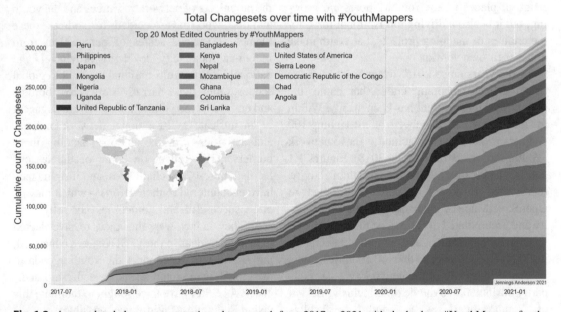

Fig. 1.2 Accumulated changesets over time show growth from 2017 to 2021 with the hashtag #YouthMappers for the top 20 most edited countries in OpenStreetMap. (Credit: J. Anderson, 2022)

muse upon the state of youth engagement – wins, lessons, challenges, struggles, and potential, alike – but authored largely by non-youth even if written for changemaker youth audiences (Figueres and Rivett-Carnoc 2020; Reimbers 2017; Sato and Dunn 2019). A few key inspira-

tional books by individual youth activists are the exception; but a collective voice of a movement has not been adequately documented and is missing from the literature. Some works may address an individual youth activist or one specific SDG (Yousafzai and Lamb 2013; Thunberg 2019), but a compendium of this sort that ranges across all of them, in the voice of many youth, has not been found. Moreover, this book features concrete science and technology solutions, centered on geospatial open data and knowledge skills for action, retaining a forward-looking perspective to make a tangible difference.

3   Who This Book Is For

We envision that this book will serve three audiences, prioritized in this order. First, we hope that this volume serves as a stellar example of youth writing for youth. The problems of the present and the futures that the SDGs seek to navigate, and open geospatial data seeks to inform, will require collective knowledge and collective action propelled by the next generation. Given our emphasis on the voices and perspectives of young leaders, we imagine that the university student peers of the thousands of YouthMappers around the world would be a key core audience to reflect on their own actions and make meaning of a movement to engage global issues. These readers are pursuing educational goals and leadership development, typically undergraduate or master's degree audiences, age 18–25, in the realm of where the United Nations defines youth.

Many of the authors are university students whose life and academic experiences draw from languages beyond English. Cognizant of the colonial legacy of this language, we hope it may be utilized nonetheless to the end of making greater more, profound connections. We preserved original phrases and analogies, rejecting typical editorial tendencies to standardize terms or overcorrect, in order to maintain the character of the stories.

We also hope to speak directly to the audience who makes up the pool of talented and dedicated faculty mentors, academic professors at universities and colleges, and even high school teachers in the 60 countries and beyond of the network who support YouthMappers specifically and youth generally in their professional and personal lives. We anticipate and hope that you find value in this book to supplement your curriculum,

engage your own students (whether they are YouthMappers or not), and use these experiences as case studies in seminars and symposia as a way to amplify their voices. This is a broad interdisciplinary audience that is interested in science for good, humanitarian and development studies, the fields represented by the UN SDGs, and globally minded social science audiences. May these chapters provide a creative resource for continuing your noble work to put youth first.

Finally, we hope that this collection also broadens and deepens the awareness of international development and environment, intergovernmental secretariats of global oriented organizations that work on various topics of the SDGs, and humanitarian professionals at UN agencies, the World Bank, NGOs, and activists who are stakeholders in engagement with youth. This tertiary audience may also include stakeholders across the OpenStreetMap community, who are involved in the fabulous idea to create a people's map but may not yet fully grasp why we seek to elevate the next generation in the broader movement. We trust you all will find the book inspiring, informative, and accessible, as we have.

4   The Framework

The authors of this book each chose the subjects they present and were then encouraged to identify a primary SDG and a secondary SDG to help orient the reader to the specific contributions of their efforts and as a framework to organize the set of cases. Both SDGs are indicated in the abstract, and the chapters of this book are ordered by the first goal, in succession, and then by the second goal for those with the same primary SDG. Authors were encouraged to present their experiences from the most authentic style of their work and aims and to include all co-authors who contributed, be they from one local chapter or multiple across countries and continents. Some chapters include mentors as co-authors as well. Many include illustrations such as maps and photographs that help to convey the meaning they wish to portray. Students were invited to participate in all aspects of the book, including writing the prospectus, recruited from among the alumni of YouthMappers leadership and research fellows, the regional ambassadors, at-large blog contributors, and other highly active youth across the network so that authors clearly had knowledge not only of their own work but some awareness of the movement as a whole. As editors, we aimed to balance the regional representation, thematic representation, and gender-inclusive set of contributions. Given the broad, global, diverse set of YouthMappers across the network, this did not require much effort to ensure.

Part I presents cases that primarily address the first seven SDGs (1–7), written by youth who envision a world with no poverty, zero hunger, good health and well-being, enjoying quality education, gender equality, clean water and sanitation, and affordable and clean energy for all. Taken collectively, they represent a set of voices advocating for fundamental elements necessary for each individual human being to live in dignity and to develop their livelihoods and households, able to meet basic needs. They appeal to the vital responsibility of providing essential resources across all regions of the world. By paying attention to spatial data and open mapping, they also remind us of the elementary role of place, underscored in the very first chapter about informal settlements emerging from rapid urbanization (Ruzindana, Gaspari, Ntakobangize, Ponti, Carlo Biraghi, Kilsedar, Tadi, Muindi, Agenga, and Mugeha, Chap. 2) and through considering the control of land at the nexus of water and energy (Mora-Mussio, Chap. 10), including reminding us of the need for universal access to electricity to power places (Charles, Chap. 11; Jolaiya, Akintola, and Nafiu, Chap. 12). We gain insights about the forces underlying the basic needs in both rural and urban places through the lens of food insecurity (Odame and Boateng, Chap. 3; Antwi, Lyford and Solís, Chap. 4). The ensuing consequences upon human health are narrated, especially in light of the shock of the global COVID-19 pandemic (Birkenes, Yang, Bachman, Ingraldi, and Diallo, Chap. 5; Andal, Islam, Shaheen, and Anderson, Chap. 7) and in the context of long-term stresses of endemic diseases (Montani, Cattaneo, Tourè, Diallo, Mari, and Casagrandi, Chap. 6). YouthMappers insist on inclusive solutions to these problems, ones that rectify gender inequality (Sharma, Clark, Dhakal, and Nepal, Chap. 9) and serve as the basis for universal education (Binte Mohiuddin and Jabot, Chap. 8).

Part II shifts to evoke a greater attention to the socio-economic infrastructure that underpins development in the places where YouthMappers live and work, addressing primarily the following three SDGs (8–10). These cases center on mapping a world that offers better and more equitable opportunities for youth participation in meaningful work and economic growth, industry (especially the geospatial industry), and innovation that stems from mapping within our cities and communities. To our students, this first and foremost means putting youth leadership and skills, especially geospatial competencies, front and center in service to the SDGs, with sustained hand-off to ensure continuity (Gautam, Aman, Ojha, and Parajuli, Chap. 13; Boateng, Manyungwa, and Anderson, Chap. 15). It means getting the most innovative tools and technologies into their hands like drones and GeoAI (Mkuwu, Mtambo, and Manyungwa, Chap. 14; Charles, Chap. 11), while at the same time honoring a place for fieldwork and collaborative mapping within their communities to make critical socio-economic realities visible through the map (Naim Elias, Bortolini, Alves, Vestena, da Silva, Damas, and Philippi, Chap. 16; Walachosky, Gómez, Martínez, Amaya, Rodríguez, Centeno, and Anderson, Chap. 17).

Here we also take a moment to examine the role of YouthMappers across three major world regions, utilizing metrics on participation, presence, and statistics about the open spatial data they have created on OpenStreetMap. In particular, we assess YouthMappers contributions in Africa (Boateng, Manyungwa, and Anderson, Chap. 15); we visualize YouthMappers contributions in Latin America (Walachosky, Gómez, Martínez, Amaya, Rodríguez, Centeno, and Anderson, Chap. 17); and we generate understanding about YouthMappers contributions in Asia (Andal, Islam, Shaheen, and Anderson, Chap. 7). These data and the narratives summarizing regional action that appear with them offer a set of voices that seek not only a future with greater sustainability but also one that is characterized by resilience.

Part III looks toward building resilience through open mapping, particularly focused on the next set of six SDGs (11–16), which attend to the dynamic systems and landscapes that form the stage for youth action. This section touches upon primary objectives to track and advance responsible systems, like that of sustainable production and consumption through the proper movement and siting of solid waste (Blevins, Karanja, Omojah, Chishala, and Oniosun, Chap. 20). It also includes advancing climate-responsible transportation, such as the movement of people for sustainability via cycling (Arruda, González, and Hermida, Chap. 19) and for resilience during disasters (Akter and Tasnim, Chap. 22; Agaton, Chap. 23). Resilience of life on land and our oceans includes open mapping to mitigate climate impacts, like flooding (Kintu and Bulley, Chap. 18; De los ríos Olarte, Peña, Arruda, and Blanco, Chap. 24), hurricanes (Agaton, Chap. 23; Akter and Tasnim, Chap. 22), and deaths from heat (Charley, Varfalameyeva, Alsanad, and Solís, Chap. 21). Mapping these systems as a stage for equitable action, YouthMappers are also attentive to justice and the need for strong institutions (SDG 16). They tell of their efforts to build this from the neighborhood scale, mapping together in participatory ways with local communities (Teixeira, Ventorini, Pôssa, Rabelo, Rocha, Figueiredo, and Silva, Chap. 25; Peña, Arruda, and Blanco, Chap. 24) and in ways that inform official cartographies (Arriaga, Birkenes, Council, Jones, Lay, McCarley, Wulf, Zhang, Wintemute, Aguirre, and Solís, Chap. 26). YouthMappers challenge us to consider how mapping can help cities of the future to be smart and just (Nakacwa and Manieson, Chap. 27).

Part IV considers the final SDG, number 17, emphasizing the importance of collective action and building partnerships for sustainability. This section features voices from the communities of support to YouthMappers, providing some additional context and insights that empower the network. This includes the importance of mentors and mentoring under best practices (Gidudu, Adames, Little, Brovelli, and Coetzee, Chap. 28). We also explore the shape of the open mapping ecosystem of partners where YouthMappers are able to thrive (Carney-Nedelman and Clark, Chap. 29), especially within the dedicated sister community that is known as HOT, the Humanitarian OpenStreetMap Team (Radford, Kateregga, Machmud, Redhead, and Mwanja, Chap. 31), and the broad-based, larger community of communities that create and use OpenStreetMap (Maron and Leson, Chap. 30). This section both celebrates successes and reviews ongoing challenges of a student-centered movement.

Part V takes a look ahead and reflects on what these experiences mean for the future of youth and of the SDGs. The organizers and sponsors of YouthMappers reflect first on the general current state of engagement with youth for global leadership (McCabe and Gale, Chap. 32). Throughout the book, we have incorporated quotes from well-known youth leaders that eloquently capture some of the sentiment we witness within this movement as well (i.e., Wathut, Yousfazai, Bastida, Gómez-Colón, and Thunberg). We end with a collection of thoughts on the path ahead particularly for this YouthMappers movement offered by the committee of organizers who continuously work to provide and steer resources to sustain the movement (Anderson, Blevins, Cowan, Carney-Nedelman, Clark, Crino, Engstrom, Hinton, Mann, McCusker, Nealon, Solís, and Zeballos, Chap. 33). We advise readers and remind ourselves to be good ancestors.

5 In Their Own Words

As we introduce you to this book, we would like to reiterate the emphasis of listening to the voices of YouthMappers. We urge you to consider the potential impact now and in the future of their open data for advancing the SDGs in the next decade and advancing global goals that they may have a hand in determining in the following decade. We encourage you to see what they have created through their mapping actions and we invite you to learn from their experiences.

When we began planning for this collection to be created, we curated some of the most salient messages that our authors hoped to ultimately convey in the pages that follow. We finish introducing you to YouthMappers, using their own words:

City planning authorities should take social-spatial structures into consideration when drawing plans to support proper living conditions in all corners of their jurisdiction.

Settlement planning must be given an important focus since it is often a key challenge that many countries face; also improving informal settlements (built environment) and improving the wellbeing of communities that live in them, which includes youth residents.

To feed the future: locate farmlands and offer dedicated services to improve agricultural output in Ghana, engaged with the next generation.

Even in a global pandemic, solutions for cities can be contextualized and localized as well through humanitarian mapping, geospatial and communications technologies used by youth.

Community projects and collaborative geospatial technologies answer the need for data and information while engaging students, researchers and citizens in the technical process, defining a learning journey that enriches both people and locations.

Local communities' engagement is fundamental to local action for SDGs. Youth can provide new ways to think about environmental health through participatory technologies that engage local communities.

The study and management of health issues as endemic diseases can be pursued through OpenStreetMap data, community engagement, capacity building and new technologies and young people already have significant expertise and inclination for team science to accomplish this.

Girls should be empowered to take part in geospatial technology and gender equality must be taken seriously.

Female leaders can serve as role models resulting in greater participation of females and decreasing the gender gap. Their involvement even in small numbers at first can provide significant contributions ensuring better results.

Innovation is critical for mapping projects that are for the social good. This can be achieved by regularly introducing youth and the community to open source tools and linking them with public and private partners.

Scientific policy advice from youth is valuable in the building of sustainable urban communities and just urban governance structures.

Young people are a group that are initiating their contributions in society and will do so throughout their lives. Ethics, methodologies and tools used in YouthMappers, such collaborative mapping, perfectly fits this new epoch of the Anthropocene, where it is not possible to contribute to sustain-

ability without recognizing the linkages between different environmental and social systems from a multidimensional social-ecological perspective.

Collaborative and open mapping of buildings, facilities and services put people and communities on the map, making them visible while addressing quantitatively the need for strategies for an equal and sustainable development.

Climate action and disaster risk reduction planning should take into account how women are affected differently during a disaster than men, and how young people can move this knowledge forward.

We, as organizations, institutions, communities, individuals, alike need to implement female-focused, driven, and led initiatives and programs to bridge the existing digital gender gap.

Continental scale organizations such as NGOs, local governments and schools should partner with YouthMappers to contribute more quality geospatial data to purpose-driven applications, given the track record of the movement.

For achieving education for sustainable development, students must not always keep themselves buried inside textbooks; they need experience to acquire knowledge, technical skills, values, and attitudes to enable a more sustainable society for all. The YouthMappers movement and experience overcomes many of the limitations that have been seen around education in that we typically teach about and not with the SDGs.

Youth seeing themselves as agents of change are pivotal to present and future development.

References

Arruda N (2021) Why is a day important to celebrate the space of women in science. Available from YouthMappers blog, 11 February. https://www.youthmappers.org/post/why-is-a-day-important-to-celebrate-the-space-of-women-in-science. Cited 14 Feb 2022

Brovelli M, Ponti M, Schade S, Solís P (2020) In: Guo H, Goodchild MF, Annoni A (eds) Citizen science in support of digital earth. Manual of digital earth, Springer, Singapore. https://doi.org/10.1007/978-981-32-9915-3_18

Carney-Nedelman D (2020) Happy 5th anniversary YouthMappers. Available from YouthMappers blog, 25 November. https://www.youthmappers.org/post/happy-5th-anniversary-youthmappers. Cited 14 Feb 2022

Chishala C, Suleiman Y (2020) How YouthMappers are contributing to 2030 agenda for Sustainable

Development. Available from YouthMappers blog, 6 August. https://www.youthmappers.org/post/2020/08/06/how-youthmappers-are-contributing-to-2030-agenda-for-sustainable-development. Cited 14 Feb 2022

Coetzee S, Minghini M, Solis P, Rautenbach V, Green C (2018) Towards understanding the impact of mapathons: reflecting on YouthMappers experiences. Int Arch Photogram Rem Sens Spat Inf Sci XLII-4/W8:35–42. https://doi.org/10.5194/isprs-archives-XLII-4-W8-35-2018

Figueres C, Rivett-Carnoc T (2020) The future we choose: surviving the climate crisis. Penguin Random-House, London

Herfort B, Lautenbach S, Porto de Albuquerque J, Anderson J, Zipf A (2021) The evolution of humanitarian mapping within the OpenStreetMap community. Sci Rep 11:3037. https://doi.org/10.1038/s41598-021-82404-z

Hite R, Solís P, Wargo L, Larsen TB (2018) Exploring affective dimensions of authentic geographic education using a qualitative document analysis of students' YouthMappers blogs. Educ Sci 8:173. https://doi.org/10.3390/educsci8040173

Larsen T, Gerike M, Harrington J (2021) Human-environment thinking and K-12 geography education. J Geogr. https://doi.org/10.1080/00221341.2021.2005666

Mugeha L (2020) 3 years, 3 stories, 3 lessons learnt. Available from YouthMappers blog, 4 June. https://www.youthmappers.org/post/2020/06/04/3-years-3-stories-3-lessons-learnt. Cited 14 Feb 2022

Price M, Berdnyk A, Brown S (2019) Open source mapping in Latin America: collaborative approaches in the classroom and field. J Latin Am Geogr. https://doi.org/10.1353/lag.0.0118

Rees A, Hawthorne T, Scott D, Spears E, Solís P (2020) Toward a community geography pedagogy: a focus on reciprocal relationships and reflection. J Geogr. https://doi.org/10.1080/00221341.2020.1841820

Reimers F (ed) (2017) Empowering students to improve the world in sixty lessons. CreateSpace Independent Publishing Platform, p 286

Sato M, Dunn P (2019) Legacy: The Sustainable Development Goals in action: 52 Changemakers. Dean Publishing, Macedon

Solís P, DeLucia P (2019) Exploring the impact of contextual information on student performance and interest in open humanitarian mapping. Professional Geogr 71(3):523–535. https://doi.org/10.1080/00330124.2018.1559655

Solís P, Mccusker B, Menkiti N, Cowan N, Blevins C (2018) Engaging global youth in participatory spatial data creation for the UN sustainable development goals: the case of open mapping for malaria prevention. Appl Geogr 98:143–155. https://doi.org/10.1016/j.apgeog.2018.07.013

Solís P, Anderson J, Rajagopalan S (2020a) Open geospatial tools for humanitarian data creation, analysis, and learning through the global lens of YouthMappers. J Geogr Syst 23(4):599–625. https://doi.org/10.1007/s10109-020-00339-x

Solís P, Rajagopalan S, Villa L, Mohiuddin MB, Boateng E, Wavamunno Nakacwa S, Peña Valencia MF (2020b) Digital humanitarians for the Sustainable Development Goals: YouthMappers as a hybrid movement. J Geogr Higher Educ:1–21. https://doi.org/10.1080/03098265.2020.1849067

Thunberg G (2019) No one is too small to make a difference. Penguin, London

USAID (2018) 100 chapters in 100 weeks. Available from United States Agency for International Development. https://www.usaid.gov/sites/default/files/documents/15396/YouthMappers-Infographic.pdf. Cited 14 Feb 2022

Yousafzai M, Lamb C (2013) I am Malala: the girl who stood up for education and was shot by the Taliban. Weidenfeld & Nicolson, London

YouthMappers (2016) About us. Available from YouthMappers website. https://www.youthmappers.org/about-us. Cited 14 Feb 2022

YouthMappers (2018) Welcoming our 100th YouthMappers Chapter: The Nature Club of Karatina University, Nyeri, Kenya. Available from YouthMappers blog, 8 February. https://www.youthmappers.org/post/2018/02/08/welcoming-our-100th-youthmappers-chapter-the-nature-club-of-karatina-university. Cited 14 Feb 2022

Mapping for the Goals on Poverty, Hunger, Health, Education, Gender, Water, and Energy

Open Data Addressing Challenges Associated with Informal Settlements in the Global South

Ernest Ruzindana, Federica Gaspari,
Erneste Ntakobangize, Chiara Ponti,
Carlo Andrea Biraghi, Candan Eylül Kilsedar,
Massimo Tadi, Zacharia Muindi, Peter Agenga,
and Laura Mugeha

Abstract

The United Nations estimates that 3 billion people living in urban contexts will need adequate and affordable housing by 2030. We urgently need alternative perspectives and methodologies for urban development that are environmentally sustainable and inclusive of the local community. This chapter illustrates the design and results of projects carried out by YouthMappers in Rwanda, Italy, and Kenya, focused on informal settlements in the Global South and the value of geospatial data for addressing SDG 1 No Poverty and SDG 11 Sustainable Cities and Communities.

Keywords

Poverty · Informal settlements · Urbanization · Rwanda · Italy · Kenya

E. Ruzindana (✉) · E. Ntakobangize
College of Science and Technology, School of Architecture and Built Environment, University of Rwanda, Kigali, Rwanda

F. Gaspari · C. Ponti · C. A. Biraghi · C. E. Kilsedar
Department of Civil and Environmental Engineering, Politecnico di Milano, Milan, Italy
e-mail: federica.gaspari@polimi.it; chiara1.ponti@mail.polimi.it; carloandrea.biraghi@polimi.it; candaneylul.kilsedar@polimi.it

M. Tadi
Department of Architecture, Built Environment and Construction Engineering, Politecnico di Milano, Milan, Italy
e-mail: massimo.tadi@polimi.it

Z. Muindi
Map Kibera Trust, Nairobi, Kenya

P. Agenga
Department of Geography, University of Nairobi, Nairobi, Kenya

L. Mugeha
Department of Geomatics Engineering and Geospatial Information Systems, Jomo Kenyatta University of Agriculture and Technology, Juja, Kenya

© The Author(s) 2023
P. Solís, M. Zeballos (eds.), *Open Mapping towards Sustainable Development Goals*, Sustainable Development Goals Series, https://doi.org/10.1007/978-3-031-05182-1_2

1   Introduction

The research, methods, activities, and stories captured here center the innovative and interdisciplinary work carried out by university students and faculty to address the local impact of global issues. YouthMappers, a global university consortium, has created a space for students to participate in the creation of geospatial data while positioning them not only as contributors but also as leaders with valuable local knowledge and experience. Students are making vital contributions to filling critical data gaps and, in the process, leading a movement toward social change with youth at the forefront.

We present three different experiences by local YouthMappers Chapters in Rwanda, Italy, and Kenya focusing on mapping activities within informal settlements. Crucial to the success of each activity were the partnerships created with local entities and across the YouthMappers network, the tools and platforms selected to implement projects, and the design of the mapping activities. The authors discuss the potential of approaches based on open data created collaboratively in critical geographical and social contexts.

2   Urbanization in the Global South

Nowadays, 56% of the world's population lives in urban areas, and this number is rapidly growing. The share of people living in cities is expected to reach 68% in 2050 (UN 2019), reversing the numbers of one century ago when rural settlements were predominant (71%). This growth will mainly happen in the countries where their share of the population in urban areas is lower than the world average. Countries with high population and urbanization rates often have very low values of the Human Development Index (HDI), an index calculated by combining human development aspects such as life expectancy, education, and per capita income indicators (UNDP 2020). In this ranking, 18/20 and 38/50 of the lowest positions are occupied by African countries.

These are low-income countries that face serious structural obstacles to achieve sustainable development. They are highly vulnerable to economic and environmental shocks and possess low levels of human resources (Maksimov et al. 2017).

Future urbanization will likely take place in unfavourable conditions with a high probability of creating informal settlements on unclaimed land in the periphery of existing cities. The rapid diffusion of slums and other forms of informal dwellings, characterized by inadequate basic services and infrastructures, makes cities more vulnerable to disasters (Rosa 2017). Life in informal settlements is precarious as they are usually overcrowded and congested. Additionally, they lack social and community networks, present stark inequalities, have crippling social problems, are particularly vulnerable to health problems, economic shocks, and the risks related to climate change and natural disasters (Habitat 2020). Improving the living conditions of vulnerable slum dwellers is a crucial challenge for making cities and informal settlements sustainable, resilient, inclusive, and safe.

Because of their spontaneous development, informal settlements are hardly manageable by authorities and institutions, and they often do not exist in official datasets. Their complicated social and physical condition is an obstacle for gathering data about them, including geospatial data. As a result, they are usually neglected, which leads to a limited amount of information about them with low quality. The lack of data limits the awareness of the dynamics of these areas and the possibility of interventions for improvement.

Luckily, open mapping can be a valid alternative solution to support the different actors involved in planning activities around the world (Chakraborty et al. 2015). The effective use of open data sources and open software gives a significant contribution to urban studies in general and for addressing the problems of informal settlements in the Global South. Initiatives of collaborative mapping are important not only for producing up-to-date geospatial data for planning purposes but also for their participatory nature, involving local citizens in bottom-up activities that can benefit the whole community (Abbott 2003).

3 Rwanda YouthMappers Chapter Experience

Rwanda is a small landlocked territory in Central–East Africa where informal settlements exist in physically congested spaces and the quality of and access to information about them are usually limited. Facilitating open data provision and accessibility helps to bridge the limited data gap by providing an alternative source of information on open platforms (Chakraborty et al. 2015). The long-term goal of Rwanda YouthMappers, the YouthMappers chapter at the University of Rwanda, is to put Rwanda on the map. In collaboration with Kigali Geospatial Development and Research Hub (GeoDR), Rwanda YouthMappers conducted a study on the challenges associated with informal settlement upgrading in the Agatare area of Kigali city (Fig. 2.1). Open mapping was used in this project to support effective planning on behalf of the government, non-governmental organizations, and the private sector to address the problems of informal settlements in the country (Fig. 2.2).

3.1 Activities in Kigali

Informal settlements are unevenly distributed among the three districts that make up Kigali city. The study was conducted in the Agatare informal settlement area in Nyarugenge district, which is considered the home of unplanned dwellings as it is where 76% of informal settlements are located (Hitayezu et al. 2018). The area is located close to the central business district (CBD) and is dominated by limited plots, rudimentary buildings, and inferior living conditions.

Kigali GeoDR partnered with the Rwanda YouthMappers with the intention of providing up-to-date and reliable mapping responses so that the government and humanitarian actors can respond quickly to the crisis in the Agatare area (Fig. 2.3). Kigali GeoDR Hub and Rwanda YouthMappers Chapter mapped building footprints, roads, water lines, drains, rainwater collection points, markets, facilities, and power lines in OpenStreetMap (OSM) (Fig. 2.4). Data generated by this activity would support local officials' efforts to design and implement inter-

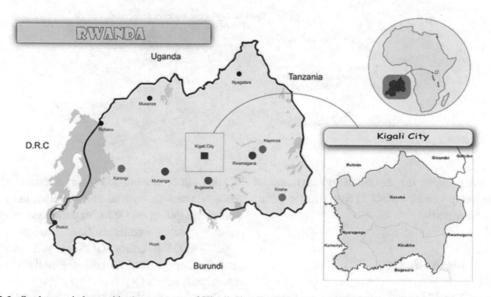

Fig. 2.1 Study area is located in Agatare area of Kigali City, Rwanda

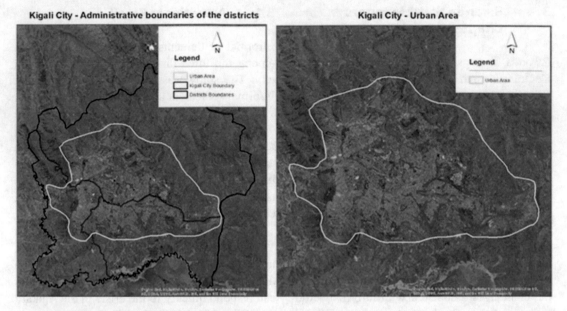

Fig. 2.2 Administrative boundaries outline the districts (red) in Kigali City, Rwanda (black), demarcating the urban settlement concentrations (yellow)

Fig. 2.3 An iterative project management cycle serves as a framework for the mapping study

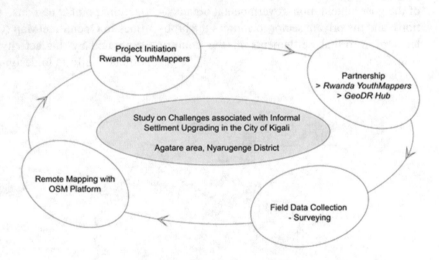

ventions to make this area safe and sustainable, in the perspective of SDG 11 Sustainable Cities and Communities.

3.2 Methodology

Rwanda YouthMappers contributed to the production of geospatial data using various resources, tools, and techniques, including the application KoBoCollect downloaded on android smartphones used to collect data on the field. For remote mapping on OSM, experienced mappers used made contributions using Java OpenStreetMap editor (JOSM) and inexperienced mappers used iD editor. Finally, the data were analyzed in QGIS.

The Rwanda YouthMappers team collected data to determine the number of homes to be demolished. As part of the project, students sup-

Informal settlement in Agatare area

Fig. 2.4 Example of features mapped by the Kigali GeoDR Hub and Rwanda YouthMappers Chapter indicates building footprints, roads, and water lines in OpenStreetMap

ported the deployment of a survey for collecting non-spatial data such as the owner of the houses, the members of the households and their number, and their daily activities for survival and tried to ask them about their living conditions. This fieldwork was carried out in conjunction with office workers in order to provide accurate spatial and non-spatial data per household. This information on buildings and services within Agatare will help the city of Kigali improve decision-making related to future planning and development of the area.

The processing of data collection took six days, and the data was organized according to its use. Non-spatial data were organized to produce tables and diagrams to show the real situation and living conditions in the Agatare area. The spatial data were organized and arranged in the following way: the points data were arranged in an Excel spreadsheet before making the shapefiles; the polygons, points, and lines were imported in QGIS to produce the geodatabase that combined all kinds of data collected. Maps were produced and printed, and the final results were presented in a report.

3.3   Project Results

A total of 6.64 km of roads, streetlights, and 8.2 km of water pipelines have been mapped to serve over 15,000 residents in the Agatare area. A total of 5.3 km of drainage channels, mainly the Mpazi and Rwampala water channels and their sub-channels, have been mapped. These water channels cross 2080 residences affected by flooding and severe overflows. A total of 6.2 km of footpaths were mapped that serve over 3000 residents in their daily business activities. A total of 24.154 building footprints were mapped using JOSM and iD editor and were uploaded to OSM.

The data created through this project was used to improve roads with lighting, footpaths, electricity, water pipes, drainage, and other facilities such as public toilets, rainwater harvesting, market, and recreational facilities (Fig. 2.5). Thus, the result of the project has been assisting the city of Kigali in decision-making and enabling the upgrading of the Agatare informal settlement with the support of the World Bank. Future improvements will contribute to making the area

Fig. 2.5 Street view of data collection sites shows examples of utility poles (left) and water harvesting points (right)

safer, more resilient, and sustainable and contribute to changing the dwellers' living conditions.

4 Italian YouthMappers Chapter Experience

Italy, with an HDI of 0.88, above the threshold of very high human development, is among the first 30 countries in the ranking of 2020 (UNDP). It has a very good availability of both authoritative and volunteer data and presents very limited cases of slums. For this reason, PoliMappers (https://polimappers.github.io/), the first Italian YouthMappers chapter, since its foundation in 2016, is very active in projects located outside national borders supporting humanitarian activities in the Global South. Indeed, in the past years, the group took an active part in a wide variety of campaigns located around the world and promoted by the Humanitarian OpenStreetMap Team (HOT), Crowd2map Tanzania, and more.

Following the participation of PoliMappers in the "Architecture for Smart City" course given by Professor Massimo Tadi at Politecnico di Milano, Piacenza campus (Gaspari 2020), at the end of 2020, we took part in a new project, in partnership with IMMdesignlab (http://www.immdesignlab.com/), led by Professor Tadi.

The project is part of the master's thesis "Slums upgrading: An integrated design approach for the environmental performance and social inclusion based on IMM methodology in the border between Bogotá and Soacha, Colombia." The slum upgrading process based on strong interdisciplinarity defines a system of connected actions and strategies involving different fields of knowledge, such as architecture and urban design, systems theory, data science, energy efficiency, mobility and transportation, geospatial information analysis, people engagement strategies, water and waste management, and food policies. This case study will allow to build a benchmark development plan for slum upgrading in the framework of the SDGs. In fact, integrated modification methodology (IMM) (Tadi et al. 2020) proposes a systemic interpretation of the SDG 11 Sustainable Cities and Communities, which suggests that local-based actions are firmly linked with SDGs and their indicators and able to trigger simultaneous improvements in environmental performances, social inclusion, and urban metabolism.

In consideration of the fact that IMM is a model-based approach, it is able to define the state of a system and its performance through a rigorous qualitative and quantitative representation. It is evident that having a dataset able to support such a model is necessary for proceeding with the work. Observing the lack of informal settlement data in the authoritative datasets, the thesis proposes to adopt open and collaborative methods for data acquisition, which could then be replicated in contexts with similar characteristics.

4.1   Collaborative Activities

PoliMappers coordinated surveying and mapping activities in the study areas of Ciudad Bolívar and Cazucá, on the border of Bogotá and Soacha, developing a workflow based on the "Architecture for Smart City" course experience (Fig. 2.6). Later, San Humberto was included in the study area. Soacha is located within the metropolitan area of Bogotá and is included in the project area for Bogotá in the rest of this chapter.

In October 2020, PoliMappers officers met with the group from IMMdesignlab to define the purpose of the collaboration, training requirements, and coordination of the process for creating geo data at the ground level to perform urban diagnostic activities in the study area.

Partnering with a local NGO, TECHO Colombia became a crucial step of the process, involving their local knowledge and expertise gained through their humanitarian campaigns in the area. Indeed, their volunteers were looking for a community approach to perform street-level surveys of urban settlements, especially in the area of San Humberto. Together, by sharing personal and professional experiences, we shaped a crowdsourced paradigm structured at both local and international scales. Following a few virtual meetings, Mapillary resulted to be the most suit-

able tool for our purpose of capturing street-level imagery. Being already familiar with the application, PoliMappers shared solutions and suggestions to perform the survey, providing training on the use of the application for TECHO volunteers. Then, a survey plan was defined together, identifying the major roads to take photo sequences of in the study area (Fig. 2.7).

Figure 2.7 shows routes surveyed for obtaining images to identify commercial activities, housing dynamics, land use, and patterns of informality in Municipality 4 (Cazucá) and Municipality 6 (San Humberto) of Soacha and in Ciudad Bolívar in Bogotá.

The COVID-19 pandemic delayed the timing of the survey, as restrictions imposed by the local authorities did not allow free movement. In February 2021, the group of volunteers from TECHO Colombia were finally able to carry out the survey, using Mapillary installed on three mobile devices, one of which mounted in a frontal position and the other two facing the right and left sides of the road, respectively, moving by car, both to reduce the time needed to cover the established routes and to minimize safety risks (Figs. 2.8 and 2.9).

PoliMappers encountered difficulties in uploading the images due to a Mapillary server problem; the images did not appear in the application database for a long time. The initial plan

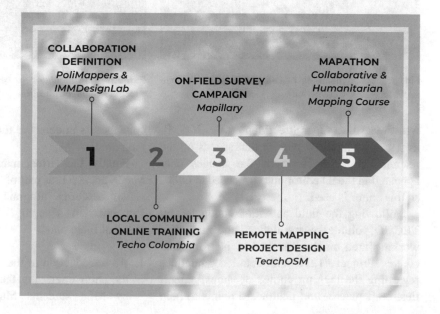

Fig. 2.6 The project workflow designed by PoliMappers serves as a framework for data collection

COLLABORATION DEFINITION
PoliMappers & IMMDesignLab

ON-FIELD SURVEY CAMPAIGN
Mapillary

MAPATHON
Collaborative & Humanitarian Mapping Course

1   2   3   4   5

LOCAL COMMUNITY ONLINE TRAINING
Techo Colombia

REMOTE MAPPING PROJECT DESIGN
TeachOSM

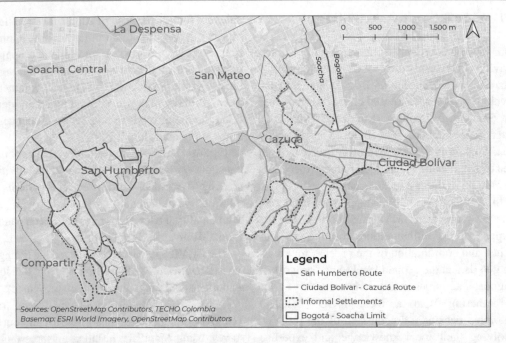

Fig. 2.7 Routes indicated serve as the basis for image collection in Cazucá and San Humberto, Bogotá, Colombia

Fig. 2.8 Front mounting of cell phone enables mobile street view image collection

Fig. 2.9 Side mounting of cell phone enables mobile street view image collection

was to use Java OpenStreetMap editor (JOSM), as it has a Mapillary plug-in that speeds up the mapping process. However, the images only appeared in the iD editor once uploaded to OSM, so this editor is used.

Following the field activities conducted by TECHO volunteers, we defined the mapathon working area on the TeachOSM platform and created Project #1270 with detailed guidelines regarding the tagging choice of shops, amenities, craft places, and healthcare points to help

contributors understand the context of the mapping area.

Finally, the virtual mapathon took place on March 26, 2021, as part of the PoliMappers innovative teaching program "Collaborative and Humanitarian Mapping." The mapathon consisted of three parts:

1. An introduction to the concept and design of the slum upgrading thesis project illustrated by MSc students Maria Alejandra Rojas

Bolaño and Silvia Raviscioni, with a focus on the collaboration with TECHO Colombia

2. A tagging and mapping tutorial carried out by PoliMappers officers that gave suggestions and highlighted possible issues

3. A guided mapping session during which attendants followed the required steps to display and explore Mapillary image sequences on iD editor, evaluate the position of points of interest by using both street-level and satellite imagery, and create or edit the targeted features on OSM

The event was held on the Zoom platform and broadcasted live on the PoliMappers Twitch channel, involving participants from both Politecnico di Milano and other universities and institutions. Throughout all of its phases, the project benefited from constant support and widespread social media promotion by partner networks: YouthMappers, IMMdesignlab, TECHO Colombia, and the Colombian YouthMappers chapter at the University of Antioquia in Medellín, Semillero Geolab UdeA.

4.2   Results, Impressions, Possible Future Activities

Collaborations at various scales made it possible to reach interested people from different areas of the world. There were 39 contributors who took part in the mapping activity and created 244 changesets, in which 508 shops, 177 amenities, and 17 healthcare places were mapped (Fig. 2.10). The contributions were then validated by the PoliMappers team using the OSMCha tool, providing constructive comments in case of errors.

Most of the participants in the mapathon had OSM editing knowledge from beginner to intermediate levels. The mapping activity presented some major challenges. The lack of knowledge of the Spanish language made it difficult to find informa-

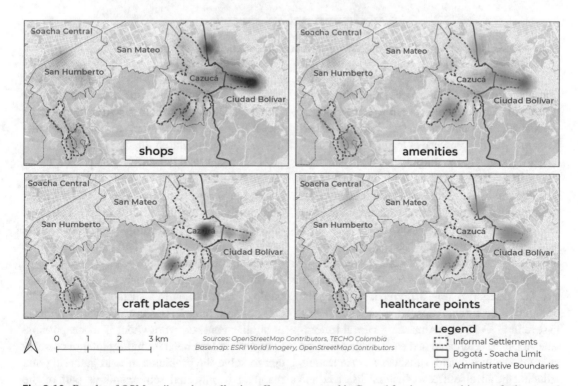

Fig. 2.10 Results of OSM attribute data collection efforts are mapped in Cazucá for shops, amenities, craft places, and healthcare points

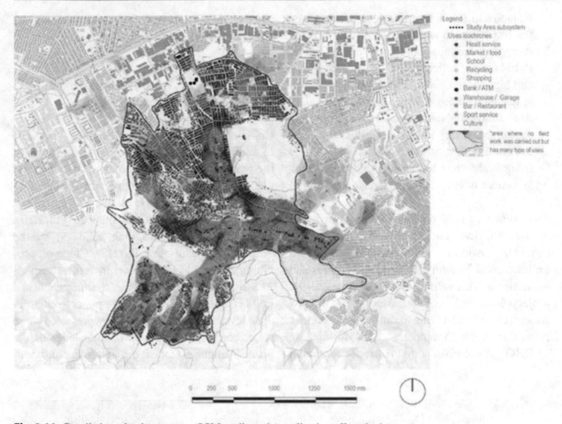

Fig. 2.11 Detailed results demonstrate OSM attribute data collection efforts by key category

tion about the use of the spaces by reading sign-posts. Another difficulty, which emerged during the mapathon in conversation with some of the participants, was the presence of features with characteristics specific to the study area, which were hard to tag in OSM with the available options. Using Mapillary images also required an additional effort to identify the features' correct location, combining satellite images and topographic layers.

The data gathered have been used to produce urban diagnostic maps. In particular, the analysis of IMM proximity (Fig. 2.11), accessibility, and diversity key categories was performed, enabling the design of a better slum upgrading project, relating to several SDG 11 targets including safe and affordable housing, affordable and sustainable transport systems, inclusive and sustainable urbanization, reducing the adverse effects of natural disasters, and the environmental impact of cities, among others.

Figure 2.10 shows densities of map features added by OSM contributors within the area of the mapathon. Data refers to the map edits tracked between March 26, 2021, and April 09, 2021. Dataset was harvested and parsed using overpass turbo. Fig. 2.11 depicts the urban diagnostic maps on proximity relative to key functions at the system level built on the collected data and Fig. 2.12 as a comparison between the proximity levels before and after the slum upgrading project.

The slum upgrading project using IMM methodology aimed to build a sustainable, adaptable, and scalable model for slum upgrading projects in similar contexts worldwide; it made possible to understand the need for a fully integrated strategy to solve the problems of data gathering and mapping in informal settlements as part of the project design approach.

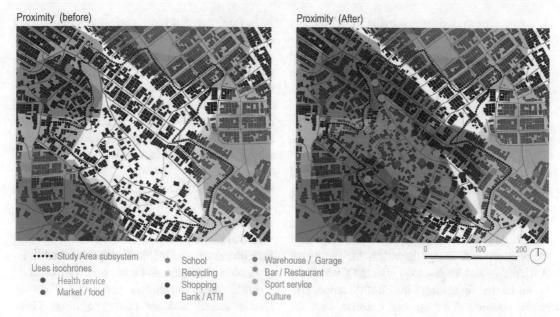

Proximity (before)   Proximity (After)

•••• Study Area subsystem   • School   • Warehouse / Garage
Uses isochrones   • Recycling   • Bar / Restaurant
   • Health service   • Shopping   • Sport service
   • Market / food   • Bank / ATM   • Culture

Fig. 2.12 Urban diagnostic map results depict local proximity to services

5   Kenyan YouthMappers Chapter Experience

5.1   Partnership with Map Kibera Trust

Kibera is the largest slum in Africa. Situated in Nairobi, Kenya, Kibera is one of the most well-known, well-researched, and well-serviced slums worldwide. Despite this focus, Kibera was literally a blank spot on the map: its patterns of traffic, scarce water resources, limited medical facilities, etc., remained invisible to the outside world and residents themselves. Without the basic knowledge of the geography of Kibera, it was impossible to have an informed discussion on how to improve the lives of its residents.

In November 2009, Map Kibera produced the first complete free and open map of Kibera, thanks to local motivated young people who learned how to create maps using OSM techniques. This included surveying with handheld GPS devices, digitizing satellite imagery, and using paper-based annotations with Walking Papers. Individuals from the blossoming Nairobi tech scene helped train and make connections with the larger community and created a sustainable group of map maintainers beyond the initial three-week effort. Data consumers were consulted for their needs to help add direction to feature types collected and immediately make use of the map data. Map Kibera has grown into a complete interactive community information project.

5.2   Joint Initiatives Between Map Kibera and YouthMappers

In 2016, Map Kibera helped launch the first YouthMappers chapter in Kenya at the University of Nairobi by providing capacity-building training to new chapter members. This led to a series of joint initiatives including:

1. The mapping of water and sanitation facilities in Mathare, a slum in Nairobi, to support the initiatives of the United States Agency for International Development (USAID) in that area
2. Hosting a joint mapathon with the Young African Leaders Initiative (YALI) during the 2017 high-level meeting for data and development in Africa (ODW 2017)

3. Launching a program to provide YouthMappers members with internship opportunities to get hands-on experience in mapping informal settlements and the methodologies used in the projects

5.3   Methodology

During the national elections in Kenya, informal settlements in Kenya are most affected by election-related violence. With the then-upcoming national elections in 2017, Map Kibera with the help of three students from the University of Nairobi and Jomo Kenyatta University of Agriculture and Technology (JKUAT), who were on an internship through the YouthMappers program, conducted a mapping exercise with the help of local community ambassadors. The main aim of this exercise was to update the security map created during the 2013 national elections.

This mapping exercise relied on information provided by peacekeeping missions (both government and non-governmental agencies) to identify the hotspots (insecure areas), i.e., places in the community known for criminal activity and had a high propensity for violence in case of social turmoil. Additionally, some areas were identified as safe places (safe havens) where community residents could run to seek humanitarian assistance in case of social unrest where they would be provided with basic needs, shelter, and medical assistance.

Before the field mapping exercise, survey questionnaires were developed and training sessions were held on the data collection process, which involved filling these questionnaires and the collection of GPS coordinates using GPS-enabled devices. A team of thirteen mappers representing all thirteen villages in Kibera was then dispatched to map the safe and unsafe places, which were characterized as follows:

Hotspots (unsafe places): These included areas that had non-functional streetlights and places that are prone to experience violence when social distress erupts.

Safe places: These included areas that had functional streetlights and areas that were close to police posts and showgrounds.

Gender-based violence (GBV) centers: These were also mapped where women and girls could seek medical and psychotherapy assistance in the case of abuse due to social unrest.

The mappers were equipped with handheld GPS gadgets and questionnaires where they would fill in the coordinates of each facility and additional information describing the condition of that facility, e.g., for streetlights, they would map if they were functional or non-functional. Additional features that were mapped included health centers and other community institutions. The entire mapping exercise took approximately one month and ran between June and July of the same year (Fig. 2.13).

Fig. 2.13 Volunteer mappers use handheld GPS and paper questionnaires for detailed data collection in Kibera, Nairobi, Kenya

5.4   Results

Afterward, the data was edited using the JOSM and later uploaded to OSM. Print maps were then designed, developed, and distributed to partner organizations under the peacekeeping missions and the police and administrative officials (Figs. 2.14 and 2.15).

The security map was also painted on the wall at a strategic point in Kibera where the community members and visitors would interac with it. The purpose of the mural was to enhanc public awareness and provide community mem bers with an offline platform for the publi (Fig. 2.16).

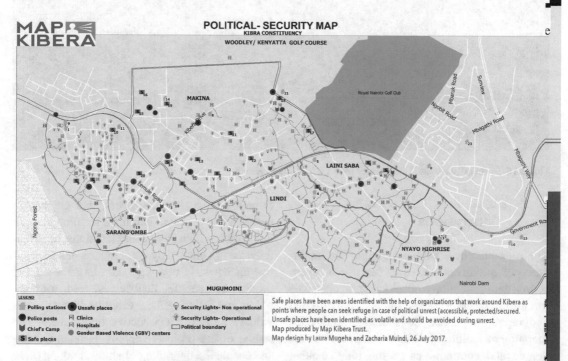

Fig. 2.14 One of the final maps of points-of-interest are distributed to partner organizations

Fig. 2.15 Administrative officials are presented the print maps developed by the project

Fig. 2.16 A mural showcases the security map via painting in Kibera

6　Conclusions

The presented case studies highlight the importance of initiatives related to the use of open-source data collection technologies and open data in informal settlements studies in different geographic contexts. Collaborative and open mapping of buildings, facilities, and services put people and communities on the map, making them visible while addressing quantitatively the need for strategies for an equal and sustainable development, especially considering the increasing number of slums around the world. Indeed, each project made evident the importance of a settlement being represented on a map as a human right, because "being on the map means being part of the city, therefore being recognized as city dweller, and then claiming certain rights as citizens" (Choplin and Lozivit 2019).

The projects carried out by the YouthMappers chapters in Rwanda, Colombia, and Kenya led to interesting insights about the adaptability of tools from the open data movement in different fields and contexts and the involvement of a wide range of expertise and backgrounds. Community projects and collaborative geospatial technologies answer the need for data and information while engaging students, researchers, and citizens in the technical process, defining a learning journey that enriches both people and locations. Map features are not created by corporations or authoritative agencies; in this way, data is gathered and owned by people themselves, redistributing power and fostering the awareness of every individual's impact in a collective environment. The results of the projects support decision-making and enable the upgrading of slums, making these areas safer and more resilient. Open culture needs to be promoted implementing strategies for communicating the obtained results, while sharing public data with local people, in order to support a larger set of data users and producers enabling the realization of a more efficient data gathering workflow.

Acknowledgments We would like to acknowledge the work done by Silvia Raviscioni and Maria Alejandra Rojas Bolaños, authors of the master's thesis "Slums upgrading: An integrated design approach for the environmental performance and social inclusion based on IMM methodology in the border between Bogotà and Soacha (Colombia)" whose data gathering phase was developed in collaboration with PoliMappers. Furthermore, we extend a special thank you to the TECHO Colombia's volunteers, who collected photos on the field with Mapillary application in the case study area. PoliMappers also would like to thank Professors Maria Antonia Brovelli and

Ludovico Giorgio Aldo Biagi, respectively, coordinator and responsible of the PoliMappers Passion in Action course, during which the mapathon for Bogotá took place virtually.

References

Abbott J (2003) The use of GIS in informal settlement upgrading: its role and impact on the community and on local government. Habitat Int 27(4):575–593. https://doi.org/10.1016/S0197-3975(03)00006-7

Chakraborty A, Wilson B, Sarraf S, Jana A (2015) Open data for informal settlements: toward a user's guide for urban managers and planners. J Urban Manag 4(2):74–91

Choplin A, Lozivit M (2019) Mapping a slum: learning from participatory mapping and digital innovation in Cotonou (Benin). Cybergeo. https://doi.org/10.4000/cybergeo.32949

Gaspari F (2020) PoliMappers at the Architecture for Smart City Course. Available via YouthMappers Blog, October 22. https://www.youthmappers.org/post/polimappers-at-the-architecture-for-smart-city-course. Cited 31 Jan 2022

Hitayezu P et al (2018) The dynamics of unplanned settlements in the City of Kigali. Laterite and International Growth Center, Kigali. http://www.theigc.org/wp-content/uploads/2019/02/Hitayezu-et-al-2018-final-report-v2.pdf

Maksimov V, Wang SL, Luo Y (2017) Reducing poverty in the least developed countries: the role of small and medium enterprises. J World Bus 52(2):244–257

ODW (2017) High level meeting: data for development in Africa. Available via Open Data Watch, June 29. https://opendatawatch.com/past-events/high-level-meeting-data-for-development-in-africa/. Nairobi, Kenya. Cited 31 Jan 2022

Rosa W (2017) Goal 11. Make cities and human settlements inclusive, safe, resilient, and sustainable. A new era in global health: nursing and the United Nations 2030 Agenda for Sustainable Development, p 339

Tadi M, Zadeh MH, Biraghi CA (2020) The integrated modification methodology. In: Masera G, Tadi M (eds) Environmental performance and social inclusion in informal settlements. Springer, Cham, pp 15–37. https://doi.org/10.1007/978-3-030-44352-8_2

UN HABITAT (2020) World cities report 2020. ISBN 978-92-1-132872-1. https://unhabitat.org/sites/default/files/2020/11/world_cities_report_2020_abridged_version.pdf

United Nations, Department of Economic and Social Affairs (2019) World population prospects 2019. https://population.un.org/wpp/Publications/Files/WPP2019_Highlights.pdf

United Nations Development Program (UNDP) (2020) Human Development Report 2020: the next frontier. Human development and the Anthropocene. ISBN: 978-92-1-126442-5. http://hdr.undp.org/sites/default/files/hdr2020.pdf

Leveraging Spatial Technology for Agricultural Intensification to Address Hunger in Ghana

3

Prince Kwame Odame
and Ebenezer Nana Kwaku Boateng

Abstract

YouthMappers are using open geospatial tools in support of initiatives seeking to achieve SGD 2 Zero Hunger and SDG 1 No Poverty in Northern Ghana. Students and researchers designed survey questions and a field data collection workflow using simple but cost-effective technology to catalogue a database of farmers, properly demarcate farm sizes, and give farmers, in particular impoverished women, the opportunity to project farm yields and increase the efficiency of their output.

Keywords

Hunger · Agriculture · Spatial analysis · Fieldwork · Ghana · Poverty

P. K. Odame (✉)
University of Education, Winneba, Winneba, Ghana
e-mail: pkwameodame@uew.edu.gh

E. N. K. Boateng
University of Cape Coast, Cape Coast, Ghana
e-mail: ebenezer.boateng@stu.ucc.edu.gh

1 Intersections of Food, Agriculture, Hunger, and Poverty

Access to food is critical to human survival, and in this case, food security only exists when all people have both economic and physical access to sufficient, nutritious, and safe food that does not only meet one's dietary needs but also offer options to meet their food preference (Ministry of Food and Agriculture 2007; Quaye 2008). In view of the aforementioned, poor food security is not only manifested in the failure of a country's agricultural sector to produce sufficient food but also witnessed in terms of an individual's failure to ensure access to sufficient food at the household level (Clover 2010). Indeed, food security has been cemented in the second goal of the Sustainable Development Goal (SDG), which aims at ending hunger, achieving food security and improved nutrition, and promoting sustainable agriculture. To the United Nations, this goal can be achieved through increased productivity and incomes of small-scale farmers, particularly through the provision of financial aid, farm inputs,

P. Solís, M. Zeballos (eds.), *Open Mapping towards Sustainable Development Goals*, Sustainable Development Goals Series, https://doi.org/10.1007/978-3-031-05182-1_3

knowledge, and necessary agricultural practices needed for sustainable food production.

Measures to ensure we stay on track in achieving the SDGs 1 and 2 must be put in place, as the attainment of these goals contributes to the achievement of other SDGs given a positive relationship between good nutrition and one's ability to attain positive outcomes in health, education, and economics (Tandoh-offin 2019). Another reason to ensure we reconsider the need to address food security to achieve SDG 2 is the recent rise in malnutrition and stunted growth in Northern Ghana where 33% of undernourished children under 5 years are considered to be stunted. Compared to global rates, Ghana's stunt rate is 12% higher than the reported 21.3% by the United Nations. Aside from Ghana, the prevalence of malnutrition has particularly affected many countries in the Global South, particularly South America and most countries on the African and Asian continent. In terms of global ratio, the number of children found to be undernourished was estimated to have reached 821 million and that translates to one malnourished child out of every 9 (UNICEF 2018).

In Ghana, over 41% of the economically active population aged 15 years and older are into agriculture, and this statistic is even higher for the three northern regions even though these regions still record the highest rate of child malnutrition (Ghana Statistical Service 2012). Being rural oriented, Hjelm and Dasori (2012) further noted that farming is the chief source of livelihood with about 80% of all households in the Northern Regions in it. Despite the contribution of agriculture to people's livelihood, Ghana's Ministry of Food and Agriculture (2013) stipulates that farmland in Ghana is mostly smallholder farms with most farm sizes being less than 2 hectares, hence the difficulty of engaging in commercial or sustainable food production.

While Ghana's agricultural sector is largely considered to be dominated by smallholder farmers, the Ministry of Food and Agriculture (2007) ranks Ghana as one of the few African countries to attain some level of parity on food security but

also cautioned that this revelation was just momentarily as further assessment on a regional basis indicates acute food insecurity in some part of the country, especially the three Northern Regions. Reasons for this caution include the fluctuations in the output of food production due to heavy reliance on rain, poor infrastructure, and use of traditional farming tools and practices. Unfortunately, data on a wide range of agricultural activities in Ghana is near absent, hence the inability of government agencies to make concrete projections or reveal the true state of Ghana's food security. Of course, investment in research by the state has always been a challenge given the stiff competition from other critical sectors of the economy. Even if such data exist, they may only reflect aggregates and be obsolete, and in-depth details may not even be available from the state (Ministry of Food and Agriculture 2007).

The need for data in the agricultural sector, in recent years, has gained attention due to its capacity to propel new knowledge, improve practices, and also improve the effectiveness of governments' policies in increasing food production. The relevance of data in the agricultural sector is seen at every stage of the food value chain where data on farms, farmers, and final consumers is always gathered for various interventions and projects. In fact, the importance of data cannot be withheld as accessibility to it becomes the game-changer in the agricultural sector, hence the need for its openness.

Open data in this regard refers to all data sources that are accessible and shareable and can be used by anyone. It has been proven to be beneficial to ensuring sustainable agriculture for smallholder rural communities mostly in developing countries. When reliable and accessible to farmers, open data can influence food security by offering famers the ability to make precise projections on farming output, battle climate change, promote higher food production, and improve access to markets. In fact, open data can also generate sector changes to provide innovations to benefit all (Lohento et al. 2017; Musker and Schaap 2018).

Despite the numerous benefits attached to open data, Jellema et al. (2015) identify the African continent to be the least inclined to open data due to limited internet access, limited funding for research or data creation, and scarce mediators who play a vital role between accessing open data and use of that data. Fortunately, Ghana has made a bold step in signing up to the Open Government Data initiative in 2014, which establishes Ghana's clear indication to harness the benefits of open data in all sectors of the economy including the agricultural sector.

Under the auspices of the USAID, the International Institute of Tropical Agriculture (IITA) engaged the University of Cape Coast YouthMappers chapter to facilitate a farmer profiling project as a way to examine the efficacy of various farming technologies that were introduced to small-scale farmers in the three Northern Regions of Ghana. Aside from the expectations of improving the output of smallholder farmers, these interventions also aimed at improving the living conditions and economic prospects of these farmers given an observed positive relationship between asset accumulation and poverty alleviation.

2   YouthMappers and the International Institute of Tropical Agriculture

From its inception in 1967, the International Institute for Tropical Agriculture (IITA) has positioned itself as a research-for-development (R4D) organization that focuses on providing tailor-made solutions to end poverty, hunger, and the destruction of natural resources across the African subregion. To achieve this, the IITA engages various national and international agencies like Ghana's Ministry of Agriculture and other allied agencies to enhance food security, improve livelihood, preserve natural resources, and increase employment. On the back of the aforementioned, the IITA is funded and constitutes one of the 15 research labs under

Consultative Group on International Agricultural Research (CGIAR).

As a research lab, the IITA has executed over 70% of the CGIAR's programs in sub-Saharan Africa and also hopes to lift about 11.5 million young people from poverty through an intensive program to rejuvenate 7.5 million hectares of farmland by 2020 (Consultative Group on International Agricultural Research 2020). Unfortunately, the evidence to evaluate the IITA's ability to deliver on this audacious goal is not available to the researcher. Among the key thematic areas that border on the IITA's activity are biotechnology and genetic improvement, natural resource management, social science, and agribusiness as well as plant production and plant health. It must be noted that each thematic area is limited to a section of the African continent and keenly related to the agricultural needs of the identified subregion.

With its footprint in 30 countries in Africa, the focus of activities in West Africa is crop improvement and biotechnology, and this centers on the improvement of crops like roots, tubers, and bananas. Fortunately, the identified crops are mostly found in the three Northern Regions of Ghana, hence the IITA interest in introducing various farm technologies to farmers in the identified communities. At present, the IITA Ghana office has an administrative facility and a demonstrational lab in all the three Northern Regions, and it is in these labs that farmers receive firsthand orientation of modern farming technologies. On the whole, the IITA is represented in 12 West African countries and has also undertaken projects that offer additional support like mechanization, agribusiness, and capacity development. On the other hand, Sudan is the only country in Northern Africa with an IITA office despite efforts to champion policies for agribusiness, nutrition, and health.

Aside from focusing on farmers' needs, the IITA also initiates programs to end poverty by addressing the increasing rate of unemployed youth in sub-Saharan Africa. By this, the IITA establishes a Youth Agripreneurs program as a conduit to create job opportunities by encouraging the youth to consider the agricultural value chain as a viable livelihood option. Such programs are executed with various business incubation programs that offer the needed environment and resources for various brainstorming activities.

3 First Engagements and a Research Method Design

The search for an outfit with the needed experience and dedication to undertake a farmer profiling project necessitated the collaboration between UCC YouthMappers and IITA. To most of our mappers, news of this potential collaboration was like winning a lottery since it did not only validate our efforts but also offered an opportunity to solve real-world problems. Additionally, location of the study site also added up to the euphoria, especially when most of our mappers appeared to originate from the South or middle belt of Ghana. We could see the smiles and excitement on their faces as everyone appeared to show interest in this project. Unfortunately, logistical constraints and project requirements did not permit us to engage over 80 mappers on campus for this project.

As the acting advisor, the choice of who to select for this project became a headache given the diversity and entrenched commitment of our chapter members. To select 5 mappers for this project, the selection criteria focused on students who required more than a year to complete their program. This was needed to ensure a smooth propagation of knowledge and lessons acquired from the project. The pool of eligible students was still around 30 students, and each mapper was poised to make the cut. Finally, a ballot was cast, and the 5 members of this project include Kwame Odame, Ebenezer Boateng, Daniel Osei Acheampong, Confidence Kpodo, and Sabina Abuga.

Typically, a bus ride from Cape Coast to Tamale is about 15 hours, but the IITA was inclined to sponsor our flights from Accra to Tamale and that trip took less than an hour.

In fact, that was a relief to us since it spared us long sitting hours. In our first meeting with staff and officials of the IITA, the chief scientist for the program mentioned that their decision to collaborate with us was heavily based on reports of our activities and their observation of our enthusiasm using open data and tools to address real-world issues. Here, we took the chance to inform our hosts of our intention to utilize open data platforms, namely, Kobo Toolbox, OSMtracker, and Java OpenStreetMap editor (JOSM). The choice of these platforms was informed by the numerous advantages that come with the use of open data and extensive experience gained from the use of these platforms in previous projects.

Guided by the objectives of IITA, our first engagement began with a demonstration of Kobo Toolbox, which was the primary data collection platform for the project (as seen in Fig. 3.1). For most of the staff of IITA, open data tools, like Kobo Toolbox, were a new phenomenon, and to make the interaction more practical, the survey questions were used in developing the research instrument since it offered the advantage of learning this tool and familiarizing ourselves with the research instrument.

The target group for this project comprised small-scale farmers under the Africa RISING program. Small-scale farmers in this regard refers to farmers with not more than 3 acres of land and grow crops for basic subsistence since their yearly output could only cater to their family needs. This includes farmers in Tingoli, Duko, Tibali, and Cheyohi of the Northern Region; Goli, Goriyiri, Zanko, and Guo in the Upper West Region; and Gia, Bonia, Nyangua, and Samboligo in the Upper East Region (map of study area is shown in Fig. 3.2). While selecting 4 farming communities in the aforementioned regions, the entire data collection exercise lasted for 10 days covering 216 farmers. It must be noted that the Africa RISING program offers support for farmers of both sexes, and though the study captured data from both sexes, emphasis was also placed on examining the efficacy of farming technologies on women's livelihood since asset acquisition is deemed critical to people's ability to escape poverty (Fig. 3.3).

Fig. 3.1 A training session on the main platform of the project, Kobo Toolbox is given in the laboratory

Fig. 3.2 A field practical session with demonstration of Kobo Toolbox launches the project in the study area

While employing Kobo Toolbox as the primary platform for data collection, the survey instrument was further divided into different themes, which reflect different focus areas for this project. Aside from the farmers' biodata, other themes on the instrument ranged from farmers' previous practices, which formed the basis to compare the efficacy of IITA's interven-tion on farmers crop yield and livelihood. Other themes include financial literacy and alternative support livelihood programs for farmers.

Having completed an engagement with a respondent (farmer), OSMtracker was employed to demarcate the boundary of their farm. This was deemed critical since most farmers did not have an objective boundary of their farmland,

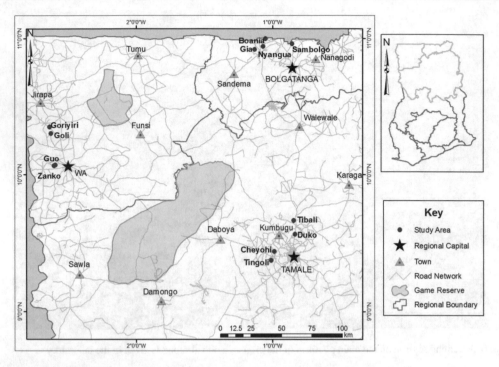

Fig. 3.3 Study areas in Northern Ghana are shown. (Credit: UCC YouthMappers)

hence their difficulty in projecting crop yield and estimating the value of their lands. At the time of the data collection, tree stumps, footpaths, stones, wooden pegs, and soil mounds were among some of the local tools that were used to separate one farmland from another. While correcting this anomaly, android phones with OSMtracker were offered to farmers while they walked round their farmlands. Figure 3.5 portrays an accompanied walk with a farmer who used OSMtracker to map the boundaries of his land.

The data collected from the field was readily available for analysis since the team relied on Kobo Toolbox. Before data analysis, data cleaning was first executed by downloading the project data in SPSS file format. This was done to eliminate or correct incomplete entries and ensure consistency in output. After the data cleaning, IBM SPSS version 21 was used for the analysis. The first analysis focused on the demographic characteristics of the respondents, while the rest of the analysis centered on the objective of the study, which employed both descriptive and inferential statistical tools to make sense of the data.

Having cited the use of the aforementioned open-source platforms, none of these 12 communities appeared to be on any map, and in this regard, the team resorted to OpenStreetMap (OSM) as a way to increase the visibility of these farming communities to the world. This situation resulted in the use of Java OpenStreetMap editor (JOSM) and OSMand, which were employed to remotely map and collect field data, respectively. The full extent of this mapping exercise is still available on OpenStreetMap and can be downloaded for any geospatial activity (Figs. 3.4, 3.5, 3.6, and 3.7).

4 Challenges from the Field

In carrying out this project, the following constraints were encountered: While all but one project member resided in Southern Ghana, a major challenge encountered in the field was language barrier. This was an issue as the major languages spoken in Southern Ghana differed significantly from the dialects spoken in the northern part. In fact, almost all of the local farmers could not

Fig. 3.4 The author speaks with male farming elders in one of the study communities

Fig. 3.5 UCC YouthMappers interact with a famer for field mapping of farm boundaries

speak or read English or any local languages spoken by the UCC YouthMappers team. To remedy this situation, the team engaged the services of translators who were familiar with the local terrain. One challenge envisaged here was the tendency of these translators to properly offer the needed service, and to this view, the team offered suitable incentives and also took time to rehearse the survey a number of times. Upon obtaining satisfactory feedback from the translators, we proceeded to the field.

The high temperatures associated with dry and dusty winds from the Sahara Desert were dominant in this area, and unlike the south, the air in the north felt thinner and warmer in the day. With desert winds, daytime temperatures were as high as 35°. This saw most team members developing cracked lips, dried nostrils, and blisters on their feet as a result of wearing boots for longer hours in the scorching heat. While enduring these conditions, both the IITA staff and the local farmers seemed fine as they went about their duties

Fig. 3.6 Female farmers gather with the mapping volunteers during a break in training

Fig. 3.7 YouthMappers accompany a farmer along a walk to deploy OSMtracker to map the boundaries of his land

without any constraints. To shield ourselves from the effect of the weather, we resorted to the use of local remedies like the shea butter, which offered a great deal of relief and insulated our lips and nostrils from the harsh weather.

The bad nature of roads and long routes to farms also posed some challenges to us. Though the team members lodged in the respective regional capitals, which had relatively good roads, the roads leading to the various farming communities were not as good as the ones encountered in the regional capitals. In some cases, the team had to endure long travelling hours, which made trips very exhausting and unpleasant. Aside from the long distance to the communities, the use of OSMtracker to map farmlands also came with some discomfort. This was typically so as most farmers lived far from their farmlands. When the team asked the farmers to estimate the distance to their farms, common responses included "the farm is not far" or "the farm is just here." Unfortunately, "just here" was found to be about a 45-minute walk from the community, and due to exhaustion, team members had to resort to a motorbike. Even with the motorbike, a team member endured about a three-and-a-half-hour ride to complete mapping of all farmlands in a community. All of these completed on a day when temperatures were between 30 and 35 °C.

Finally, with the IITA staff being our local representatives in the three Northern Regions, the teams also encountered low turnouts in some isolated cases. This stems from late communication to farmers, and in some cases, farmers showed up late although information about the exercise had been received earlier. Also, some farmers had travelled outside the communities to attend to other commitments and were not reached. During moments of low turnout, the team took time to tour the communities since life in Northern Ghana was different from what we knew in the South. This provided the opportunity to learn more about local culture as the team interacted and perfected some phrases in their local dialect. On occasions where farmers had travelled outside the community, necessary transport was provided to get in touch with these farmers though this situation prolonged the time for engagements. For those who could not be reached, attempts were made to reach out to them after the survey period.

5 Findings from Our Work

From the 12 communities engaged, a total of 182 (which translates to an 86% response rate) smallholder farmers were found to be under the Africa

Table 3.1 Percentage of farmers reached, by community, gender, education level

Variable and category	Percent
Community	
Bonia	2.2
Cheyohi	2.7
Cheyohi No. 2	4.9
Goriyiri	7.7
Zanko	7.7
Guo	7.7
Samboligo	8.2
Gia	8.8
Tingoli	8.8
Duko	9.3
Tibali	9.3
Goli	10.4
Nyangua	12.1
Gender	
Female	48.8
Male	51.6
Level of education	
No education	72.0
Basic education	15.4
Junior high school	5.5
O/A level	2.2

RISING program. It is worth mentioning that the number of farmers found in each community was also subject to the population of local farmers in each community and farmers' ability to meet the IITA criteria for inclusion into the Africa RISING program. As seen in Table 3.1, Nyangua in the Upper East had the highest population (12%) for local farmers who were reached for this study. On the other side, Bonia in the Upper East recorded the least number of farmers (2.2%), which may be attributed to the relatively low population in this community.

The gender distribution of smallholder farmers revealed that there were 51.6% males and 48.2% females. This shows a relatively low disparity among males and females, although the land tenure system in Northern Ghana is entrusted in the hands of males, which perpetuates a patriarchal dominance and also limits women's access to land for farming or any other purpose. It can be said that IITA's Africa RISING program in Northern Ghana promoted gender equality, which would trickle down to ensuring food security, thus ensuring equal access to their program for both males and females. In terms of respondents' level of education, 72% had no formal education, and this finding is in a tandem with the Ghana Statistical Service, which stipulates that 44–54% of the people in the North have never been to school before despite a strong belief in the decline of this statistic given government policy to make basic education free. On the back of this, the remaining 28% of respondents appeared to have some level of education, but this does not include any tertiary education.

Obtaining spatial information of farmlands was regarded as important for this study. On that note, all farmlands enrolled in the Africa RISING program were mapped out and presented on a regional basis as seen in Figs. 3.8 (Northern Regions), 3.9 (Upper East), and 3.10 (Upper West). We expected that farmers can use this data to make precise projections about their activities.

6   Assessing Results Relative to Other Interventions

This section presents findings on existing programs enjoyed by smallholder farmers under the Africa RISING program. This was deemed important since it offers a fair assessment of the farming intervention programs, which may also serve as a basis to compare its efficacy on farmers' livelihoods. Having presented a catalogue of interventions to farmers, the results showed that the majority of the farmers practiced the maize leaf stripping technology (52.7%). This was based on the relatively shorter gestation period of maize, being a staple food and always in high demand. On the other hand, cowpea living mulch was identified to be the second most practiced intervention (36.8%). This intervention was practiced since it provides nitrogen for other crops such as maize (Table 3.2).

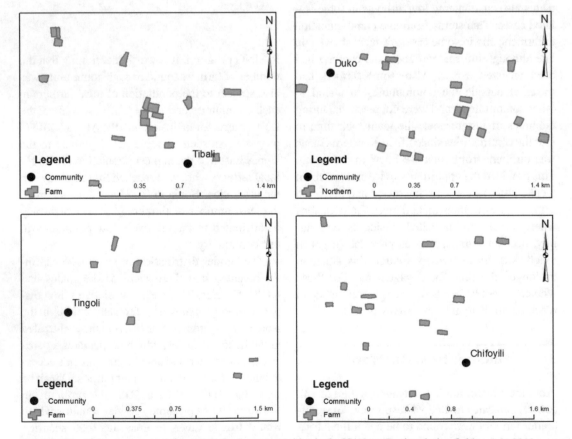

Fig. 3.8 YouthMappers map out farmlands for the communities in the Northern Region during fieldwork in 2019

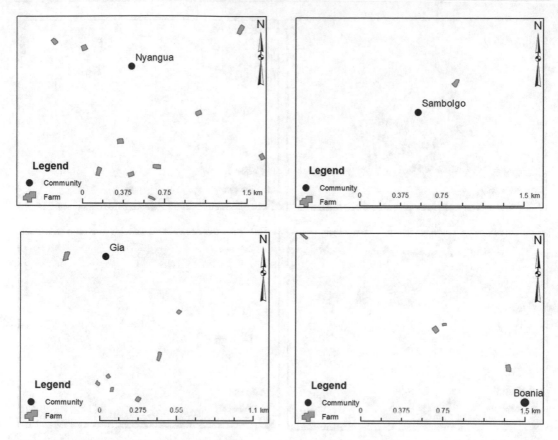

Fig. 3.9 YouthMappers map out farmlands for the communities in the Upper East Region during fieldwork in 2019

Having identified the existing programs practiced by the farmers, the perspectives of farmers were sought on which of the Africa RISING program they preferred. It was found that the maize leaf stripping (72.41%) seemed to be the obvious choice of most farmers. The reasons for this selection were no different from the ones enumerated above. Aside from maize, cowpea (68.97%) and groundnut (34.48%) also ranked second after maize (Table 3.3).

Having identified farmers' current crop program, the study further sought to identify the farm management practices executed by farmers to increase their crop yield. Maize leaf stripping was the most preferred (42.9%) farm management practice. In simple terms, maize stripping involves the removal of lower leaves from the maize plant at anthesis or post-anthesis. In fact, this is done to increase radiant energy penetration to the understory crop and also boost crop output. When farmers also engage in the rearing of rumi-

nates, the strips from the maize also serve as feed to the animals. Also, other leading management practices include cowpea living mulch (24.7%) and cowpea living mulch, maize leaf stripping (11%) (Table 3.4).

To assess the impact of the program, the yields of the smallholder farmers were measured (refer to Table 3.5). It was observed that there had been a significant increase in the number of smallholder farmers who cultivated maize as other farmers who preferred cultivating other crops changed and aspired to enjoy the numerous benefits associated with maize cultivation. From this data, this appeared to be the presence of female farmers who also considered the short gestation period for maize as an avenue to increase household food supply and sell excess for income. With such surpluses, female farmers may have enough to procure some necessities of life, and this was evident when most women did not own mobile phones, which could cost as low as $5.

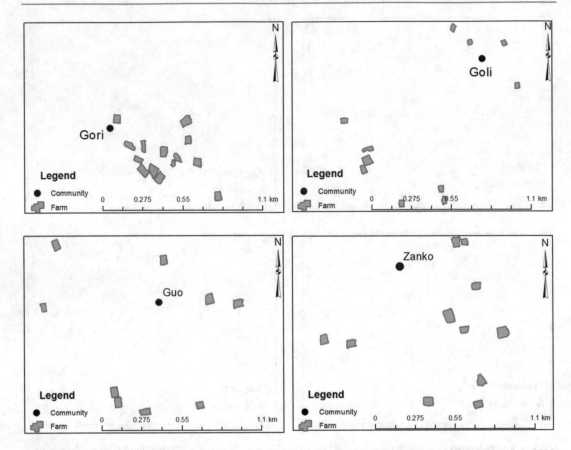

Fig. 3.10 YouthMappers map out farmlands for the communities in the Upper West Region during fieldwork in 2019

Table 3.2 Intervention program practiced

Name	Frequency	Percent
Cowpea living mulch	67	36.8
Groundnut spacing	19	10.4
Maize leaf stripping	96	52.7
Total	182	100.0

Table 3.3 Preferred program (among the previous intervention)

Intervention	Frequency	Percentage
Maize leaf stripping	42	72.41
Cowpea living mulch	40	68.97
Fertilizer trials	20	34.48
Groundnut spacing	20	34.48
Strip cropping	20	34.48
Livestock feed improvement & livestock housing	10	17.24
Vegetable maize intercropping	6	12.07
Integrated soil fertility management	9	15.52
Cowpea spraying regime	17	29.31
Nutrition intervention among women	5	8.62
Irrigation	7	12.07
Natural resource management	3	5.17

Table 3.4 Management practices

Variable	Frequency	Percent
No response	3	1.6
Cowpea living mulch	45	24.7
Cowpea living mulch, groundnut spacing	11	6.0
Cowpea living mulch, groundnut spacing, maize leaf stripping	10	5.5
Cowpea living mulch, maize leaf stripping	20	11.0
Fertilizer trials (nitrogen rate on maize), groundnut spacing, maize leaf stripping, livestock feed improvement, & livestock housing	1	.5
Fertilizer trials (nitrogen rate on maize), livestock feed improvement, & livestock housing	1	.5
Groundnut spacing	8	4.4
Groundnut spacing, maize leaf stripping	5	2.7
Maize leaf stripping	78	42.9
Total	182	100.0

Table 3.5 Yield of smallholder farmers, by number of bags

	N	Minimum	Maximum	Mean	Std. deviation
What was your yield last year (maize)?	83	.00	8.00	2.8223	1.59435
What was your yield last year (cowpea)?	73	.0	14.0	2.167	2.4869
What was your yield last two years (maize)?	73	.0	10.0	2.397	1.9947
What was your yield last two years (cowpea)?	72	.0	10.0	1.882	2.2083

Indeed, with the absence of mobile phones, the only way to contact these women was through other family members or friends who owned mobile phones. This situation also posed a challenge for the IITA staff, in making access to information on weather and prevailing market prices of crops a difficult task, and all these further reinforced women's reliance on men.

7 SDGs and Lessons Learnt

As the world seeks to eradicate hunger (SDG 2) and poverty (SDG 1), organizations and partnerships like YouthMappers and IITA have equally played an active role in providing various support services to smallholder farmers in Northern Ghana. Despite reports of lower farm fields from various research across the country, these interventions from the IITA were found to significantly boost crop yield across the farming seasons. Indeed, the success stories were found to have a greater impact

Fig. 3.11 An elder woman farmer enjoys the training sessions with YouthMappers

on female farmers since higher farm yields implied some level of economic independence and a general improvement of food security, which influences malnourishment and hunger. It was expected that the IITA would offer more opportunities to increase support for female farmers in Northern Ghana (Figs. 3.11, 3.12, and 3.13).

Engaging YouthMappers students in this project offered us a rare opportunity to explore the breadbasket of our country, its people, and

Fig. 3.12 A woman farmer with her child attends the trainings

their culture. We gained firsthand account of what it takes to be a farmer, and more importantly, we understood the daily living situations or struggles of female smaller-scale farmers. While this project was limited to selected farmers, and to selected students, the potential to properly demarcate farm size using open tools offers women the opportunity to project farm yields and employ simple but cost-effective technology in making the best of their land. From this project, we appreciate the potential of the Kobo Toolbox in developing a database of all farmers under this program. Our contributions to this data and the findings thus offer a solid backbone to understand the unique needs of female farmers since this will be crucial in improving Ghana's food security and economically empower them, addressing both hunger and poverty through leveraging spatial technologies for agricultural intensification (Figs. 3.14, 3.15, 3.16, and 3.17).

Fig. 3.13 One of the study communities gathers under tree shade to commemorate trainings with a photo

Fig. 3.14 The author interviews farmer families in Northern Ghana

Fig. 3.15 YouthMappers team members intake and validate data using Kobo Toolbox

Fig. 3.16 Farming
families provide critical
data for the mapping
activities

Fig. 3.17 Women
farmers in Ghana
especially stand to gain
from policies that the
data can inform

Acknowledgments We are grateful for the support received from Chad Blevins, the entire YouthMappers community, and the Africa RISING project.

References

Clover J (2010) Food security in sub-Saharan Africa. Afr Security Rev 12(1):5–15. https://doi.org/10.1080/10246029.2003.9627566

Consultative Group on International Agricultural Research (2020) International Institute of Tropical Agriculture (IITA). https://www.cgiar.org/research/center/iita/. Cited 29 Jan 2022

Ghana Statistical Service (2012) 2010 population and housing census – summary reports of final results. Ghana

Hjelm and Dasori (2012) Comprehensive Food Security & Vulnerability Analysis - Focus on Northern Ghana, United Nations World Food Programme Headquarters, Rome, Italy.

Jellema A, Meijninger W, Addison C (2015) Global open data for agriculture and nutrition open data and smallholder food and nutritional security of GODAN (the Global Open Data for Agriculture and Nutrition initiative). www.facebook.com/CTApage. Cited 29 Jan 2022

Lohento K, Janssen S, Ørnemark C, Adieno D, Atz U (2017) Open data benefits for agriculture and nutrition

Gender data for gender equality ICT Update 84:24. http://ictupdate.cta.int/. Cited 29 Jan 2022

Ministry of Food and Agriculture (2007) Food and Agriculture Sector Development Policy (FASDEP II) (Issue August). https://www.fao.org/faolex/results/details/es/c/LEX-FAOC144957/. Cited 29 Jan 2022

Musker R, Schaap B (2018) Global Open Data in Agriculture and Nutrition (GODAN) initiative partner network analysis. F1000Research 7(47). https://doi.org/10.12688/f1000research.13044.1

Quaye W (2008) Food security situation in northern Ghana, coping strategies and related constraints. Afr J Agric Res 3(5):334–342

Tandoh-offin P (2019) Ghana and Global Development Agendas: the case of the Sustainable Development Goals. AJPSDG 2(1):49–69

UNICEF (2018) Global hunger continues to rise, new UN report says. https://www.unicef.org/rosa/press-releases/global-hunger-continues-rise-new-un-report-says. Cited 29 Jan 2022

Rural Household Food Insecurity and Child Malnutrition in Northern Ghana

4

Kwaku Antwi, Conrad Lyford, and Patricia Solís

Abstract

Close to 750 million or nearly one in ten people in the world are exposed to severe levels of food insecurity, and 2 billion people do not have regular access to safe, nutritious, and sufficient food. A critical but overlooked question is how to properly locate the most food insecure and malnourished households within geographical areas that have been generally identified as food insecure and malnourished – especially where such areas are poorly mapped. The YouthMappers approach served as a baseline to inform our spatial analysis of food insecurity and underpinned household surveys on malnutrition. The resulting maps paint a powerful picture of the spatial variation of factors affecting different places, knowledge which can better inform interventions that are tailored to households in need and ultimately help to meet the goals of SDG 2, Zero Hunger.

K. Antwi (✉)
Regional Department of Food and Agriculture, Northern Region, Ghana

C. Lyford
Department of Agricultural and Applied Economics, Texas Tech University, Lubbock, TX, USA
e-mail: conrad.lyford@ttu.edu

P. Solís
Knowledge Exchange for Resilience, Arizona State University, Tempe, AZ, USA
e-mail: patricia.solis@asu.edu

Keywords

Food insecurity · Rural development · Spatial analysis · Ghana · Nutrition · Hunger

1 What We Need to Know About Hunger

Food insecurity and malnutrition continue to persist as relevant development challenges due to the growing number of the world's population that suffer from these two undesired conditions. Despite the reduction in the prevalence of hunger worldwide, pockets of places persist where increases in hunger levels have been observed. Currently, the realization of the 2030 targets (SDG 2) by many African countries seems unattainable because many developing countries have been thrown out of track to achieving zero hunger in the next decade. If recent trends continue, the number of people affected by hunger would surpass 840 million by 2030. Food production and consumption are not aligned with where hunger

P. Solís, M. Zeballos (eds.), *Open Mapping towards Sustainable Development Goals*, Sustainable Development Goals Series, https://doi.org/10.1007/978-3-031-05182-1_4

is happening, implicating SDG 12 as well. Consequently, an increase in food insecurity and hunger are directly linked with rising incidence of malnutrition levels in sub-Saharan African countries including Ghana where incidence of malnutrition is currently above 20% (FAO 2020).

A lot of research has been undertaken to address the problem of food insecurity and malnutrition, particularly among households in Africa and in Ghana. Most of these studies have dwelt on the determinants of household food insecurity and malnutrition, and data has been analyzed using econometric models, accompanied by recommendations to inform policy decisions. However, the missing middle in most of these research studies has been how to properly locate the most food insecure and malnourished households in geographical areas that have been generally identified as food insecure and having malnourished populations.

2   Seeking Answers with YouthMappers

The question of "where" inspired me to seek answers through my doctoral dissertation, which was focused on agricultural economics, but I had very little background in spatial science. Through my mentors, I got introduced and became part of the YouthMappers Chapter of Texas Tech University, one of the co-founding universities of the network. Serendipitously, the University of Cape Coast in Ghana, my home country, became the first international inaugural chapter to join that year.

Through this experience, I also met members of other YouthMappers chapters in many countries including Kenya, India, and Columbia. That summer, we visited the YouthMappers chapter at the University of Cape Coast in Ghana where we enlisted local support of geographers and others for mapping and input to the research. I then learned more about mapping and spatial analysis, taking a service-learning course that falls to better understand not only how to create

but also how to use the open spatial data from OpenStreetMap (OSM).

2.1   Locating Food Insecure and Malnourished Regions

In this research, we explore how a combination of both econometric analysis and spatial technologies brings out the status of household food insecurity and malnutrition in some selected communities in the northern region of Ghana. The study communities are found in the Central Gonja (8°57′N 0°13′W), East Mamprusi (10°26′N 0°37′W), Gushiegu (9°55′N 0°13′W), Mion (9°44′N 0°00′W), Tolon (9°26′N 1°4′W), and Zabzugu (9°17′N 0°22′E) Districts in the northern region. According to the Ghana Ministry of Food and Agriculture (MoFA), it is estimated that about 70% of the population in these communities live in rural communities and their main livelihood activity is agriculture, which mainly involves the production of crops and livestock.

2.2   Laying Baseline Data in Poorly Mapped Food Insecure Regions

Using OSM procedures, my OSM study group members and I at Texas Tech University were able to map out some agricultural-related infrastructure such as roads, houses in farming communities, and water bodies in the northern region of Ghana. I used this approach during my dissertation research to provide geographical location to the households that were found to be food insecure and malnourished in the northern region of Ghana. Figures 4.1 and 4.2 demonstrate the extensive mapping performed with the work of many mappers from Texas, Ghana, and beyond, depicting food insecure, poorly mapped Northern Ghana before and after this work, and the close example of mapping done in rural Gushiegu, respectively. The study presented

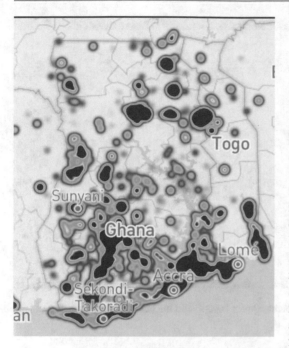

Fig. 4.1 A heat map of new edits demonstrates the extensive contributions of YouthMappers to OSM in Ghana

Fig. 4.2 Prior to this collaborative project, the extent of edits on OSM from YouthMappers throughout the country of Ghana were limited

here takes that baseline information as a starting point to locate the important factors for this particular region (Fig. 4.3).

3   Building the Case to Measure and Map Hunger

TheUnited States Department of Agriculture (USDA) defines household food as having the means to always access enough food by all members for an active, healthy life (USDA 2006). The Food and Agriculture Organization (FOA) also defines food security as a state that exists when all people, at all times, have physical and economic access to sufficient, safe, and nutritious food that meet their dietary needs and food preferences for an active and healthy life (World Food Summit, 1996). The achievement of food security among households, particularly those in rural communities, directly contributes to the attainment of SDG 2 and indirectly to SDG 12. It is noteworthy that across many regions, the overall prevalence of hunger and food insecurity has fallen since the millennium from 14.8% in 2000 to 10.8% in 2018. Despite these reductions in the prevalence of hunger worldwide, increases in hunger levels in particular areas remain high or have increased, where currently the realization of the 2030 targets (SDG 2) by many African countries seems far from being reached.

3.1   Tracking Hunger Through the SDGs

The world is not on track to achieve zero hunger by 2030. If recent trends continue, the number of people affected by hunger would surpass 840 million by 2030. The patterns of food insecurity and hunger are directly linked with rising incidence of malnutrition levels in sub-Saharan African countries, including Ghana, where incidence of malnutrition is currently above 20% (FAO 2020).

Malnutrition or malnourishment is a condition that results from continuous eating of diet in which nutrients are either not enough or are too much, which causes health problems and manifested in conditions such as stunting and obesity, particularly among children. The majority of the world's undernourished 381 million are still found in Asia, while more than 250 million live in Africa, where the number of under-

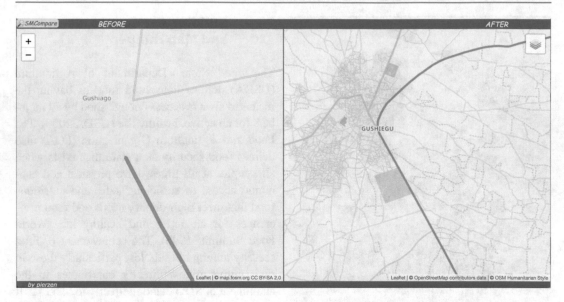

Fig. 4.3 One example of the detailed contributions of YouthMappers is visible in the before image of Gushiago [sic] (left) compared to Gushiegu (right)

nourished is growing faster than anywhere in the world. In 2019, close to 750 million or nearly one in ten people in the world were exposed to severe levels of food insecurity. An estimated 2 billion people in the world did not have regular access to safe, nutritious, and sufficient food in 2019.

According to the World Health Organization (WHO), lack of access to highly nutritious foods is the main cause of malnutrition, a condition that is estimated to contribute to 45% of child deaths. The WHO estimates that 162 million children under 5 years around the world are identified as stunted, 51 million are underweight, while 44 million are overweight or obese because of poor feeding. According to the Ghana Health and Demographic Survey (GHDS), in Northern Ghana, 33% of children are stunted, 11% are underweight, and 6.3% are wasted as compared to the national averages of 19%, 11%, and 5% for stunting, underweight, and wasting, respectively.

Poor nutrition has implications for a child's development, since a lack of adequate calories and nutrients to sustain normal growth puts chil-

dren at a greater risk of being vulnerable to diseases and has adverse effects on their physical, cognitive, and mental development. There has been a global effort to achieve food security and improved nutrition, particularly among rural households in developing countries. Despite these efforts, rural household food insecurity and undernourishment have been major developmental challenges to governments of many developing countries including Ghana. In Ghana, the northern region has been identified as the home of many of Ghana's food insecure and malnourished households, and research has shown that more than 70% of rural households in the northern region of Ghana are food insecure and malnourished.

3.2 Current Indicators of Causes of Food Insecurity and Malnutrition

The measurement of household food insecurity status has been done over the years by food security experts using different indicators.

Commonly used indicators of household food insecurity status include the food consumption score (FCS), household dietary diversity score (HDDS), coping strategies index (CSI), household hunger scale (HHS), and household food insecurity and access scale (HFIAS). The HDDS and the FCS are metrics that assess the number of different types of food or food groups that individuals or households consume and the frequency at which they are eaten. The idea here is that the more diverse a meal is, the more likely it is to contain essential calories and nutrients. The difference between HDDS and FCS has been found to be the reference time. While the HDDS uses a 24-h recall for the measurement of food security, the FCS uses 7-day recall for food security measurement. The CSI is an indirect measure of food security and measures the consumption behaviors of households while assessing the frequency and severity of households' behaviors during the period they do not have enough food or enough resources to purchase food. The HFIAS and the HHS are measures that assess households' behaviors that show a lack of sufficient and quality food and anxiety over future insecure access to food.

The body mass index (BMI), which is defined as the ratio of the weight (kg) and the square of the height (m), BMI = weight (m)/height^2 (m), has been used by the WHO and other health organizations as the main anthropometric measure of humans. This anthropometric measure is used to indirectly assess the nutritional status of children between the ages of two years and five years. It is used as an indicator of child malnutrition within the sampled households. Research has shown that the assessment of the nutritional status of any population using anthropometric data is the most popular indirect approach since the BMI is mainly sensitive to changes in the food security situation of a household. Furthermore, it is less subject to systematic measurement errors, it can be disaggregated to provide individual-level information, and it is well

suitable for monitoring and evaluating program interventions.

The weight-for-height, height-for-age, and weight-for-age outcomes when determined are compared with the World Health Organization (WHO) standards to determine the level of *wasting*, *stunting*, and *underweight* in the sample. The level of wasting within a population provides an indication of acute malnutrition, stunting provides information on chronic malnutrition, and the level of underweight provides indication of both acute and chronic malnutrition, and the three indicators provided information on the prevalence of malnourishment in the population.

3.3 Going Beyond Goals and Indicators Through the Map

Proper identification and targeting of malnourished and food-insecure households in any population are an important step in solving the problem of malnutrition and food insecurity in order to achieve SDG 2. But there is the need to go beyond identifying the causes of food insecurity and malnutrition to properly identify the physical locations and distribution/spread of the food insecure and undernourished households to effectively target them with intervention strategies that make the most sense for the context where they are happening.

This study explores and demonstrates the need to add a spatial dimension to complement the already-existing quantitative and econometric analysis to efficiently tackle the problem of food insecurity and undernutrition. The development and presentation of a spatial visualization of food insecurity and undernourishment are vital to providing insights including how these phenomena are geographically distributed, which otherwise would have been missed by policymakers, development partners, and implementers of food security intervention programs without spatial visualization of these patterns. Spatial visualization of food insecurity and undernourishment using Geographical Information System (GIS) technology is important for the advancement and sustainability of food production and distribution.

Spatial patterns of food insecurity and malnutrition have been widely used in numerous policy and research applications ranging from targeting emergency food aid and food security intervention programs to the assessment of causes of food insecurity and malnourishment (Davies 2003). In addition, analyzing food insecurity and undernourishment using spatial visualization provides information that can be more accessible to a wider range of users. A combination of food insecurity estimates and GIS tools to analyze food insecurity and undernourishment is a way of displaying implicit information that may not be apparent from conventional statistical tables.

4 Assessing Household Food Security Status

In this study, we used the household dietary diversity score (HDDS) to measure rural households' food security status. We used this indicator because the HDDS provides us with a direct outcome food security indicator, which is related to food and nutritional security. In determining the household dietary diversity score, we categorized the food items that households responded to have eaten over the last 24 h into ten food categories including cereals and grains; roots and tubers; vegetables; fruits; meat; eggs; fish and seafood; legumes, nuts, and seeds; milk and milk products; and oils and fats. We scored one for a particular food group if members of that household had consumed any food item belonging to that food category. We scored 0 for a particular category if members of that household had not consumed any food item belonging to that food category during the past 24 h. In this study, we categorized the sampled households into *food-secure households* and *food-insecure households* based on the household dietary diversity score obtained by the household.

We used a multistage sampling approach, which included random stratified sampling combined with probability proportionate to size procedure to select households that were used for data collection. At the first stage of sampling, we randomly sampled six out of the ten districts for the data collection. At the final stage, we used the probability proportionate to size procedure to select representative households in each of the district. We used this procedure to select 504 households from the selected communities for the data collection. We collected the data through personal interviews using a semi-structured questionnaire in two periods. We did that to ensure that seasonal variations in household food consumption did not influence data analysis and interpretation. The head of the household and/or partner and the person responsible for preparing meals for the household were the main respondents to the survey questions.

We collected household demographic data and data on households' meal consumption a day before the household was interviewed. Furthermore, we collected data on the weight and height of children between the ages of two years and five years. Anthropometric measurements on everyone including height and weight were repeatedly measured, and the average values were used for the data analysis. In situations where a household had more than one eligible child, following the WHO guidelines and standards, the average or mean anthropometric Z-scores were calculated and used during the analysis. To ensure that the human rights of the respondents were protected and respected during the data collection, the entire study was reviewed and approved by the Institutional Review Board of Texas Tech University, Lubbock, Texas, with protocol number IRB2017-646. Per the guidelines established, we sought respondents' consent by getting approval from respondents after we read the respondents' consent forms to them.

4.1   Determining Malnutrition Status

We used the anthropometric measurements of children between the ages of two and five years as an indicator of the undernourishment status of children of sampled households. Anthropometric indicators are widely used to understand the demographic dynamics of nutritional status within a household, particularly of mothers and infants as they are among the most vulnerable in society. We derived the Z-score for weight-for-age (WAZ), height-for-age (HAZ), and weight-for-height (WHZ) for the sampled children using the 2006 WHO growth standard with ENA software. We used the Z-scores we obtained to determine the prevalence of underweight, stunting, and wasting among the sampled households. In this study, we classified stunting prevalence rate of <20 as low, 20–29 as medium, 30–39 as high, and > = 40 as very high. Similarly, we classified underweight prevalence rate of <10 as low, 10–19 as medium, 20–29 as high, and > = 30 as very high. Similarly, we classified wasting prevalence rate of <5 as low, 5–9 as medium, 10–19 as high, and > = 20 as very high.

4.2   Data Analysis

Based on the household food insecurity estimates (HDDS), we grouped all households into 2 categories including *food secure* and *food insecure* households. We made this categorization because HDDS, unlike the other indicators of household food security status, defines a household as being either food insecure or food secure. We further categorized the dependent variable (HDDS) as 0 for any household that was found to be food insecure and 1 for households that were categorized as food secure. The relationship between the categorical variables including sex of household head, marital status of household head, whether the household owns livestock, whether the household has access to land, and whether the household has other farms, and food insecurity status was determined. We employed the chi-square test to determine the level of significance of the relationship between household food security status and associated categorical variables. We visualized the spatial distribution of food insecurity and malnutrition using ArcGIS 10.1. The household points regarding latitude and longitude that we obtained during the data collection and information of the central point of each household were created. Food insecurity and malnutrition (stunting, wasting, and underweight) were matched to the point of the household.

4.3   Findings

The overall prevalence of food insecurity in the study area was 82%, which signifies that more than half of all the households in the study area were found to be food insecure (Table 4.1).

Figure 4.4 shows the spatial distribution of food insecurity among households in the selected communities where we collected the data.

We found the overall prevalence of stunting, wasting, and underweight among the children is 30.2%, 5.8%, and 21.1%, respectively, which is an indication of high prevalence rate of child malnutrition in the study (Table 4.2).

The results further show that the sex of the household head, household access to land, livestock ownership status of the household, years of formal education of household head, and household size were found as significant variables associated with household food security status in the study area. The results in Table 4.3 show that we found a positive significant relationship between household access to agricultural land and household food security ($p = 0.043$)

Table 4.1 Prevalence of food insecurity

Degree of food insecurity (HDDS)	Frequency	Percentage
Food secure (HDDS > 6)	91	19.1
Food insecure (HDDS < 6)	413	81.9

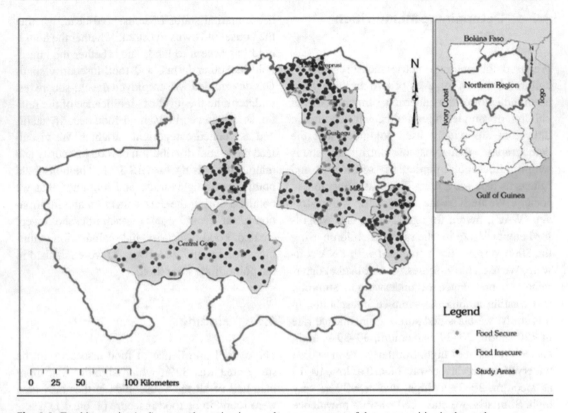

Fig. 4.4 Food insecurity (red) across the region is prevalent across most of the communities in the study

Table 4.2 Degree of malnutrition across the study area

Prevalence of malnutrition	Frequency	Percentage
Stunting (%)		
<20	78	16.6
20–29	142	30.2
30–39	169	35.9
> = 40	82	17.3
Underweight (%)		
<10	33	7.1
10–19	172	36.5
20–29	192	40.7
> = 30	74	15.7
Wasting (%)		
<5	67	14.2
5–9	215	45.6
10–19	121	25.8
> = 20	68	14.4

suggesting that households with enhanced access to agricultural lands are more likely to be food secure. Further, the results show that households that own livestock are more likely to be food

secure ($p = 0.031$). The chi-square test results further show significant relationship between household food security status and stunting, but there is no significant relation between household food security status and underweight and wasting.

The food insecurity map (Fig. 4.4) shows that households in the study area are generally not food secure, which can easily be seen from the map without any complex statistical interpretations. Though the situation is generally similar across the study area, the map shows that food insecurity varies from one geographical area to another. Hence, planning for interventions using aggregated data at the district level may reduce the effectiveness of such programs. Spatial analysis of food insecurity can improve food insecurity intervention coverage effectiveness through the identification of specific geographical locations that needs assistance. The map generated can provide valuable

Table 4.3 Household-level categorical variables stratified by food security status

Variable	Food secure (%)	Food insecure (%)	χ^{2*}p
Sex			0.037**
Male	35.3	64.7	
Female	28.1	71.9	
Marital status			0.726
Married	13.3	86.7	
Not married	0	100	
Land access			0.043**
Access	6.2	93.8	
No access	18.5	81.5	
Livestock ownership			0.031**
Owned	25.3	74.7	
No livestock	30.8	69.2	
Other farms			0.614
Yes	24.7	75.3	
No	22.7	77.3	
Stunting (%)			0.043**
<20 [1]	34.6	65.4	
20–29 [11]	28.2	7.18	
30–39 [111]	33.1	66.9	
> = 40 [1111]	21.9	78.1	
Underweight (%)			0.216
<10 [1]	27.3	72.7	
10–19 [11]	39.5	60.5	
20–29 [111]	36.0	64.0	
> = 30 [1111]	20.3	79.7	
Wasting (%)			0.178
<5 [1]	31.3	68.7	
5–9 [11]	30.2	69.8	
10–19 [111]	24.8	75.2	
> = 20 [1111]	16.2	83.8	

χ^{2*}p is the calculated chi-square values, ** with statistical significance
[1] = Low, [11] = medium, [111] = high, [1111] = very high, per categorization of public health significance (WHO 1995)

information about spatial disparity of household food insecurity that may be relevant to policymakers, development partners, and institutions that work to reduce the incidence of food insecurity among households. Similarly, the level of stunting, underweight, and wasting is high among the East Mamprusi, Tolon, and Zabzugu Districts. This is because household food insecurity has been found to be highly related to the nutritional status of household members.

5   Conclusion

We found that the overall prevalence of household food insecurity and malnutrition is high in places that matter. The results of the analysis show that rural households' food security status in the northern region of Ghana is negatively influenced by contextual factors including household size. The negative effect of these factors on households in the short term is not in doubt, and in the long run, the life and survival of rural households could be threatened if measures are not put in place to curb the negative effects of these factors.

We further found that spatial variations of food insecurity and malnutrition exist in the study area. Designing food insecurity intervention programs and plans using regional-level evidence and government administration units might mask the true picture of spatial distribution of the problem in local context as shown by the district variation in the food insecurity status revealed by this study. It is, thus, important that program-level planning should consider district-based microlevel variation in allocating resources for intervention to address food insecurity.

A next step in this type of work would be to actively utilize spatial mapping in food security and malnutrition research, refining the methods we innovated here, based upon what is most effective to generate new insights for the region in question. Further study on spatial distribution of food insecurity and malnutrition using time series data at a microlevel is recommended. This will bring out the temporal variation in spatial disparity of household food insecurity under different seasons that will be relevant for further research on the topic and an even greater refinement of policies that promise to reduce hunger and align production and consumption to the needs and demands.

Finally, additional research that treats food security as a continuum rather than the binary model of HDDS that is in wide use by researchers and practitioners would shed further light on additional questions that this work raises. Studying the varied spatial patterns over time would further shed light on the dynamics of food security in this region, both drawing from and building upon both the open spatial data created through this work. In the end, these results will provide more information to recommend how to best tailor food security policies to the varying needs of different households and communities across different districts, generating a missing link between global goals like SDG 2 and SDG 12 to local realities.

References

Davies B. (2003). Choosing a Method for Poverty Mapping. FAO-Rome, Italy

FAO (2020). The State of Food Security and Nutrition in the World 2020: Transforming Food Systems for Affordable Healthy Diets. Rome-Italy

USDA (2006). Household Food Security in the United States, 2006. Economic Research Report, 2006. United States Department of Agriculture

WHO (1995). WHO Global Database on Child Growth and Malnutrition. World Health Organization Geneva

World Food Summit (1996). Rome Declaration on World Food Security and World Food Summit Plan of Action: World Food Summit, 13–17 November 1996, Rome, Italy

Where Is the Closest Health Clinic? YouthMappers Map Their Communities Before and During the COVID-19 Pandemic

5

Adele Birkenes, Siennah Yang,
Benjamin Bachman, Stephanie Ingraldi,
and Ibrahima Sory Diallo

Abstract

YouthMappers chapters are both locally and globally situated, fostering a confluence of community input, GIS skill sets, subject-matter expertise, and creativity that drives progress toward the Sustainable Development Goals (SDGs). Examining Hudson Valley Mappers' and Gaston Berger University YouthMappers' community mapping projects in parallel offers rich insights on open mapping for SDG 3 Good Health and Well-being and highlights the value of SDG 4 Quality Education. Both YouthMappers chapters collaborated with local partners to map health facilities and other community resources; the resulting maps and data filled critical information gaps during the COVID-19 pandemic. The project outcomes underscore the unique niche that the YouthMappers network occupies in the open mapping world.

A. Birkenes (✉) · S. Yang · B. Bachman · S. Ingraldi
Vassar College, Poughkeepsie, NY, USA
e-mail: abirkenes@alum.vassar.edu; siyang@alum.
vassar.edu; bbachman21@alum.vassar.edu;
singraldi@alum.vassar.edu

I. S. Diallo
Gaston Berger University, Saint-Louis, Senegal
e-mail: diallo.ibrahima-sory@ugb.edu.sn

Keywords

Health · Community assets · Pandemic · GIS ·
New York · Senegal

1 Amid a Global Health Crisis

In the wake of the COVID-19 pandemic, access to reliable healthcare location data is more vital than ever. Both prior to and during the pandemic, YouthMappers chapters around the world strengthened their communities by providing residents and local stakeholders with accurate and up-to-date information on healthcare facilities and other resources. They designed these mapping projects with community input and in collaboration with local and international partner organizations. In doing so, YouthMappers chapters contributed to the third Sustainable Development Goal (SDG): Ensure healthy lives and promote well-being for all at all ages. This chapter demonstrates how open mapping efforts can "strengthen the capacity of all countries, in particular developing countries, for early warning, risk reduction and management of national

P. Solís, M. Zeballos (eds.), *Open Mapping towards Sustainable Development Goals*, Sustainable
Development Goals Series, https://doi.org/10.1007/978-3-031-05182-1_5

and global health risks" (SDG Target 3.d). Additionally, the chapter illustrates how quality education (SDG 4) equips university students with the knowledge and skills needed to promote sustainable development (SDG Target 4.7).

This chapter discusses two community resource mapping projects led by YouthMappers chapters at Vassar College in Poughkeepsie, New York, and Gaston Berger University in Saint-Louis, Senegal. At the onset of the COVID-19 pandemic in the spring of 2020, Hudson Valley Mappers at Vassar College created a shareable online web map of local resources, including healthcare facilities, COVID-19 testing sites, and food pantries, to help residents of Dutchess County navigate life during the pandemic. In 2019, Gaston Berger University YouthMappers added over 100 healthcare facilities in Saint-Louis to OpenStreetMap using the Healthsites OpenDataKit (ODK) collection tool. By the time the pandemic hit, this comprehensive dataset was already available in OpenStreetMap for public use. Both of these projects serve as models for how students can execute volunteer-led community Geographic Information Systems (GIS) projects in preparation for and during times of crisis. The success of the two initiatives is rooted in the chapters' commitment to engaging local stakeholders and community organizations and incorporating their needs into mapping processes and products.

2 Community Resources in Dutchess County, New York

Whereas some YouthMappers chapters in the United States may focus solely on global humanitarian and disaster relief efforts, Hudson Valley Mappers prides itself on our commitment to community geography and working with local organizations to help achieve their vision of community engagement and development. Our ability to build and maintain relationships with community partners in our home locations of Poughkeepsie and Dutchess County in New York

has been invaluable to our COVID-19 mapping efforts. Our community partners provided essential feedback throughout multiple stages of the project and helped disseminate the map to as many eyes as possible. With a combination of community input and dedication from a passionate student volunteer base, we were able to produce a tool that was immediately available to community members in a time of crisis.

2.1 Poughkeepsie and the Hudson Valley Mappers

Hudson Valley Mappers is a YouthMappers chapter at Vassar College in Poughkeepsie, New York. Poughkeepsie is located in Dutchess County in the Hudson Valley, midway between the state capital, Albany, and New York City. The city and town of Poughkeepsie have a total of 74,558 residents, and Dutchess County has 294,218 residents (American Community Survey 2019). The city of Poughkeepsie, which neighbors the town, shares a similar history with other smaller rustbelt cities that have an industrial past. In the 1940s, the arrival of IBM and other manufacturing and design firms brought an influx of jobs to the area, which led to a rise in housing and commercial development in the city (Flad and Griffen 2009). However, in the 1960s, the city experienced the effects of suburbanization and white flight, and major industries relocated outside of the city. Federal investment in urban renewal projects led to the construction of highways and arterials that cut through and demolished existing neighborhoods, severely impacting the quality of life of the residents. The processes of urban renewal, suburbanization, and deindustrialization spurred spatial and socio-economic changes that resulted in a higher concentration of lower-income communities of color and a cityscape of abandoned factories and buildings.

In the late twentieth century, the city introduced a myriad of strategies to revitalize

Poughkeepsie and enhance its reputation. The non-profit sector grew in response to glaring socio-economic inequality and the needs of its residents. As many cities in the nation transitioned from an industrial to a service economy, newer businesses in the service sector also came to Poughkeepsie. In recent years, Poughkeepsie has undergone rapid changes, such as the construction of new apartment buildings, the opening of new businesses and restaurants, and the expansion of anchor institutions.

However, revitalization and rising property values in the city have increased the risk of pricing out current residents. To address the deepening socio-economic inequality in the city, the city has instituted a variety of policy interventions and programs. For instance, an anti-blight task force and the Dutchess County and City of Poughkeepsie Land Bank were created to return vacant and abandoned properties in the community to productive use. The city updated its comprehensive plan in 2021; the plan will provide a long-range vision for future growth and development (pk4keeps.org). The city also has a robust network of community-based organizations that creatively address the legacies of urban renewal and the impacts of systemic racism and promote equity and enhance the quality of life for Poughkeepsie residents.

Vassar students, through coursework and internships, have served as a key volunteer labor force for these community organizations. At the helm of fostering local collaborations is the Office of Community-Engaged Learning (OCEL), which facilitates semester-long community-engaged learning (CEL) internships for students at various organizations in Poughkeepsie as part of their coursework. Vassar geography students have frequently participated in CEL internships focused on providing GIS support to local organizations. Due to the individualized nature of CEL placements (i.e., one to two students placed at an organization each semester), partner organizations lacked a way to formally engage large groups of students in projects involving extensive field data collection or longer-term analytical work. This type of volunteer effort is critical to the operations of local nonprofits that require up-to-date geospatial data on on-the-ground conditions but do not have the financial resources to hire paid GIS specialists. Thus, Vassar's YouthMappers chapter, Hudson Valley Mappers, was born out of a need to formalize and consolidate the volunteer geographical support that Vassar students provide to the college's community partners in the city of Poughkeepsie.

Hudson Valley Mappers was founded in September 2018 as a joint venture between the OCEL and the Department of Earth Science & Geography. The OCEL created a part-time community geographer work-study position to help ensure adequate resourcing for the chapter. In the year and a half between Hudson Valley Mappers' inception and the onset of the COVID-19 pandemic, the chapter conducted several sustained mapping projects in close collaboration with local nonprofits and the city of Poughkeepsie government. These organizations were part of the OCEL's existing network of community partners. The projects included mapping sites for the city government's tree plantings, businesses and vacant buildings along Main Street for the nonprofit Hudson River Housing, and waste disposal infrastructure as part of a partnership with the environmental education program No Child Left Inside. Consequently, by March 2020, Hudson Valley Mappers had already cultivated long-term relationships with local partners that contribute to community health and well-being in Poughkeepsie (Fig. 5.1).

2.2 Immediate GIS Response

At the start of COVID-19 in March 2020, Hudson Valley Mappers began exploring how Vassar students, faculty, and staff, as well as Poughkeepsie community members, might use their GIS interests and backgrounds to meaningfully engage in COVID-19 response efforts. While most collective mapping efforts related to COVID-19 were primarily dedicated to mapping COVID-19 cases or healthcare-related facilities, Hudson Valley Mappers decided to create a web map of community resources and services in Dutchess

Fig. 5.1 Hudson Valley Mappers identify tree planting sites in collaboration with the City of Poughkeepsie government and other community partners at their inaugural community mapping event in November 2018

County. There were numerous new community efforts and changes beginning in mid-March 2020, and various news sites and social media circulated information about the different types of community resources available in Dutchess County. It was becoming difficult to keep track of the services available, given that information was constantly updating and changing. Our YouthMappers chapter concluded that making a web map to consolidate the wealth of information about community resources that lived on disparate websites and social media could help residents easily visualize where resources were located and how to access them. We also viewed this project as a way for the Vassar community to continue to engage with the local community virtually when the college transitioned to remote learning.

Centering community priorities is a core practice of Hudson Valley Mappers and one that we carried into this project. Therefore, our first step was to solicit input from our community partners on what data they believed should be included in the map. We contacted nonprofit staff, city government officials, and school district administra-

tors, who provided lists of categories of resources to prioritize and raw data to represent spatially. Data included locations and hours of operation of food services (such as food pantries, meal programs, and school meal distribution sites) and health services (such as coronavirus testing facilities, hospitals, and pharmacies). We loaded the data layers in a prototype web map on ArcGIS Online and circulated it to our partners. They responded enthusiastically with their approval of the map and suggestions for improvement. At that point, with such strong support from our partners, we were confident that the tool filled a niche in terms of resources available to County residents at the onset of the pandemic; consequently, we decided to launch the project on a larger scale.

We created the COVID-19 Community Resources in Dutchess County Map, an interactive map using online software that enabled community partners, including those without prior experience with GIS, to update the map with information related to their areas of expertise. We accomplished this by linking the web map to Google Sheets spreadsheets so that the map automatically reflected all changes that community partners made to the spreadsheets. We also recruited a team of student and faculty volunteers who assisted with additional data input tasks to get the map up and running, such as transferring information on community resources from PDFs and webpages to the Google Sheets spreadsheets.

The final stage of the first phase of the project was distributing the map to residents of Dutchess County through the leveraging of a range of platforms, including social media, e-mail, and the city of Poughkeepsie's weekly e-newsletter. Community partners also included the map on their websites and in informational handouts.

After the map was distributed, several community partners and Vassar faculty members requested that we add additional layers to the map, such as restaurants offering delivery and curbside pickup and locations with free public Wi-Fi. Faculty and students from Bard College in Annandale-on-Hudson also collaborated with us to add locations of various community services in Northern Dutchess County. A team of students and faculty from Vassar and Bard, as well as a few community

partners, continued to update the map throughout the spring of 2020 (Figs. 5.2 and 5.3).

2.3 Longer-Term GIS Support to the Community

Updates to the COVID-19 Community Resources Map paused in the summer and fall months as cases significantly dropped and remained low in New York following the April 2020 spike. However, cases were on the rise again by November 2020. With renewed COVID-19 restrictions, Hudson Valley Mappers decided it was time to reopen the project. In January 2021, we updated each layer on the map using the most recent information available, most of which we gathered from Dutchess County's COVID-19 Response website. We prioritized food pantries and COVID-19 testing sites, and a core group of student volunteers tackled the restaurants.

On December 11, 2020, the FDA issued emergency use authorization for the first COVID-19 vaccine developed by Pfizer (Hinton 2021), and on January 13, 2021, New York opened up its

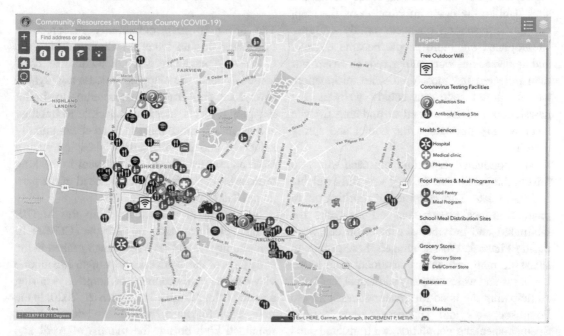

Fig. 5.2 City of Poughkeepsie community resources are shown in this interactive map at the end of the first phase of the project, June 2020

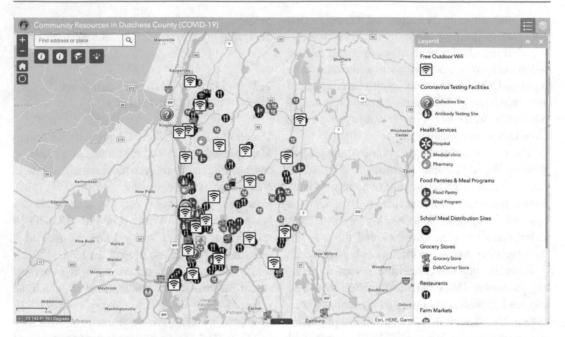

Fig. 5.3 Dutchess County community resources are shown in this interactive map at the end of the first phase of the project, June 2020

first statewide vaccination sites (Governor's Press Office 2021). Thus, in addition to updating existing layers, we created a new layer for vaccination sites. Unlike the other information on the map, vaccine information changed rapidly as more sites opened up, more people became eligible, and new vaccines were authorized. To ensure our map reflected the most up-to-date information, we monitored state and county websites and newsletters and committed to updating the site layer weekly throughout the winter and spring months.

We conducted a second outreach effort to inform the public that the map was still available and up-to-date. This consisted of reaching out to contacts in local government and nonprofits that facilitated and provided feedback on our work during Phase 1. We encouraged these groups to share the map with their constituents and member bases and welcomed any additional feedback. Anticipating the need for greater accessibility to Poughkeepsie's and Dutchess County's large Spanish-speaking population, we translated parts of the web map prior to the second outreach campaign.

2.4 Local Perspectives on Challenges and Opportunities

The effort to map community resources in Poughkeepsie and Dutchess County during the COVID-19 pandemic was not without its challenges. Indeed, the mapping process highlighted several key concerns that are common to similar mapping projects, including a lack of technology access among those most in need of community resources and the difficulty of keeping such a large dataset up-to-date and relevant in the long term, even when the heightened need of the pandemic had largely subsided.

According to usage data from the ArcGIS Online platform that hosted the COVID-19 Community Resources Map, usage peaked in the early days of the pandemic, with up to 296 unique views of the map occurring on some days during a period from March 25 to March 31, 2020. In the time that followed, viewership dropped but remained high during the months of April and May, then declined to a steady rate around three or four views per day. We suspect viewership was higher in the early days of the pandemic because people were struggling to adapt to the rapidly evolving COVID-19 restrictions and were in need of a reliable, centralized information source. This viewership pattern could have also indicated the need for a stronger outreach campaign in Phase 2 of the project. Early in the pandemic, heightened need, increased communication across community networks, and widespread uncertainty over how individuals, businesses, and service providers would respond to the pandemic created a niche in which the COVID-19 Community Resources Map was able to connect community members with resources amid rapidly changing public health measures. Later in the pandemic, this role became less clear. Organizations and businesses adapted to communicate pandemic response measures to their patrons. Vaccination providers, including pharmacies and the state government, developed their own platforms to assist the public in finding appointments. Student mappers could not guarantee the same level of timeliness and accuracy in the COVID-19 Community Resources Map.

One overarching challenge for community mapping projects is keeping datasets current as volunteer capacity changes over time. Student volunteers were not always able to keep up with rapidly changing information, meaning that some of the information in the map may have been inaccurate or out of date. The map included a field to indicate when information was last updated, but that required viewers to use discretion when interpreting the map and made it less appealing to government entities that were committed to providing the most accurate and up-to-date information at all times. We also learned that some local organizations were already offering services similar to the one provided by our map. For example, the United Way of Dutchess-Orange developed a 211 service in which users could call the number 211 to be connected with a number of services in the community.

Accessibility is a major concern for community mapping projects, and the COVID-19 Community Resources Map was no exception. First, even though the map was developed with the Spanish-speaking community in mind,

English was used for most of the map text, aside from key labels. Second, lack of access to technology to view the map is a limitation that became clear over time. Many of those with the highest need for community resources do not have access to computers/smartphones and Wi-Fi that make viewing the map possible in the first place. New approaches to improve accessibility should be developed through software capabilities and map design. One of Hudson Valley Mappers' main community partners, Hudson River Housing, informed us that it would be difficult to share the map with residents of their affordable housing sites due to lack of internet access and technology. Many residents receive information primarily through a hard copy newsletter. Even though a QR-code link could be included in the newsletter, those without a smartphone could not read the code or access the internet to reach the map website. Moreover, because the ArcGIS Online interface is optimized for computer browsers, smartphone users may not find the interactive map sufficiently intuitive and easy to browse.

In spite of these limitations, the COVID-19 Community Resources Map served as an unparalleled tool for connecting Dutchess County residents with local services, especially during the first few months of the pandemic. Due to our existing relationships with community partners and close monitoring of local news sources, our YouthMappers chapter was able to identify an information gap within days of the onset of pandemic restrictions and design a platform well before other groups like medical providers and government entities developed their own systems.

2.5 Challenges and Opportunities for Open Mapping from a Youth Perspective

While designing and creating the community resource map, Hudson Valley Mappers connected with other YouthMappers undertaking similar COVID-19 response mapping projects. We presented our interactive map at a YouthMappers network-wide webinar and published a blog post that detailed the steps of our project. We were also in close contact with María Fernanda Peña Valencia, a YouthMappers regional ambassador at the University of Antioquia in Medellin, Colombia. María Fernanda launched the Mappers4Med initiative to map community food and health resources available to residents of Medellín. Unlike our project, which used online Esri products, Mappers4Med's utilized OpenStreetMap, which better aligns with the open mapping principles of the YouthMappers community. Any member of the public can easily find, edit, and download the data that the Mappers4Med team added to OpenStreetMap, whereas with our project, members of the public would first have to know about our project and then contact us for editing access to the raw data (although it is downloadable from our web map) (Fig. 5.4).

Hudson Valley Mappers' decision to work with Esri software is a reflection of the students' and faculty members' expertise in proprietary GIS software. An open-source geospatial curriculum is far from the norm at US universities, and when we mapped in response to the crisis sce-nario of the COVID pandemic, we used the tools we knew best from coursework, even though they may not have facilitated maximum data sharing and transparency for the general public. At the same time, our approach did have advantages, such as allowing volunteers without geospatial expertise to update the map by editing our shared Google Sheets file.

Ultimately, Hudson Valley Mappers is com-mitted to providing Vassar students with expo-sure to open-source mapping software via integration in GIS curricula and extracurricular training sessions. For instance, in the 2020–2021 school year, our group hosted a mapathon to train Vassar and local high school students in editing OpenStreetMap, with a focus on digitiz-ing buildings in the city of Poughkeepsie. We also pivoted from employing Esri field survey-ing tools to using the open-source tool KoBo Toolbox. One of our long-term goals is to trans-fer data from the community resource map to OpenStreetMap so that our data can become part of a collective repository that is of use to community actors beyond the duration of the pandemic.

Fig. 5.4 Hudson Valley Mappers president Adele Birkenes presents the project at a virtual YouthMappers webinar in April 2020

3 Health Facilities in Saint-Louis, Senegal

Meanwhile, another engaged local mapping effort during the pandemic was happening as part of a pilot project to map health facilities in the department of Saint-Louis, Senegal. The work was cooperatively organized by Healthsites.io, CartONG, Geomatica, and the OpenStreetMap Senegal community.

We, YouthMappers from the UGB YouthMappers chapter at Gaston Berger University, participated in data collection trainings and field mapping activities. This work, similar to the one in New York, worked to make community resources known and visible.

3.1 Gaston Berger University YouthMappers

We are 10 enumerators who participated fully in the data collection phase on health facilities in our home locations. This entailed remote work and fieldwork and created spatial data in the form of attributes, tags, and descriptions, in addition to siting of features onto OSM. This was a critical phase especially since we collected exhaustive amounts of data, making it open through OSM and following the global community guidelines for quality.

In July 2019, we underwent extensive training on the use of Healthsites' open-source data collection tool. The mobile app ODK allows users to collect data and save it directly to OpenStreetMap through the healthsites.io platform (Fig. 5.5).

3.2 Mobilizing to Collect Health Asset Data in Saint-Louis

During the fieldwork period, it took us 48 hours to collect data on 104 health facilities located throughout the entire department of Saint-Louis. The data collection form generated by ODK was composed of several questions. It allowed us to collect as much information as possible from the

Fig. 5.5 The data collection team consists of (left to right) Ndeye Khady Ndoye and Ibrahima Sory Diallo of UGB YouthMappers, Mohamet Lamine Ndiaye of OSM Senegal

health facilities, among the questions we can mention, and Table 5.1 outlines these data.

To start the pilot phase of the project, Mark Herringer, Healthsites project manager, and Mohamet Lamine Ndiaye of the OpenStreetMap Senegal community conducted several meetings with health stakeholders during the month of April 2019. The objective of these meetings was to create a synergy around the collaborative mapping of health structures in the district of Saint-Louis. Other meetings were planned to present the results to the various stakeholders.

The facilities collected included 41 private, 58 public, 3 community, one NGO, and one combination. Twenty-four facilities were not wheelchair accessible. Four facilities were powered by solar electricity (Healthsites.io). This information, stored on OpenStreetMap, is also shared directly with the Senegalese Ministry of Health (Fig. 5.6).

Table 5.1 Description of data collected

Tag	Description
Amenity	For describing useful and important facilities for visitors and residents, e.g., identifying the health site was a clinic, doctor's office, hospital, dentist, pharmacy.
Healthcare	A key to tag all places that provide healthcare services or are related to the healthcare sector, e.g., doctor, pharmacy, hospital, clinic, dentist, physiotherapist, alternative, laboratory.
Healthcare: speciality	A key to detail the special services provided by a healthcare facility, e.g., biology, blood_check, clinical_pathology.
Name	The primary tag used for naming an element.
Operator	The operator tag is used to name a company, corporation, person, or any other entity who is directly in charge of the current operation of a map object.
Operator: type	This tag is used to give more information about the type of operator for a feature, e.g., public, private, community, religious, government, NGO, combination.
Addr: full	Used for a full-text, often multi-line, address for buildings and facilities.
Contact: phone	The contact tag is the prefix for several contact: * keys to describe contacts.
Operational_status	Used to document an observation of the current functional status of a mapped feature.
Opening_hours	Describes when something is open or closed: days/times of opening.
Beds	Indicates the number of beds in a hotel or hospital
Staff_count: doctors	Indicates the number of doctors in a health facility.
Staff_count: nurses	Indicates the number of nurses in a health facility.
Healthcare: equipment	Indicates what type of specialty medical equipment is available at the health facility, e.g., ultrasound, MRI, x_ray, dialysis, operating_theater, laboratory, imaging_equipment, intensive_care_unit, emergency_department.
Dispensing	Whether a pharmacy dispenses prescription drugs or not.
Wheelchair	Used to mark places or ways that are suitable to be used with a wheelchair and a person with a disability who uses another mobility device (like a walker).
Emergency	This key describes various emergency services.
Insurance: health	This key describes the type of health insurance accepted at the health site.
Water_source	Used to indicate the source of the water for features that provide or use water, e.g., well, water_works, manual_pump, powered_pump, groundwater, rain.
Electricity	Used to indicate the source of the power generated, e.g., grid, generator, solar, other, none.
URL	Specifying a URL related to a feature, in this case, the wiki page if it is linked to an organized mapping effort, both through surveying and importing.
Source	Used to indicate the source of information (e.g., meta data) added to OpenStreetMap.

3.3 Validating Data to Identify User Stories

To validate the data, the team held additional meetings with health stakeholders. The Ministry of Health's regional manager for the Saint-Louis district appreciated the geolocation data of health facilities and the up-to-date information on the status of services present. Together, we identified several user stories enabled by the data, such as routing maps to determine travel times to health services and their accessibility. A meeting was also held with the team from the Centre Opérationnel d'Urgence Sanitaire du Sénégal to validate the data.

4 Social Impact of Data in a Pandemic for the SDGs

Many should benefit from the availability of the data created and visualized, both in New York and in Senegal, because the purpose of making open spatial data available as a public good is to allow decision-makers to better respond to the needs of the population. Healthcare industries could use this data to streamline supply chains or identify opportunities for innovation. This data on healthcare facilities could permit public health service providers to plan responses and awareness campaigns. As a public good, OSM data created by the YouthMappers could allow the

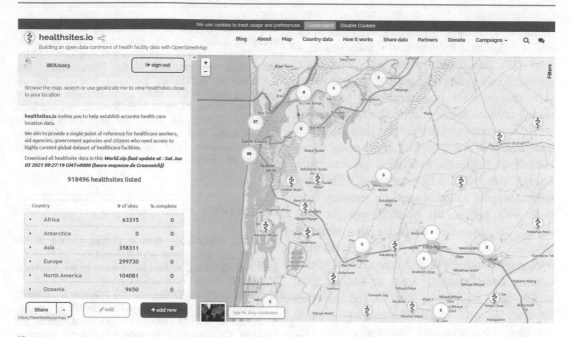

Fig. 5.6 A map features the contributed data on health facilities in Saint-Louis, Senegal. (Credit: www.healthsites.io)

Senegalese Ministry of Health and Dutchess County in their own ways to have a clear picture of the state of the healthcare system and identify gaps that may endanger the lives of the populations that they serve. Entrepreneurs and civil society could also use this data to develop innovative solutions to health crises. And, finally, the public, in general, and OpenStreetMap users alike can always build on this work and further develop user stories alongside response entities, public health agencies, and humanitarian aid organizations to optimize their support during health crises, in ways that advance Sustainable Development Goals. Yet it is not enough to simply create the data. In order for these projects to truly make a difference, stakeholders must recognize the value of the data as a public good for decision making.

Clearly, examining Hudson Valley Mappers' and Gaston Berger University (UGB) YouthMappers' community mapping projects in parallel offers rich insights on open mapping for SDG 3 for good health and well-being. Consistent and inclusive collaboration with local stakeholders and community organizations is key to the success and sustainability of these projects. Both YouthMappers chapters built these relationships

long before the COVID-19 crisis hit. In the case of Hudson Valley Mappers, the chapter's existing community geography connections in the city of Poughkeepsie enabled student volunteers to immediately solicit input from community partners at the onset of the pandemic and design a mapping tool that best fit local residents' and nonprofits' needs. In the case of UGB YouthMappers, the chapter collaborated with partner organizations Healthsites.io, CartONG, Geomatica, and OpenStreetMap Senegal to map health facilities in Saint-Louis. Their mapping process involved input from both local healthcare stakeholders and the Senegalese government (Centre Opérationnel d'Urgence Sanitaire du Sénégal). As an outcome of these partnerships, UGB YouthMappers successfully added a comprehensive health facility dataset for Saint-Louis to OpenStreetMap even *before* the pandemic began.

These two examples of YouthMappers collaborating with local partners also underscore the unique niche that the YouthMappers network occupies in the open mapping world. On a global scale, YouthMappers is the premier bridge between academia and the development and humanitarian assistance community. Student members of YouthMappers chapters listen to the needs articulated by community members, connect with local organizations and international partners, and mobilize groups of student volunteers to contribute geographic data before and during crises. Their specialized knowledge about their home communities enables them to be this bridge, and in doing so, they are simultaneously contributing directly to SDG 4 on education. During the COVID-19 pandemic, with many YouthMappers students mapping remotely from their own homes, the YouthMappers network served as connective tissue between chapters; through webinars, blog posts, and informal exchanges, students shared ideas on how best to employ open mapping tools to address local health and well-being needs. YouthMappers chapters are both locally and globally situated, fostering a confluence of community input, GIS skill sets, subject–matter expertise, and creativity that drives progress toward the SDGs.

Acknowledgments The Hudson Valley Mappers team would like to acknowledge Dr. Lisa Kaul, former director of the Vassar College Office of Community-Engaged Learning, and professors of Geography Dr. Mary Ann Cunningham and Neil Curri for their unending support. We would also like to thank the student volunteers who helped create the COVID-19 Community Resources Map, including Jamie Greer, Zsa Zsa Toms, Claire Kendrick, Anik Parayil, Thomas Tomikawa, and Zoe Turner-Debs. We extend our gratitude to the Vassar, Bard, and Poughkeepsie community members who assisted us, including Amanda Goodman, Lydia Hatfield, Professor Yu Zhou, Professor Brooke Jude, Sarah Salem, and Ezra Weissman. Finally, we would like to thank our fellow Hudson Valley Mappers student leaders, past and present: Mariah Caballero, Dina Onish, Hannah Benton, Hanqi Wu, and Phoebe Murray.

References

American Community Survey (2019) United States Bureau of the Census: Washington, DC

Flad HK, Griffen C (2009) Main street to main frames. State University of New York Press

Governor's Press Office (2021) Governor Cuomo announces expanded vaccination network to accelerate distribution of COVID-19 Vaccine. Available via New York State, 8 January. https://www.governor.ny.gov/news/governor-cuomo-announces-expanded-vaccination-network-accelerate-distribution-covid-19-vaccine. Cited 29 Jan 2022

Healthsites.io (2019) Healthsites Senegalese Pilot. Available via Medium, July. https://medium.com/healthsites-io/healthsites-sengalese-pilot-d7bd6f5a0b16. Cited 15 June 2021

Hinton DM (2021) Letter to Pfizer Inc. [Letter, 10 May]. Available via US FDA. https://www.fda.gov/media/144412/download. Cited 29 Jan 2022

Cross-Continental YouthMappers Action to Fight Schistosomiasis Transmission in Senegal

Michael Montani, Fabio Cattaneo,
Amadou Lamine Tourè, Ibrahima Sory Diallo,
Lorenzo Mari, and Renato Casagrandi

Abstract

The authors detail the design of an innovative and cooperative approach to ground truthing geospatial data through cross-continental YouthMappers coordinated action. This effort provided key geographic information to design control actions and served as a powerful, active tool to disseminate awareness about the importance of neglected tropical diseases in remote regions of the planet in support of SDG 3 Good Health and Well-being and SDG 6 Clean Water and Sanitation.

Keywords

Sanitation · Pathogens · Habitat · Global collaboration · Senegal · Italy · Health

M. Montani (✉)
Polimappers, Politecnico di Milano, Milan, Italy

United Nations Global Service Center, Brindisi, Italy
e-mail: michael.montani@un.org

F. Cattaneo
Polimappers, Politecnico di Milano, Milan, Italy

A. L. Tourè
DEIB, Politecnico di Milano, Milan, Italy

African Institute for Mathematical Sciences (AIMS),
M'bour-Thies, Senegal

I. S. Diallo
UGB YouthMappers, Université Gaston Berger,
Saint-Louis, Senegal

L. Mari · R. Casagrandi
DEIB, Politecnico di Milano, Milan, Italy
e-mail: lorenzo.mari@polimi.it; renato.casagrandi@polimi.it

1 The Complex Transmission Cycle of Schistosomiasis

In many regions of the world, the relationship between humans and the environment is troublesome, as it can give both access to natural resources and exposure to pathogens. Human intervention associated, for instance, with expanding urbanization and the construction of infrastructure can destabilize ecological equilibria, causing at the same time easier access to resources and higher exposure to environmentally transmitted diseases. Examples of this twofold relationship can be found in many regions of the world, including Saint-Louis, located in northwest Senegal, West Africa, where an increase of schistosomiasis transmission has occurred between 1996 and 2016, according to

P. Solís, M. Zeballos (eds.), *Open Mapping towards Sustainable Development Goals*, Sustainable Development Goals Series, https://doi.org/10.1007/978-3-031-05182-1_6

data at the Ministry of Health. This event followed the construction of the Diama dam, which was designed mainly to prevent saltwater intrusion into the Senegal River.

Schistosomiasis belongs to the group of the so-called neglected tropical diseases (NTDs), a diverse group of infections that in 2021 was still globally affecting more than 1.7 billion people according to the World Health Organization (Neglected Tropical Diseases progress dashboard, WHO 2021). The adjective "neglected" qualifies diseases whose burden is overlooked by the relevant decision-making bodies, leading to continuing and possibly widespread transmission in developing regions and in the poorest groups within developed countries. In particular, the 2030 UN Agenda for Sustainable Development Goals (SDGs) identifies the target of ending NTDs as SDG 3.3, whose associated indicator (3.3.5) is the number of people requiring interventions against them. Just to have an idea of the numbers involved to date at the global scale, the progress dashboard of WHO (2021) also reports that in 2019 alone, 230 million people were affected by schistosomiasis globally, 90% of

whom lived in Africa. Action items to try solving this issue consider the increase of political commitment to sustain domestic financing, development of new drugs, and creation of mapping systems to target treatments and monitor drug resistance. In this chapter, we show how cooperative YouthMappers action can at the same time (*i*) provide key geographic information to design control actions (such as identifying the most vulnerable villages that may be afflicted by schistosomiasis) and (*ii*) serve as a powerful, active tool to disseminate awareness about the importance of NTDs in remote regions of the planet.

Schistosomiasis is a water-based disease with a complex transmission cycle (Fig. 6.1). Human exposure is determined simply by contact with freshwater where the intermediate host (snails of genera *Biomphalaria* or *Bulinus*) of the parasite (worms of genus *Schistosoma*) lives. Humans are definitive hosts of schistosomes, which sexually reproduce in the bodies of infected people, who then release worm eggs with their excrements. Eggs hatch into miracidia, a free-living stage of the parasite that can infect snails. Schistosomes undergo asexual replication in infected snails,

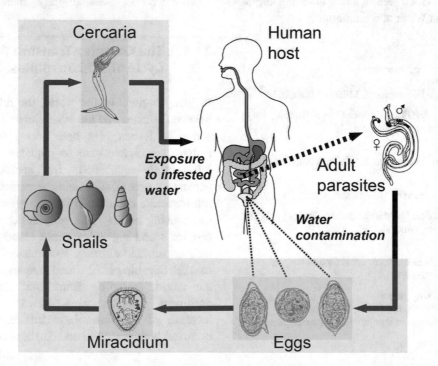

Fig. 6.1 The complex transmission cycle of the water-based disease named *schistosomiasis*. (Mari et al. 2017)

which then release cercariae, another free-living form of the parasite. Cercariae close the transmission cycle by penetrating the skin of humans and other animals who enter in contact with infested water. Schistosomiasis can cause urogenital and/or enteric problems in human carriers and, in children, may determine serious impairment of body growth and mental development, leading to a run-down state and preventing regular school attendance, a first step into (or a key step to perpetuate) a disease-induced poverty trap (Bonds et al. 2010). The focal human populations affected by this disease are in fact those who (*i*) already live in extreme poverty conditions, (*ii*) need to draw and use water directly from possibly contaminated environmental sources (Fig. 6.2), and (*iii*) may also release their wastes directly into the environment, thereby contributing to onward transmission and maximizing the probability to become infected.

In the Saint-Louis region of Senegal, anthropogenic interference with the natural ecosystem was linked to an increase in the snail population size, particularly around Lac de Guiers, the largest lake in the area (Sow et al. 2002). As a consequence, there has been a rehearsal of cases of schistosomiasis within the Saint-Louis region. To help health authorities improve the resilience of vulnerable populations, a deepened, quantitative understanding of disease transmission dynamics, accounting for both socio-economic and environmental variables, is fundamental. For this purpose, geospatial information is key, as it allows mapping locations where human and snail hosts reside, which is rural villages where vulnerable people live and the snail habitats where disease vectors are potentially most abundant. Furthermore, the creation and maintenance of open data repositories allow both local and remote citizens to actively participate in these challenging activities, namely, by collaboratively extracting information, sharing knowledge, and sensibilizing the local population about the risks related to schistosomiasis. This was the main

Fig. 6.2   A mother enters natural waters possibly contaminated by *Schistosoma* parasites to wash clothes, with children assisting her and playing within water in Ndiawdoun village. (www.openstreetmap.org/#map=19/16.06791/-16.39380)

motivation for us to participate in the 2018 YouthMappers Research Fellowship program with the project "Cross-continental YouthMappers fighting schistosomiasis in the Senegal River Valley," whose activities are summarized in the present chapter. We emphasize that the spirit of international cooperation among YouthMappers that we experienced during the Research Fellowship Symposium at the George Washington University and West Virginia University provided an important burst to realize the project as described here. After that event, in fact, we directly contacted the YouthMappers chapter at Université Gaston Berger (UGB) in Saint-Louis and asked for collaboration. Their answer was so enthusiastic that we decided to form a team to simultaneously map from three connected units in three locations (and two continents, Europe and Africa). The three units were PoliMappers (the YouthMappers chapter of Politecnico di Milano in Milan, Italy), UGB YouthMappers (the YouthMappers chapter at Université Gaston Berger in Saint-Louis, Senegal), and the PhD students at the African Institute for Mathematical Sciences (AIMS) in M'bour, Senegal. This last group was coordinated by one of us (Amadou Lamine Touré), who serves since years as researcher at the AIMS and spent one year as postdoc at the Politecnico di Milano within the Polisocial project "Mapping Schistosomiasis Risk in the Saint Louis region in Senegal" (MASTR-SLS, 2016). The diversity of experience and the multicultural background of the team resulted in a formidable facilitation of activities (other than fun).

2   Data Collection to Fight Transmission

We organized our geospatial data collection campaign in the following way. On one side, we started mapping the snail habitat: we performed field surveys in Senegal to retrieve multispectral drone imagery, also involving local mappers, in order to create a spectral signature to identify *Typha*, an aquatic genus of reed that is often associated with the presence of freshwater snails

(Chamberlin et al. 2020). On the other hand, we mapped the "human habitat": OpenStreetMap (OSM) was fed with information about the villages where vulnerable people live. Many of these small and informal settlements are indeed unknown, even to the local government.

2.1   Mapping the Habitat of Snails

To identify *Typha* vegetation and possibly use that information to run eco-epidemiological models for schistosomiasis, it may be important to accurately describe the spectral signature of the aquatic reed. This would simplify habitat identification at wider spatial scales because large areas can be explored using drone or satellite images instead of field campaigns. The knowledge on the location of such vegetation is a valuable proxy to produce risk maps of schistosomiasis, as *Typha* may serve as habitat of the intermediate snail host of schistosomiasis. An accurate spectral signature of the reed can be obtained through surveys with drones equipped with multispectral sensors, typically used for precision agriculture.

In March–April 2019, thanks to the 2018 YouthMappers Research Fellowship and to the Polisocial Award by Politecnico di Milano (project MASTR-SLS, 2016), we had the opportunity to conduct an extensive ground-truth retrieval campaign. During this experience, we explored the rural region around the city of Saint-Louis to find spots of *Typha* (Fig. 6.3). Thanks to preliminary mapping of *Typha* areas in the region made by a local company (Fig. 6.4), our team had an initial hint of the places where the vegetation potentially grew. As the team was based in the city of Saint-Louis and a significant amount of time was required to reach remote areas of interest, the duration of the set of batteries of the drones was not sufficient to cover daylong survey activities, and regular trips from and to the city had to be planned. Different spots of *Typha* have been visited, in particular, near the delta of the Senegal River and the city of Saint-Louis. Rural villages and small inhabited places are dispersed all along the way. People in rural

Fig. 6.3 The YouthMappers field team consider a potentially contaminated area and stop to take a selfie via drone while crossing a waterfield with *Typha*

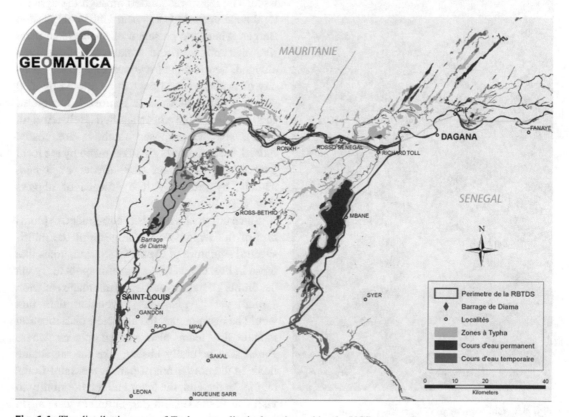

Fig. 6.4 The distribution map of *Typha* as qualitatively estimated by the UGB startup Geomatica serves as a basis for *Typha* scouting in the region

places enter rivers and water points to retrieve water for daily activities and domestic uses and to release wastewater, exposing them to high risks of being infected by schistosomiasis (see again Fig. 6.1).

During the first days of exploration, the team got in contact with people living in the study area's villages. The multicultural diversity of members in the team was very beneficial to engage local populations. Thanks to the presence of both Senegalese and Italian "YouthMappers with drones" members, the team had an empathic relation with inhabitants in the local language (Fig. 6.5). Even though the official language in Senegal is French, a large part of the population living in cities and the totality of the population living in rural areas typically speak Wolof, the most widespread language in West sub-Saharan Africa. The

Fig. 6.5 Children in a village are proud to hold our strange technical bird, the multispectral drone

team (Fig. 6.6) had the possibility to explain the goals of its work, to inform the population, and to take drone surveys after having properly asked the permission of the chiefs of the different rural settlements.

Amidst the necessary adjustments, the first week of fieldwork did not lead to significant advancements in terms of ground truthing, as we only sporadically found spots of *Typha* and those were mainly mixed with other vegetation. A huge help was given to us when we entered in contact with researchers at Université Gaston Berger, more precisely with the group led by Dr. Elhadji Babacar Ly and Dr. Diene Ndiaye. They study the possible use of *Typha*, which has been found to have good thermal and mechanical properties, as a construction material (Ly et al. 2019), similar to what is done with bamboo in Asia. Members of the local university staff could indicate clearly where *Typha* was causing serious problems in the region, independently of the epidemiological issue. The team was guided through the agricultural zone of Saint-Louis near Université Gaston Berger, where the presence of *Typha* is so massive that most of the canals and drains are clogged, and local cultivations suffer from scarcity of water, whose flow can be altered by reed swamps. Efficient ways to remove big spots of such vegetation have been studied (Hellsten et al. 1999), but the proposed methods are costly. Indeed, several efforts are undertaken by the local government to reduce the effects of *Typha*, namely, by removing it by means of diggers (Fig. 6.7).

Since our goal with this drone-guided exploration of the region was to fingerprint the multispectral signature of *Typha* vegetation, areas like those in Fig. 6.3 were very useful spots to fly via the drone. To gather multispectral images of "non *Typha*" yet "*Typha*-like" vegetation data that would be instrumental to calibrate a classification model, the team also visited places where *Typha* was naturally absent, like the sugarcane fields in the northernmost part of the Saint-Louis region. In the end, the team had the possibility to visit Université Gaston Berger and to meet again

Fig. 6.6 The drone-mapping team greets the camera, from left to right: Amadou Lamine Toure, Michael Montani, Fabio Cattaneo and Ibrahima Sory Diallo

Fig. 6.7 Reducing the risk of exposure to the disease is possible via mechanical removal of *Typha* in the region surrounding Universitè Gaston Berger. (www.openstreetmap.org/#map=18/16.03875/-16.41664)

with the UGB YouthMappers, sharing mapping knowledge and experiences.

2.2 Mapping the Habitat of Humans

This task was carried out by setting up an extensive collaborative mapping campaign to create new OpenStreetMap data for the focal areas of interest involving YouthMappers and interested citizens from different continents, thereby fostering international collaboration among experienced and new mappers.

To identify the most vulnerable villages that may be afflicted by schistosomiasis, the mapping campaign focused on the extraction of buildings over the western part of the Saint-Louis region and around Lac de Guiers. The collaborative project was posted on the TeachOSM Tasking Manager for every OpenStreetMap (OSM) volunteer to contribute to. Moreover, several mapping events were held to achieve global participation in the production of open geospatial information for our focal area. The project was officially launched at State of the Map 2018 (SotM), the annual international conference about OSM and its community, which took place at Politecnico di Milano in July. Since then, we organized and hosted several editing gatherings with a variety of mapping groups, ranging from inexperienced volunteers to geoinformatics experts, with ages ranging from high-school teenagers to university students.

The main event, which we named "Cross-continental YouthMapathon," was organized involving three different gatherings mapping on the project simultaneously yet in different geographic locations. Each of these three local meetings was led by members of the MASTR-SLS project, who shortly introduced the research problem to the participants. Specifically, the three main poles were organized as follows:

- Politecnico di Milano (Milan, Italy), with Michael Montani and Lorenzo Mari, and the participation of the members of PoliMappers
- Université Gaston Berger (Saint-Louis, Senegal), with Fabio Cattaneo and Renato Casagrandi, and the fundamental contribution and collaboration of the local chapter of UGB YouthMappers, led by Ibrahima Sory Diallo
- African Institute of Mathematical Sciences (M'bour, Senegal), with Amadou Lamine Touré leading a group of students and staff members

Other YouthMappers chapters joined the mapping activity from other locations, for instance, Semillero GeoLab UdeA, based in Universidad de Antioquia, Medellìn, Colombia. As a result, our cross-continental youth mapping activity did involve three continents: Africa, Europe, and South America.

3 Results of a Cross-Continental Effort

3.1 Insights from Drone Mapping

The field mapping campaign led us to retrieve up to 28 square kilometers of high-resolution drone imagery, which allowed us to better define the spectral signature of *Typha*, notably on the green, red, and near-infrared bands. The dataset is composed of drone images taken during surveys of spots with *Typha* plants alone, with *Typha* mixed with other vegetation, or without *Typha*, like sugarcane farmlands. Specifically, the information coming from the imagery of *Typha* spots has been used to describe the spectral signature of the aquatic reed. As a first exercise, we applied machine learning techniques, notably logistic regression, decision tree, and random forest, to train classifiers aimed to identify *Typha* presence from drone imagery in images that were not used as training data for the construction of the algorithms. We avoid

Fig. 6.8 Manually identified *Typha* are groundtruthed (in green) on a drone survey

Fig. 6.9 Pixels predict *Typha* (in orange) from a Random Forest model trained on our dataset from another drone survey

detailing here the methodology used and the details of the analysis, which is still undergoing. In some of the areas, we had quite a good result in matching the true presence of *Typha*, as shown in Figs. 6.8 and 6.9.

3.2 Insights from OSM Mapping

The OSM mapping campaign lasted from July 2018 to November 2019. During the entire mapping campaign, a total of 23,165 buildings (from now on, edits) were mapped by a total of 240 different contributors from all over the world. Approximately 80% of the edits are geographically located in the Saint-Louis region, while the remaining ones are in Louga, the adjacent region to the South. Two simple queries to the OSM database about the status of the map in the focal region of interest at the beginning and end of our project can reveal the entity of our contribution (Table 6.1).

The corresponding map snapshots show the situation before (Fig. 6.10) and after (Fig. 6.11) our mapping campaign, in terms of buildings edited in Saint-Louis within the project.

Besides the sheer increase in the number of mapped buildings, participative mapping campaigns also offer the opportunity to study the emergent collective behavior of the participants. The goal of collaborative mapping is indeed not only to build maps but also to build mappers, as brilliantly summarized in the YouthMappers vision. And that is of primary importance to all of us. The cumulative distribution of OSM edits performed by all volunteers during the mapping campaign is reported in Fig. 6.12.

It is possible to notice how the steepest ascents occur exactly at the time of our major mapping events. Typically, those are named "retaining edits", that is edits occurring from the date of the event up to a few days later, as contributors often

Table 6.1 Summary of the mapping contribution of our campaign

Area of interest	Total buildings on OSM	Buildings we mapped (our edits)	Our edits over total	Increase
Senegal	282,016	23,127	8.2%	+9%
Saint-Louis	74,773	18,791	25.1%	+33.6%

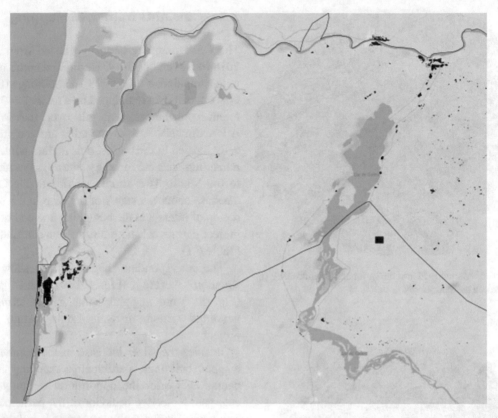

Fig. 6.10 Mapped buildings in the Senegal River Valley Region appear as black marks in the left panel before the project, shown as OSM data retrieved on 1 July 2018

kept on mapping well after the formal meetings. The statistics related to each major event are summarized in Table 6.2.

An interesting aspect of the statistics is the average number of edits per contributor at each event, which was much higher in meetings attended by expert mappers than it was in events aimed at beginners. While the expert mappers attending the SotM or NoG events were already proficient at OSM editing, the IAT event and PoliMappers@Statale gathered people who, in some cases, did not even know what mapping is. In the latter case, the expected edits-per-contributor ratio and retention are obviously lower.

We also estimated that up to 5901 mapped buildings within the project are not related to any of these major events, as they have been probably contributed by occasional mappers either on a voluntary basis or through small meetings not directly supervised by us. This unsolicited yet highly appreciated effort corresponds to 25.4%

of the total edits—a "spillover" effect that is always hoped for but never to be taken for granted when launching collaborative events.

It is possible to further describe the behavior of volunteers by classifying them with respect to their past mapping history (if any) and contributed edits within our project. To this purpose, we identified four categories of users:

- *Newcomers*: People who did not know anything about OSM and never mapped earlier in their life
- *Mayflies*: Newcomers who mapped only on the day on which they joined OSM but did not contribute further to our project
- *Regular mappers*: Contributors who already knew how to map and participated regularly in OSM events
- *Power mappers*: Expert contributors who added way more features than regular mappers

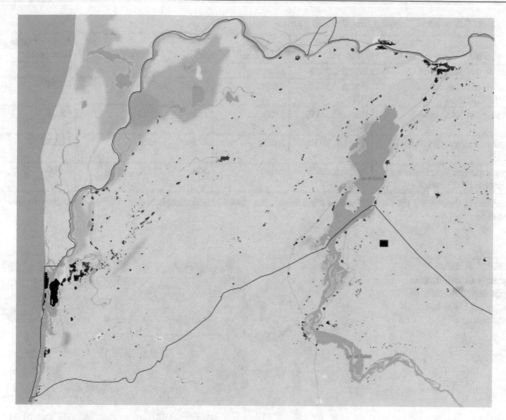

Fig. 6.11 Mapped buildings appear after the project as red marks in the OSM data retrieved on 26 November 2019

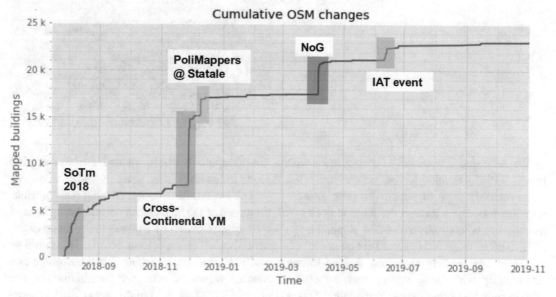

Fig. 6.12 Cumulative distribution of OSM changes performed by all volunteers during the mapping campaign show a step-wise increase in the phases of the effort

Table 6.2 Statistics of mapping contributions by event

Event name	Date	Retention[a]	Edits	Contributors	Average edits per contributor
State of the Map 2018 (SotM)	July 28, 2018	14 days	4762	15	317.47
Cross-Continental YouthMapathon	November 28, 2018	2 days	7053	71	99.34
PoliMappers@Statale	December 10, 2019	1 day	1691	26	65.04
Night of Geography 2019 (NoG)	April 5, 2019	3 days	3254	29	112.21
Ingegneria per l'Ambiente e il Territorio (IAT) event	June 13, 2019	2 days	504	18	28.00

[a]Retention is defined as the time for which there have been above-average crowdsourced OSM contributions after the event date

Fig. 6.13 Relative frequency distribution of the per capita number of changes on OSM are shown above

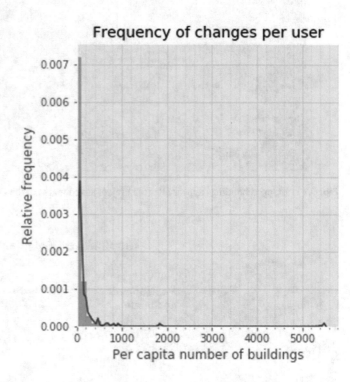

We recorded a total of 41 newcomers who joined OSM, thanks to our mapathons, via approximately 90% of which (36 people) were mayflies too. Even though the large majority of newcomers who discovered OSM stopped mapping, 10% of them did not and kept on contributing, possibly a first step into a long-lasting relationship with OSM editing and, hopefully, with introducing other people to mapping. This is the case of a Senegalese UGB student, who, after participating in the Cross-Continental YouthMapathon event, created a new mapping group in his home city, Richard-Toll.

The overall distribution of edits per contributor turned out to be far from normal, being instead well approximated by a power law with an exponential cutoff (Figs. 6.13 and 6.14). This could be easily noticed by observing that the average and median values of edits per contributor did not coincide (112 vs. 37 buildings per user, respectively). The fact that most of the mappers performed few edits while few experts mapped a disproportionately large number of features is not only related specifically to our project. It is instead a widely known effect in the OSM environment (Coetzee and Rautenbach 2016; Brovelli et al. 2020) and, more generally, in many citizen science projects (Lintott et al. 2008; Dickinson and Bonney 2012; Curtis 2014; Sauermann and Franzoni 2015) (Figs. 6.15 and 6.16).

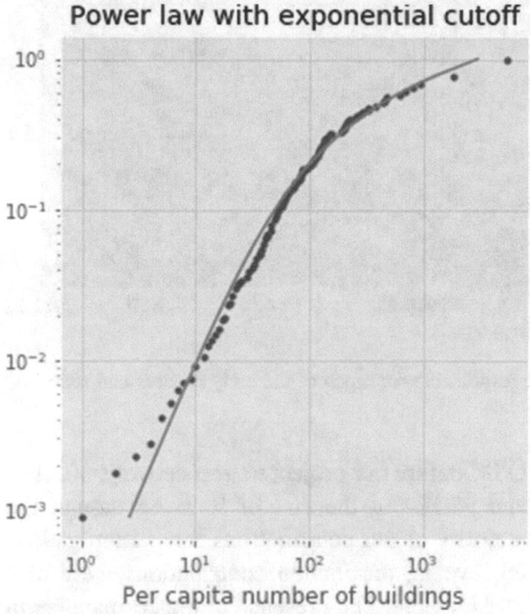

4 Conclusions for Joint Youth Action on SDGs

This research project allowed us to advance the knowledge on the use of open collaborative mapping to fight a disease, in our case schistosomiasis. We created geographic information that permitted us to improve the habitat mapping of both the intermediate (snails) and the definitive (humans) hosts of schistosome parasites. As for the former, we collected and analyzed high-resolution multispectral drone imagery of aquatic vegetation spots; for the latter, we leveraged open participatory editing on OSM to improve or, in some cases, to start the mapping of the villages of the focal region of Saint-Louis.

The field mapping experience developed through this research project has been essential to gather drone imagery that can be used to feed eco-epidemiological models. Such geographically informed data can be important to refine the description of disease dynamics. Also, our data may be important to train supervised classifica-

Fig. 6.14 Power law with exponential cutoff (in blue) fits the cumulative frequency to the per capita number of changes on OSM

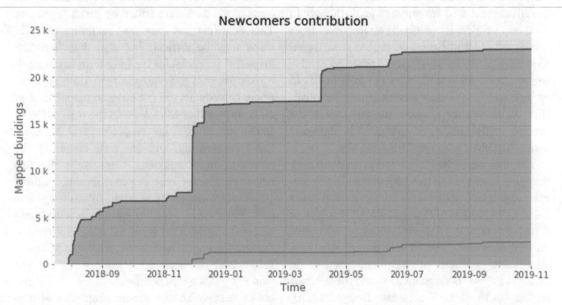

Fig. 6.15 Newcomers' share of total contributions began later in the project and remained a steady source of data

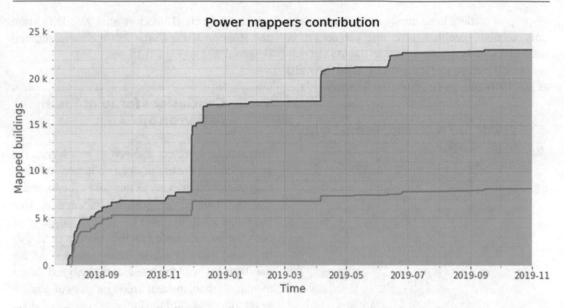

Fig. 6.16 Power mappers' share of total contributions were responsible for much of the early success and remained steady throughout the duration of the project

tion algorithms designed to perform large-scale mapping of spots of aquatic vegetation. What we also experienced in our project is that the engagement of local researchers, YouthMappers, and citizens proved to be key to performing field activities, not only in terms of logistics but also to overcome language and cultural barriers and to identify and access the best survey areas. The benefits of an expanded team were not limited to the advancement of the project's goals though. In fact, on one side, by involving contributors from the region of Saint-Louis, the project contributed to local capacity building; on the other, by involving international YouthMappers at their first field experience, it served as a crash course introduction to the reality of what mapping from the ground really entails, as compared to the most usual "armchair mapping" from the sky.

On the other hand, the open participatory mapping leveraged in this project was key to create geographical data on the human habitat in the Saint-Louis region of Senegal, allowing to put on a map vulnerable villages through collaborative editing activities. The project contributed to a significant increase of buildings mapped in Senegal (+9%), even more so in the Saint-Louis region (+33.6%). We went from 282,016 (Senegal) and 74,773 (Saint-Louis) buildings in

OSM before our project to, respectively, 305,143 and 93,564 at the end of it. Even though the majority of our contributions came from gathering events, unsolicited contributions were also considerable. The presence of expert mappers in our meetings led to events with more contributions and longer retention times. The average number of edits per contributor typically followed a power-law distribution, meaning that most of the data were edited by just a few users. This confirms not only that retaining users in OSM may be difficult but also that "building mappers" is crucial for the long-term sustainability of open data and communities. Although most of the newcomers did not keep on mapping, they all contributed at least once to the generation of spatial information for the project and the wider OSM community, and are now aware of the power of open mapping. Conversely, the few newcomers who kept on contributing (around 10%) might hopefully become future long-term contributors and possibly introduce other newcomers to OSM. That is indeed what happened to the three youngest authors of this work: Michael, Fabio, and Ibrahima were all newcomers less than six years ago; nonetheless, they were capable of leading YouthMappers chapters and running an international mapping effort (Fig. 6.17).

Fig. 6.17 A joyful meeting unites UGB YouthMappers and PoliMappers at Université Gaston Berger in Saint-Louis, Senegal

Acknowledgments Data collection activities were made possible thanks to the support provided by the MASTR-SLS project (http://www.mastr-sls.polimi.it) funded by Politecnico di Milano through the Polisocial program for responsible research (http://www.polisocial.polimi.it) and by the 2018 YouthMappers Research Fellowship (https://www.youthmappers.org/faces-voices). The authors are also grateful to several partners who provided epidemiological data (in particular to Dr. Elhadji D. Dia of the Senegalese Ministry of Health) or facilitated the field mapping campaign, such as Geomatica Services, Dr. Abdou Ka Dioungue, UGB YouthMappers, and the local population in rural settlements.

References

Bonds MH, Keenan DC, Rohani P, Sachs JD (2010) Poverty trap formed by the ecology of infectious diseases. Proc R Soc B Biol Sci 277:1185–1192

Brovelli M, Ponti M, Schade S, Solís P (2020) Citizen science in support of digital earth. In: Guo H, Goodchild MF, Annoni A (eds) Manual of digital earth. Springer, Singapore. https://doi.org/10.1007/978-981-32-9915-3_18

Chamberlin AJ et al (2020) Visualization of schistosomiasis snail habitats using light unmanned aerial vehicles. Geospat Health 15:818

Coetzee S, Rautenbach V (2016) Reflections on a community-based service learning approach in a geoinformatics project module. In: Gruner S (ed) ICT education. SACLA 2016, vol Communications in Computer and Information Science: 642. Springer, Cham

Curtis V (2014) Online citizen science games: opportunities for the biological sciences. Appl Transl Genom 3:90–94

Dickinson J, Bonney R (2012) Citizen science. Public participation in environmental research. Cornell University Press, Cornell, New York

Hellsten S, Dieme C, Mbengue M, Janauer GA, den Hollander N, Pieterse AH (1999) Typha control efficiency of a weed-cutting boat in the Lac de Guiers in Senegal, schistosomiasis transmission: a preliminary study on mowing speed and re-growth capacity. Hydrobiologia 415:249

Lintott CJ, Schawinski K, Slosar A, Land K, Bamford S, Thomas D, Raddick MJ, Nichol RC, Szalay A,

Andreescu D, Murray P, Vandenberg J (2008) Galaxy zoo: morphologies derived from visual inspection of galaxies from the sloan digital sky survey. Mon Not R Astron Soc 389:1179–1189

Ly EB, Lette M, Diallo AK, Gassama A, Takasaki A, Ndiaye D (2019) Effect of reinforcing fillers and fibres treatment on morphological and mechanical properties of typha-phenolic resin composites. Fibers Polym 20(5):1046–1053. https://doi.org/10.1007/s12221-019-1087-y

Mari L, Gatto M, Ciddio M, Dia ED, Sokolow SH, De Leo GA, Casagrandi R (2017) Big-data-driven modeling unveils country-wide drivers of endemic schistosomiasis. Sci Rep 7:489

MASTR-SLS (2016) Mapping schistosomiasis risk in the Saint Louis region, Senegal. Available via Politecnico Milano. http://www.mastr-sls.polimi.it. Cited 29 Jan 2022

Sauermann H, Franzoni C (2015) Crowd science user contribution patterns and their implications. Proc. Natl. Acad. Sci. U.S.A 112:679–684

Sow S, De Vlas S, Engels D, Gryseels B (2002) Water-related disease patterns before and after the construction of the Diama dam in northern Senegal. Ann Trop Med Parasitol 96:575–586

World Health Organisation (2021) Control of neglected tropical diseases. Available via WHO. www.who.int/teams/control-of-neglected-tropical-diseases/overview/progress-dashboard-2011-2020. Cited 13 Aug 2021

YouthMappers (2018) YouthMappers research fellowship. Available via YouthMappers. www.youthmappers.org/post/2018/04/27/introducing-the-2018-youthmappers-research-fellows. Cited 5 Sept 2021

Understanding YouthMappers' Contributions to Building Resilient Communities in Asia

Feye Andal, Manjurul Islam,
Ataur Rahman Shaheen, and Jennings Anderson

Abstract

This chapter considers the contributions of YouthMappers chapters in Asia. In addition to a regional overview, we highlight actions of students in Bangladesh and the Philippines to fill critical data gaps that support community access to information during emergencies, natural disasters, and pandemics. Lack of data leads to poor decision-making at any time, but in the context of shocks and hazards, it can have an especially profound impact on local communities. By creating open geospatial data and by advancing the geospatial capacity of university students and local community members, local governing bodies will be able to plan for the well-being of their constituents and community members will have access to the information necessary to keep their families safe. This contributes to better health and well-being (SDG 3) and a more resilient society in the face of impacts of climate change (SDG 13).

F. Andal (✉)
University of the Philippines, Philippines, PA, USA
e-mail: dhandal@up.edu.ph

M. Islam · A. R. Shaheen
Dhaka College, Dhaka, Bangladesh

J. Anderson
YetiGeoLabs, LLC, and YouthMappers,
Helena, MT, USA

Keywords

YouthMappers in Asia · Mapping contributions · Climate change · Pandemic · Health

1 Footprints of Youth and Open Data in Asia

Let us listen to what is happening in Asia where youth are making themselves heard through the map.

The voices coming from this region of the world are uniting under the banner of YouthMappers, creating data to begin to address all our development objectives, following the United Nations Sustainable Development Goals. In particular, the youth in this region are motivated to contribute toward SDG 3 (Good Health and Well-Being) and SDG 13 (Climate Action) through data collection activities and campaigns to accurately provide open and accessible information on health sites and other assets ahead of natural disasters. Bottom line is that all of the

SDGs, when focused on reduced inequalities, are able to improve our well-being as a youth population and a region as a whole.

2 Generation of Open Spatial Data in Asia

Asia is the region of the world with the highest numbers of people on earth, and it is also among the regions with a great potential to engage very large numbers of YouthMappers in terms of chapters, students, and edits. Asia is one of the continents to receive considerable mapping attention from OpenStreetMap (OSM) edits where the total edits by YouthMappers in any world region hit the highest, with 7.2 million, for more than 6 million distinct visible map features (latest edited by YouthMappers). Edits to building objects are also the highest for any region among YouthMappers, topping 6.5 million edits on about 5.7 million buildings.

These numbers are impressive considering that Asia is not the region with the largest number of university chapters (Africa has the most). But our students are very productive and engaged, and there have been early and frequent mapping campaigns that have added significantly to the map.

Chapters are present in Bangladesh, Bhutan, India, Iraq, Indonesia, Japan, Nepal, the Philippines, Singapore, and Sri Lanka, while mapping is happening throughout the entire region by students here and elsewhere in the global network. Figure 7.1 presents the total edits in the Asian region since the beginning of YouthMappers.

Looking at the cumulative count of features edited by YouthMappers in Asia, the edits in Bangladesh make up approximately half of the contributions to OSM. This is followed by Nepal and the Philippines seeing the next most editing (by feature count) among all YouthMappers. In addition, YouthMappers have created tagging for 13,507 distinct amenities, such as schools, businesses, or points of interest. As of the start of 2022, 97,287 km of roads and paths across the continent were last edited by YouthMappers. The campaigns to create this data have often directly led to the well-being of the population in our region, frequently in response to disasters and hazards like flooding, hurricanes, and earthquakes. Having this data created by and for our region to help those most in need, making this data more open and visible, certainly contributes to regional equity and the health of our residents, including ourselves.

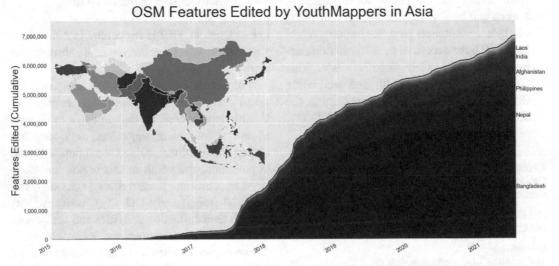

Fig. 7.1 Accumulated changesets in Asia over time show growth since the beginning of YouthMappers in OpenStreetMap. (Credit: J. Anderson 2022)

3 Resilience and the Sustainable Development Goals

Our contributions have significantly increased the amount of open geospatial data accessible by all for several purposes geared toward the attainment of SDG 3 Good Health and Well-being and SDG 13 Climate Action. Not all geospatial data are readily available and accessible to the general public, making it difficult to provide an analysis on the socio-cultural issues that we are continuously facing. Data on critical facilities such as health facilities, essential services, and residential building footprints are needed when responding to natural disasters or a global pandemic. Investing in activities and campaigns that generate data and fill in gaps also supports community-level decision-making as they navigate uncertainty by making critical information available to them. This awareness of both vulnerabilities and assets is a foundation for building resilience (Rodin 2014).

4 Chapter Contributions in Response to COVID-19

The COVID-19 global pandemic is reshaping how geospatial communities respond to the socio-cultural problems in society. The pandemic demonstrated the need for collaborative geospatial data to help local communities better understand the crisis we are facing collectively. Since the onset of the pandemic, there is an increasing demand for real-time maps, and readily available crowdsourced information plays a critical role in keeping the general public informed and prepared.

Bangladesh and the Philippines were both affected by the pandemic. In the Philippines, crowdsourced mapping accelerated to provide the general public with information relevant to the pandemic since communities in the Philippines were under a strict lockdown. Crowdsourced geospatial data with national coverage has the advantage that anyone can use the data. It makes the relevant information accessible to anyone, which can also support decision- and policy-makers in responding to the pandemic.

4.1 Mapping Health Facilities

The first mapathon, organized by the local OSM-Philippines community, was "mapagaling," which means *mapa* (map) and *galing* (to cure). This mapathon gathered volunteers in the Philippines to map health facilities and services in the country using OSM. The mapathon consisted of a short introduction about using OSM and MapContrib to train users with little to no experience of using a map to contribute missing data such as location and contact details. Once uploaded to OSM, this information became available to the general public.

The student members of University of the Philippines Resilience Institute (UPRI) YouthMappers, the YouthMappers chapter at the University of the Philippines, who joined the mapathon, enjoyed it because they got to learn more about using MapContrib, which was the platform used specifically for the project. Unlike the iD editor for contributing to OSM, it provides a quicker way to add information without having to trace the individual buildings. Moreover, the students were glad to participate in this project as they stated that they were able to contribute to the map data in the comfort of their homes (Fig. 7.2).

Meanwhile, the Bangladesh Government ran a monthlong campaign in April 2020 named "Bangladesh Challenge" with the objective of encouraging Bangladeshi youth to mark emergency and important locations known to them. Through this initiative, people contributed information on the location of pharmacies, hospitals, clinics, recharge points, and other important landmarks on the Google map and OSM and through the Bangladesh Challenge's website. During the lockdown, it helped engage people for a greater national cause and to fight the COVID-19 pandemic by getting emergency supplies using the map.

Most YouthMappers from Bangladesh contributed to update all the amenity data around health centers. Over 60 members of the YouthMappers chapter at Dhaka College contributed to this project. They played a vital role in updating data and received recognition from the organizing government authority (Fig. 7.3).

Fig. 7.2 OSM PH's first mapping initiative during the pandemic focuses on mapping the health facilities in the Philippines through OSM

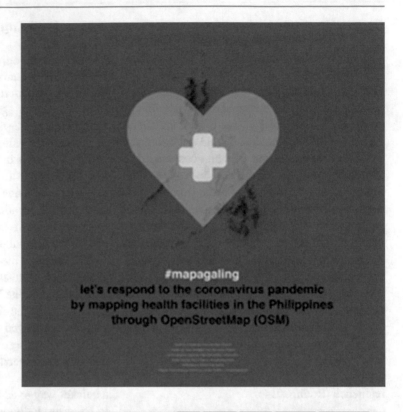

Fig. 7.3 Campaigns publicize mapping on personal devices

4.2 Mapping Essential Services

In addition to mapping health services and facilities, mapping essential services such as supermarkets and groceries, local wet and dry markets, water-refilling stations, shops, and restaurants that remained open during the lockdown was necessary as this information was not readily available to the public. Mikko Tamura, MapBeks founder, led this initiative and partnered with different organizations including Map the Philippines, Ministry of Mapping, University of the Philippines Department of Geography, Grab Map operations, MapBeks, and OpenStreetMap-Philippines to carry out this crowdsourced mapping project. YouthMappers at the University of

Fig. 7.4 A mappy birthday party aims to map the open stores and essential facilities during the lockdown while celebrating Mikko's Birthday

the Philippines and volunteers enjoyed participating and contributing to the project since this event was a combination of a mapathon and a birthday celebration despite having to participate remotely because of the lockdown (Fig. 7.4).

4.3 Mapping Residential Households

As COVID-19 active cases rise in the Philippines, specifically in Quezon City of Metro Manila, the University of the Philippines Resilience Institute (UPRI) collaborated with the Humanitarian OpenStreetMap Team (HOT)-Philippines to map the building footprints in Quezon City. The OSM building footprints were used by UPRI to develop a data-driven analysis and recommendation to the local government unit of Quezon City for the city's pandemic response. The HOT-Philippines team created a project in the HOT Tasking

Manager to make it accessible to anyone who is interested in contributing to mapping Quezon City. UPRI needed the completed building footprints before June 2020 to provide analysis and recommendations for the national government in deciding whether to downgrade the quarantine restrictions over the country's national capital region. However, it was especially difficult for the mappers and validators of this project, including the YouthMappers chapter members and volunteers, to complete it before June as the city is one of the most densely populated in Metro Manila. It took the 644 volunteers to completely map and validate the buildings in 62 days. The whole mapping and validation process ended on June 25, 2021.

From this project, we realized that it was difficult to compromise between the quality of data and meeting the deadline. As information is critical during crises like the pandemic, making decisions based on flawed or incomplete data may cause inability to accurately assess the current COVID-19 situation. Poor data input leads to poor decision-making and can have a direct impact on the state of the local communities (Fig. 7.5).

5 Chapter Contributions in Response to Climate Change

Now the world is experiencing different types of natural hazards because the climate is changing. As mappers, we cannot directly reduce climate change but can help minimize it. Unplanned urbanization and industrialization are the main reasons for climate change. Many vulnerable and disaster-prone areas in the world are unmapped. Coastal areas, islands, hill tracts, and disconnected remote areas remain incompletely mapped. For this reason, the people of communities in these vulnerable areas are always at risk, so community mapping is a must.

Disasters are a common phenomenon in Bangladesh. Among the different disasters, floods are one of the biggest threats for people who are living in coastal areas, near the rivers,

Fig. 7.5 HOT-Philippines' PhilAWARE Project helps to advance building footprints mapping in Quezon City

and on islands. With the support of the International Federation of the Red Cross (IFRC) and Bangladesh Red Crescent Society, different projects are being organized in these areas for disaster risk reduction, preparedness, and resilience. In Khasbaroshimul, Sirajgang is a flood-prone small island on the Jamuna River. Student mappers from YouthMappers Dhaka College visited this small island beside the Jamuna River, under the Sirajgang district of Bangladesh, to map that area. We visited the place and mapped the area by using GPS and field papers and collected data door to door with the support of the local community. We worked with local community leaders to form a group of community mappers and train them because when community members map areas, they provide real data and

also become aware of their blind spots and need for resources and safety areas.

We collected data about each household, its condition (based on flood lasting and floor height), material, level, and address. Community members indicated important places like schools, cyclone centers, mosques, and others. They also marked the river erosion areas and drew roads and paths and indicated where the roads and embankment were broken and specific locations where floodwater entered localities easily in a community collaborative map. Mappers tagged important points such as water entry points and also added some special tags about flood signals, community disaster response team (CDRT) members, doctor offices, veterinary clinics, and the location of datri (midwives). According to its geographical location, mappers marked this area as flood prone. Most of the people are farmers and day laborers and are struggling for their livelihood. They need more support, and after updating the data, a clearer picture of the needs and struggles of the community is visible. Because community members participated in the collection of this data, the data accuracy level was 100%. Once uploaded to OSM, as a public good, anyone is able to see and use the data.

6   Digital Innovation During a Pandemic

It is essential to develop methods, technologies, and innovations quickly in times of great pressure like disasters and the pandemic. This preparedness is part of building resilience as much as it is part of advancing toward the SDGs. The pandemic has illustrated that tools and geospatial data should always be readily available as decision-makers quickly need information for analysis and policy-making purposes. As volunteer student mappers were not allowed to conduct fieldwork to collect exact locations and additional relevant information of the health facilities and essential services because of the risk faced by the pandemic, various local communities developed creative and resourceful ways on how volunteers can contribute to OSM despite having

no prior experience on mapping and to ensure that the data being ingested in OSM are accurate and correct.

7   Future Outlook, Collaboration, and Engagement

All types of work are possible by collecting, uploading, and using OpenStreetMap data. Development work, disaster preparedness, relief distribution, rehabilitation, resilience, health, and medical support can be enhanced through the incorporation of crowdsourced, open data. Throughout the pandemic, we have observed efficient and effective collaborations between the different local mapping communities and local YouthMappers chapters to deliver critical information that can be used by the local community and the general public. This specific time during the pandemic has been a great example of what resilience means and how much we can achieve and help local communities.

References

Anderson, J. (2022). YouthMappers Activity Dashboard. https://activity.youthmappers.org. Cited 6 May 2022
Rodin J (2014) The resilience dividend. The Rockefeller Foundation, New York

Activating Education for Sustainable Development Goals Through YouthMappers

Maliha Binte Mohiuddin and Michael Jabot

Abstract

In hopes of building inclusive and sustainable societies, SDG 4, quality education, is central to helping to build a knowledge base to tackle some of the most pressing challenges faced by society. YouthMappers around the world are applying their knowledge coupled with critical reflection tools to act on and bring others along in making changes that improve the world. As such, they can be considered among a generation of "Solutionaries," students who extend their understanding beyond typical boundaries to include a systematic application of their learning. Youth, in general, and young women, in particular, can get aligned to the opportunity to learn through practical knowledge, by way of inclusive mapping communities, which sparks their passion for learning and supports SDG 5 gender equality in education, as well.

Keywords

Education · Sustainable development · New York · Bangladesh · Pedagogy · Gender · Science

1   What Does Education for SDGs Mean? Setting the Stage

Among the most foundational and crosscutting of the 17 Sustainable Development Goals set forth by the 2030 Agenda for Sustainable Development is SDG 4, ensuring quality education. At its center is the tenet of ensuring inclusive and equitable quality education and promoting lifelong learning opportunities for all. In hopes of building inclusive and sustainable societies, it is a fundamental goal that is central to helping to build the knowledge base to tackle some of the most pressing challenges faced by society – and all the other SDGs. It is a key factor in reducing the inequalities that are rampant across societies, including gender inequity; and it is the foundation for helping to expand the opportunities of youth to move

M. B. Mohiuddin (✉)
Department of Geography, West Virginia University, Morgantown, WV, USA
e-mail: mb00061@mix.wvu.edu

M. Jabot
Department of Curriculum and Instruction, State University of New York – Fredonia, Fredonia, NY, USA
e-mail: Michael.Jabot@fredonia.edu

P. Solís, M. Zeballos (eds.), *Open Mapping towards Sustainable Development Goals*, Sustainable Development Goals Series, https://doi.org/10.1007/978-3-031-05182-1_8

into high-skill jobs that will build the resource base to allow peaceful societies to exist.

But the real challenge is how best to meet this goal with those that can lead this change, that being the educators and learners themselves.

Well before the development of the Sustainable Development Goals, teachers were recognized as a key factor in the development of understanding of global issues (the Belgrade Charter [UNESCO 1976]; the 1977 Intergovernmental Conference on Environmental Education [UNESCO 1978] specifically refers to preservice teacher education and in-service teacher education and calls for teacher education to address global issues as part of teacher preparation. These early recommendations were framed around the idea that teachers need to understand the importance of a sustainable emphasis in their teaching and that this overarching idea should be included in curricula for preservice teacher education and be part of in-service teacher professional development (UNESCO 1978: 35–36).

Likewise, the role of students as the drivers of their own educational experiences has a lengthy history, evidenced by many variations of engaged pedagogies and active learning techniques. YouthMappers recognizes and honors both roles in its student-led faculty-mentored design. This chapter explores the idea of how education for SDGs can be activated through these activities, offering our perspective as two co-authors who are student and teacher, respectively, from opposite sides of the planet.

2 YouthMappers Among a Generation of "Solutionaries"

These tenets for education for the SDGs have been cast around the idea that educators have the ability to guide and empower learners to make change possible. Zoe Weil has described this as having the goal of "developing a generation of solutionaries" (Weil 2016). This term describes students specifically educated in new ways to create a more sustainable, equitable future for all. When learners participate and are empowered as

"solutionaries," we/they have the ability to apply our/their knowledge coupled with critical reflection tools to act on and bring others along in making changes that improve the world. Mapping is one of these tools that YouthMappers promotes in many of the same ways and with the same spirit.

Like teachers, students want their work to matter. It is important to most that our work matters not only to us but to others, our friends, our families, and our community. But a larger question that exists is, "Does our work matter to those outside our realm of immediate influence?"

2.1 The Spirit of Solutionaries at the University of Dhaka

The students at the University of Dhaka in Bangladesh were among the very first chapters to establish within YouthMappers. Growing from chapter local leader to regional ambassador to global leadership fellow within the YouthMappers programming, one co-author has followed an empowered trajectory, which could be characterized as one that is followed by solutionaries. Following this, she participated in one of the assignments of the National NGO Program on Humanitarian Leadership (NNPHL) workshop organized by Concern Worldwide, International Medical Corps, and Harvard Humanitarian Initiative.

These experiences led to sharing this learned ability to apply skills, ethical reflection, mapping tools, and fieldwork to act on and bring other students along in making changes that create a greater gender equity. As cultivating the inspirational entitlement of young leaders, we have shared the lessons with fellow students at the University of Dhaka by organizing interactive training, which was combined with a technical session and fieldwork. The students were introduced to the YouthMappers consortium, the use of software related to mapping, and arranging fieldwork exercises involved in interviewing internally displaced women in the Moghbazar slum area about the issues behind mental stress and stress triggers in where the students have used Kobo Toolbox and Mapillary for gathering geolocated survey data (Figs. 8.1 and 8.2).

The internally displaced women are moving from disaster-affected remote areas to the capital city of Dhaka in search of job opportunities and improved living conditions. This migration decision puts them under threat of living in new environmental territories and determinants of cultural changes. However, engaging in the mapping activities made internally displaced women more aware of safety through their experience, using spaces and place values, control over resources, the balance of rights, livelihood strategies, and knowledge production in terms of rapid economic and ecological transformation. Through participatory mapping activities, the internally displaced women are concerned with specific safety issues, sanitary facilities, gathering cultural and ecological information, and identifying unsafe areas such as insufficient lighting on the street and dark places. Students working with these projects learn firsthand and gain an education unlike many others.

In Bangladesh, the threats from climate change, natural disasters, and internal migration are treating women as vulnerable human beings. Expanding the participation of many women mappers and leaders in the YouthMappers chapters in Bangladesh has motivated and encouraged

gender-inclusive mapping communities to assist with women's needs. Completing a circle of education, the mapping activities on spatial data are openly available where schools and clinics have supported quality education, women's health services, economic productivity, and improved knowledge production. Mapping allows female mappers in Bangladesh to develop community livelihood, leadership experiences, empowered capacity building, and advance economic development by creating new geospatial data for development projects in places vulnerable to natural and human-made disasters. The involvement of female mappers is one of the most effective tools to assist in humanitarian services to build resilient communities, help address women's issues, make inclusive mapping communities, and promote education for the following generation.

2.2 The Spirit of Solutionaries at the State University of New York at Fredonia

The State University of New York (SUNY), Fredonia, is proud to have been an inaugural chapter for YouthMappers. The work at the State University of New York at Fredonia (SUNY

Fig. 8.1 YouthMappers interview internally displaced women in the Moghbazer settlement

Fig. 8.2 A group discusses mental stress of displacement in a community conversation convened by YouthMappers

Fredonia) and our YouthMappers chapter, Geoventurers, has revolved around the infusion of humanitarian mapping into the learning and teaching of science. SUNY Fredonia has a long history of teacher preparation, and this history has been enhanced by the inclusion of the incredible resources and support of YouthMappers. The focus of what is shared here is to cast a small glance at how the presence of the Geoventurers YouthMappers chapter has helped shape the academic work that the educator co-author oversees.

A question we could ask as an educator is, what if the work we encouraged in our students could serve to immediately change lives for the better in the real world? What would be the possibilities of such an approach to inspire students to even learn more? Again, we have a long-standing awareness of how we can focus on Education for Sustainable Development (ESD). The proceedings of the UNESCO World Conference on Education for Sustainable Development (UNESCO 2009) set forth recommendations for the integration of ESD into teacher preparation and practice. The recommendations encourage the development and extension of ESD to include all sectors (civil society,

public and private sectors, NGOs, and development partners) involved in the overall goal of sustainable development; to support ways in which sustainable development issues can be integrated in a systematic way across all education settings with particular focus given to effective pedagogical approaches, teaching practice, and curricular learning materials; and to reorient teacher preparation programs around sound pedagogical practices in introducing ESD. In the courses at SUNY Fredonia, the recommendations made for ESD are addressed in a number of ways with humanitarian mapping playing a central role. In the *Earth as a System* course, a general education course, the science content is focused on helping students understand the interconnected nature of natural systems.

3   Recasting Science Education for Human Solutionaries

As our unique yet connected experiences demonstrate, education for solutionaries needs to be grounded in real-world experiences, leverage fieldwork, focus on interconnected systems, and be inclusive. While an interconnected approach is common in science education, the content presented looks at the ways that the natural world interacts in the realities of humans around the world. The central tenet of the *Earth as a System* course is to focus on the science of "wicked problems" (Kreuter et al. 2004) and how the science and the cases studied play out differently depending on the reality of those impacted by the challenges studied. The use of humanitarian mapping around these topics (e.g., access to water) allows for the students to contribute to projects that are seeking to help others in need and does so in the temporal frame of issues that are impacting others right now.

Our connections to humanitarian and participatory mapping have ranged from addressing the call around natural crises that have occurred, like the flooding caused in Puerto Rico by Hurricane Sandy in 2012, where many of the students had close personal connections to the communities that were impacted, to using mapping technologies to look at issues around water in our local community. This opportunity has allowed students from across all majors to bring their disciplinary perspectives to the work in addition to the science content being studied. This opportunity directly addresses the recommendations made by UNESCO for ESD where students are allowed to extend their understanding beyond typical boundaries to include a systematic application of their learning.

While the *Earth as a System* course is a general education course open to all majors on the SUNY Fredonia campus, it is a required course for all students who are pursuing teacher certification. The ability to have this course and the experiences they have with humanitarian and participatory mapping as a common experience among these future teachers is an incredible opportunity to encourage the ideas experienced here to be put into practice in classrooms when these teachers enter their own classrooms.

The transition from the *Earth as a System* course to science methods course offerings has been shaped around the recommendations and findings around ESD discussed previously but most importantly a shift to the idea of "teaching

with and not just about ESD" (Zamora-Polo and Sánchez-Martín 2019). The idea of teaching with and not just about the SDGs allows for the development of the recommended pedagogical approaches in the context of near-real-time data and occurrences.

The work in the science methods course is typically shaped by attempting to address the incredible challenge of how best to help our students reflect on the way that they have learned science across their lives and to reflect on how their mode of learning may, in fact, need to be modified as they think about teaching science to others. The benefit of having the shared experience of the *Earth as a System* course is that the process of this transformation has begun based on their own learning, and we can continue in the science methods course.

The pedagogical strategies shared in this course have as their legacy the incredible work developed in the TeachOSM curriculum (Cowan and Hinton 2014). The inclusion of TeachOSM strategies and the Tasking Manager designed by TeachOSM allow future educators to see ways that collaborative mapping with a humanitarian focus can be seamlessly integrated into their work.

The use of these strategies is then centered broadly around the Geo-Inquiry Process (National Geographic Society 2017) where the future teachers design instructional modules that introduce their students to ways in which they can look at significant issues of interest at varying scales from local communities to the larger global perspectives. The Geo-Inquiry Process shapes students' thinking by asking questions that shape their thinking and understanding through the collection of data that helps answer these questions. It is in this data collection and visualization that humanitarian and participatory mapping come most into play.

If we return to the question of access to water, a module shaped by this process would look at how access to water on the shores of the Great Lakes is quite different than if we look at access to water in other regions of the world. What is critically important to this is the idea that through the use of participatory mapping, students can investigate if what they believe to be a "non-issue," access to clean water along the Great Lakes, is in fact a much more complicated issue. Students in this module could, in fact, map points of access to water around the community and investigate if, in fact, we should not be worried about this access. In light of the understanding of the importance of the human right of access to clean water, the students can also investigate ways that they can use humanitarian mapping tools to help others around the world have the same information to help inform their communities. It is in the focus on this type of learning that can foster their understanding that their work plays in global collaborative community.

4   Tying It All Together

Young people are considered the "solutionaries" in terms of humanitarian mapping because they are crossing the trans-boundary borders of continents to mark the meaning of their existence on the world map by creating sustainability around us. Mapping is a way of knowledge of the production of lived experience, which is preparing the youth to be solutionaries. Young people as "solutionaries" are bringing knowledge and skills to solve the development challenges through

humanitarian mapping to bring positive changes in communities.

Some of us educators have been taught that science is the foremost objective and that this objectivity is what, in fact, gives science its "power." But as we think about a shift in science education away from the process of science being driven by the scientific method to science being driven by inquiry and a greater focus on the practices and process of science, there is an increased need for welcoming a level of subjectivity that comes from investigating how science plays a role in people's lives and that this role is often different because of the variables that shape people's lives. It is in this that the application of humanitarian and participatory mapping in both the structure of the content that future teachers will develop and the use of these technologies as a pedagogical tool that empowers students to contribute to their communities, both locally and globally. It is through humanitarian and participatory mapping that future learners will develop real-world experiences that empower profound change, both locally and globally (Fig. 8.3).

Speaking as an educator, for most of my formal education, I didn't much think about who actually was responsible for the scientific advances I was learning about nor what was going on in the world at the time those advances were happening. It wasn't until later that I began

Fig. 8.3 Mapping quantitative and qualitative data reveals significant learning opportunities for university students

to truly appreciate what science as a human endeavor means. Fortunately, because of the role that YouthMappers has played on our campus, the students I work with have opportunities to explore the interactions between science and society in ways that I did not, and in the case of our future teachers, take this learning into their own classrooms to help shape this understanding in their students.

From the perspective of learners, we can share the idea that as we start to raise the awareness of the challenges faced in the real world, we begin to grow in understanding of the way that what we have learned about social issues in humanities courses interacts with what we have learned in other areas, like the sciences (Solís and DeLucia 2019). Often, we as students react passionately when we encounter information about injustices that occur around the world, especially when we see these injustices as being couched as "wicked problems" where solutions to lessen inequity are known but not applied. The debate and fervor for discussing these inequalities often end as the class comes to an end. But what if we could find ways to link the passion of teaching and learning with the opportunity to take action. This is where YouthMappers has had the greatest impact on our work.

Students don't just learn about problems faced by real people in the real world; educators don't just teach about them; both actually can contribute to do something about them. Our work with humanitarian mapping offers deep inspiration about the impact that education for the SDGs has to offer and removes the all too frequent question of "why am I learning this?" This work can serve as the inspiration to focus on innovation and entrepreneurship and to engage in creative thinking well beyond the academic classroom. It acts to empower a generation of mappers as "solutionaries," bringing together the interdisciplinarity of the social sciences and natural sciences to broaden understanding and purpose.

References

Cowan N, Hinton RA (2014) TeachOSM. Foss4G, Open Source Geospatial Foundation (OSGeo)

Kreuter MW, De Rosa C, Howze EH, Baldwin GT (2004) Understanding wicked problems: a key to advancing environmental health promotion. Health Educ Behav 31(4):441–454

National Geographic Society (2017) The geo-inquiry process. Available via NGS. https://www.nationalgeographic.org/education/programs/geo-inquiry/. Cited 20 Jan 2022

Solís P, DeLucia P (2019) Exploring the impact of contextual information on student performance and interest in open humanitarian mapping. Prof Geogr 71(3):523–535. https://doi.org/10.1080/00330124.2018.1559655

UNESCO (1978) Intergovernmental conference on environmental education: Tbilisi (USSR), 14–26 October 1977. Final report. United Nations Educational, Scientific and Cultural Organization, Paris

UNESCO (2009) World conference on education for sustainable development: the Bonn declaration. Accessed at: www.esd-world-conference-2009.org on 4 October 2009. United Nations Educational, Scientific and Cultural Organization

UNESCO-UNEP (1976) The Belgrade charter: a global framework for environmental education. International Environmental Education Workshop, Belgrade

Weil Z (2016) The world becomes what we teach: educating a generation of solutionaries. Lantern Books, Brooklyn

Zamora-Polo F, Sánchez-Martín J (2019) Teaching for a better world. Sustainability and sustainable development goals in the construction of a change-maker university. Sustainability 11(15):4224

Seeing the World Through Maps: An Inclusive and Youth-Oriented Approach

Shraddha Sharma, Courtney Clark, Sandhya Dhakal, and Saugat Nepal

Abstract

Through YouthMappers, young women students are able to design and participate in activities that collect, create, and disseminate spatial data and information to prepare spatial databases and maps for visualization, analysis, planning, and decision-making in their local communities. Because the field of geography is dominated by men, the contributions, needs, and priorities of women are frequently overshadowed and ignored. We cannot imagine a society labeled as fully mapped if it does not represent all its members. Through its innovation programming, YouthMappers aims to engage future generations of women mappers to reduce inequalities (SDG 10) and support gender equality (SDG 5).

Keywords

Gender · Equality · Women · Nepal · Everywhere She Maps

1 Maps and Their Makers Often Exclude Women's Perspectives

Oftentimes the world of women cartographers seems to be hidden, much like the so-called dark side of the moon.

Will C. van Den Hoonaard wrote this in his book, *Map Worlds: A History of Women in Cartography* (published by Wilfrid Laurier University Press), which was shared by Cathy Newman on her blog post for National Geographic, "Mapping Out the Hidden World of Women Cartographers." Many women have contributed significantly to the fields of mapmaking, GIS, surveying, geospatial technology, and more, but oftentimes those who write history and control resources fail to mention their names. The contributions and names of some

S. Sharma (✉) · S. Dhakal · S. Nepal
Tribhuwan University, Institute of Engineering, Kathmandu, Nepal

C. Clark
YouthMappers and American Geographical Society, New York, NY, USA
e-mail: cclark@americangeo.org

P. Solís, M. Zeballos (eds.), *Open Mapping towards Sustainable Development Goals*, Sustainable Development Goals Series, https://doi.org/10.1007/978-3-031-05182-1_9

women geographers may be lost entirely and their work made invisible by institutional and individual gatekeepers of geographic knowledge. Vincent Varney featured three often forgotten women geographers on his blog in a post titled "3 Women Mapmakers Who Changed the Way We View the World," including Marie Tharp, Kira B. Shingareva, and Florence Kelley. Marie Tharp was an oceanographic cartographer who created the first scientific map of the ocean floor. Florence Kelley was a social and political reformer who created a map of Chicago showing demographic information that helped improve the lives of residents living in poverty. Kira Shingareva was the first cartographer to map the dark side of the moon. They revolutionized the cartographic field and helped us change our perspective of how we conceive the world. But sadly, very few of us discuss their names when we talk about innovative and leading cartographers.

> Maps are representations of the world. Whoever is making the map is showing what they are perceiving that reality to be, they are privileging and prioritizing elements and attributes. Having a more diverse range of mappers means we actually have richer information on those maps – not just for the women to use, but for everyone.

Maps have always been symbols and instruments of power that strive to represent the reality of our world, a sentiment captured in the quote above from YouthMappers co-founder and director, Dr. Patricia Solís. Clearly, it requires great effort from engineers, cartographers, and volunteers to design maps that we use in our day-to-day life, from finding the best route during the busiest time of the day to having a parcel delivered to our footsteps. Humans have made maps from the earliest periods of our existence, but what is needed now is a much stronger focus on inclusivity. Inclusiveness in maps ranges from involving people of different races to people of different genders to people of different economic standards to people of different age groups. A global map should represent everyone, regardless of their origin, color, race, gender, ability, or ethnicity.

Maps are essential to many sectors and should contain more diverse data. For this to happen, increasing diversity among mappers is vital. Gender inclusion in maps seeks to contribute to the global community by collecting and creating

critical data to amplify evidence-based interventions that can be sustainable, impactful, and scalable (Moloney 2020).

Because the field of geography is still dominated by men, the contributions, needs, and priorities of women are frequently overshadowed and ignored. This problem is visible in both the OpenStreetMap (OSM) database and the network of communities built around OSM. OSM contributors aren't asked about their gender when they create their account and establish their username, so it is impossible to know exactly how many women or nonbinary people add data to the platform. However, based on the results of a 2021 survey hosted by the OpenStreetMap Foundation, it is safe to estimate no more than approximately 8% of OSM contributors fall in these categories. Some experts estimate that as few as 2–5% of contributors are women or nonbinary. In this era of an increased focus on inclusivity, women are still significantly underrepresented in the geospatial industry. By comparison, YouthMappers engage a much higher rate of female and nonbinary mappers: Some chapters are 100% women, and overall among the nearly 300 chapters in 60+ countries, estimates of female participation fall in the range of 35–45%. While there are many other steps to take to ensure full participation and inclusivity, YouthMappers has shown through its design and programming that it could be a mechanism to potentially reduce inequalities for mapping, OSM, the fields related to geography, and technology in general. But what is even more important are the potential outcomes that inclusive spatial data created by an inclusive community could generate.

2　Missing Data → Missing Representation → Missing Societal Development

Accurate and complete data is critical to the success of development initiatives and to all of the SDGs. Data users range from simple route navigators to policymakers and planners. But what if the data we rely upon miss critical information? Even riskier, what if we are unaware of the impact of missing data? Sadly, this is often the case of our present world.

In 2019, Melinda and Bill Gates (2019) included the following thoughts in their annual blog post:

> Data helps us create goals and measure progress. It enables advocacy and accountability. That's why the missing data about women and girls' lives is so harmful. It gets in the way of helping them make their lives better. The data we do have – data that policymakers depend on – is bad. You might even call it sexist.

We cannot imagine a society labeled as fully mapped if it does not represent all its members. The work needed to collect and analyze data can seem tedious. But what's not tedious is using this data to empower millions of women and girls. When women map, they are more likely than men to represent women's specific needs and priorities, which is key to driving changes in local policies, plans, and budgets.

Geographic attributes, details, and services that are important to one group of people might be overlooked by other OSM contributors, which ultimately results in a biased map. Data cannot be decoupled from the biases of those who create, collect, and analyze them. Services such as bus parks, childcare, women's health clinics, safe routes at night, washrooms, police stations, streetlights, and organizations that specifically help women may be overlooked by men mappers and absent from OSM due to the lesser number of women mappers involved.

Aishworya Shrestha, a research assistant at Kathmandu Living Labs, a leading civic-tech organization in Nepal, told us the following:

> Since data drives most of our world today, algorithmic bias plays a huge role in what is perceived as "real" and to what extent it is real. These are not gender specific problems, these are societal problems. It's not the battle of the sexes, it's a struggle between all of us and the power system. We need more inclusive data for the world to be more inclusive. Here, merely being NON-SEXIST is not enough, we need to be ANTI- SEXIST.

The unsafe streets of the cities and towns can be mapped using, in part, OSM data, which not only helps secure the well-being of individuals using the maps but can also provide important insight to planners and policymakers who seek to address security problems. Details shown by more gender-equal maps can help planners and policymakers identify the changes to improve public spaces, services, and facilities in order to make the community a better place for everyone.

3 Bridging the Gender Gap Through Everywhere She Maps

In 2020 YouthMappers launched the Everywhere She Maps program with the aim of increasing women's participation in the network and strengthening the inclusiveness of the geospatial community to ensure women's perspectives are represented in apps, websites, and mapping platforms. Through Everywhere She Maps, the next generation of women mappers will build their technical capacity, enhance their professional and networking skills, and contribute to mapping projects focused on adding data relevant to women's needs, with the effect of improving security, saving lives, spurring innovation, and increasing prosperity. The figure below illustrates the four interconnected core activities of Everywhere She Maps (Fig. 9.1):

3.1 Professional Development and Internship Matching

Women and nonbinary individuals who pursue careers in geography, data science, technology, and other STEM (science, technology, engineering, and mathematics (STEM) fields often face challenges due to their gender. Just a few of the many barriers to women's full participation in these fields are:

- False stereotypes about women's ability to perform and their men counterparts in STEM courses and careers.
- A lack of women in leadership roles in university geography departments and the geospatial industry, for example, only 22% of GIS executives globally are women (*GIS Lounge*).
- Paternalistic perceptions among some employers that women should not be allowed to con-

Fig. 9.1 Four interconnected core activities of Everywhere She Maps appear in the framework designed for the project

duct geographic fieldwork for the women's own safety.

- Impostor syndrome, which is especially common among women, who are more likely to doubt their own abilities and skills when they work in fields dominated by men.
- Unequal distribution of domestic and care labor between women and men: Globally, women spend about 4.2 h daily on domestic and care work, while men spend about 1.7 h. As a result, women have less time to participate in extracurricular activities – such as attending an OpenStreetMap training after class – while they are university students and to engage in networking, volunteering, and other critical career-building activities once they graduate.

To help women and nonbinary YouthMappers students overcome these barriers, we have developed an Internship Match program and series of online professional development sessions. Through the Internship Match program, Everywhere She Maps pairs qualified YouthMappers participants who identify as women or nonbinary gender with internship opportunities from YouthMappers' partners that will allow them to develop professionally and strengthen their geospatial and technology skills as they help their host organization achieve its goals.

3.2 Mapping Campaigns for Gender Equality

Everywhere She Maps is also leading mapping and geospatial data activities that invest in women and girls' security, livelihoods, prosperity, and participation in innovation and the digital economy.

One of our current mapping campaigns includes mapping rural locations in Sierra Leone, Nigeria, and Niger to increase remote communities' access to adequate electricity. Power distribution networks in these countries are scarcely mapped, especially in rural locations. Mapping buildings helps us know where residents who use electricity live, and mapping roads provides a strong inference for where power lines are likely to be. This data will be used for electrification planning by the Ministry of Energy, national utilities, and other energy sector stakeholders in the target countries. Women and girls will especially benefit from access to modern and affordable energy in remote communities.

3.3 Women in Technology Leadership Fellowship

YouthMappers' Leadership Fellowship program brings together YouthMappers student leaders, partners, and staff from around the world for two weeks for an intensive, in-person, leadership development training. Everywhere She Maps will adapt this model to design and implement a Leadership Fellowship for Women in Technology (WiT).

The Fellowship workshops will cover topics such as challenges faced by women geospatial professionals and WiT as they prepare to enter and develop professionally as members of fields that have historically been and currently are male dominated, the principles of intersectional feminism and the unique challenges faced by women of color and women from the Global South, and the opportunities available to participants as they define and work to achieve their professional and personal goals.

Everywhere She Maps will also assist Women in Technology Leadership Fellows in receiving placement for external internships, fellowships, and employment and empower them to lead efforts to increase the number of OSM edits made by women and nonbinary YouthMappers chapter members.

3.4 Regional Ambassadors

Everywhere She Maps engages eight regional ambassadors from Asia, Africa, and Latin America who conduct outreach, host trainings, and provide mentorship to YouthMappers chapters in their region of focus, with the goal of increasing women's participation in the chapters and in OSM editing activities. These ambassadors are highly accomplished and respected members of the YouthMappers network and serve as inspiring role models for other YouthMappers students who are women or nonbinary (Fig. 9.2).

Airin Akter serves as an Everywhere She Maps regional ambassador in Bangladesh. She is a well-recognized leader in YouthMappers and OpenStreetMap and serves as a senior research associate for Capacity-Building Service Group in Bangladesh. Below, she shares her experience as part of Everywhere She Maps (Fig. 9.3):

Fig. 9.2 Team members take a drone selfie

Fig. 9.3 Everywhere She Maps Regional Ambassador participating in a training in women empowerment

Being a small-town girl, I have always dreamt to be a changemaker. After getting admitted to the University of Dhaka, Bangladesh, I started participating in different projects, unaware that women were considered less productive in various sectors of job and education. In the course of exposing myself to available opportunities, I initiated the mapping in OpenStreetMap. In 2016, we started YouthMappers chapters at the university. Afterward, I began to organize activities like mapathons with some of my fellows. It was the first time I experienced sexism when many of the male students started humiliating me, 'Oh!! That girl!! What would she do? This program seems like a disaster to them!' Rather, I took those comments as the fuel to expand more, to prove of being capable.

Everywhere She Maps Project. I am overwhelmed to get selected as one of the regional ambassadors under this project. I have learned many things about Gender Equality and Equity. What allures me is "Engaging Men in Women's Empowerment" which is a new thing to me. When I completed the training, I was fascinated by thinking about the dimension of the idea. If you want to empower women, you need to engage men in this campaign as women and men are essential parts of society.

4 Breaking Stereotypes: Mapping Modi Gaupalika

Shraddha, co-author of this chapter, first learned about OpenStreetMap (OSM) during Open Data Day 2016. Open Data Day is an annual celebration of open data with events around the world. Since then, she, along with her friends Kshitiz, Sovas, Susmina, and Sandhay frequently heard about the applicability of OSM in post-disaster response activities and humanitarian activities. However, they hadn't had the opportunity to use it themselves, until one of the group members came up with the idea to start a project called

During my Master's, I took the course 'Gender Development and Environment' in which I got to know more about gender equality, and it attracted me a lot. At that time, YouthMappers published their Research Fellowship Program, where I applied under the supervision of one of my professors. She instructed me rigorously, and I happened to get selected as one of the YouthMappers Research fellows where my study was about Gender Equality and Awareness of Climate Change. And then comes the

"*Mapping Modi Gaupalika*." Here, Shraddha tells the story of the project from her perspective.

The main objective of this project was to collect, create, and disseminate spatial data and information to prepare spatial databases and maps for visualization, analysis, planning, and decision-making in Modi Rural Municipality. We officially started "*Mapping Modi Gaupalika*" at the end of September 2018, and we collaborated with the municipal bodies of Modi for the smooth execution of the project.

Kshitiz was the leader of the project and a resident of the municipality. At the end of September, we received a call from Kshitiz that we were ready to begin with a field survey as the first step of the project. We were informed that it would take us approximately a month to collect the data and were supposed to travel the whole municipality on foot. "*How are we going to be treated?*", "*Are the routes even safe to travel?*", and "*What if we don't get proper food and a warm place to sleep?*" were just a few of the questions that lingered for days. With all the doubts and confusion, we asked our parents for permission. Surely, being overprotective Nepali parents, they were skeptical about our travels. They were concerned about our safety; however, we were able to convince them. Being female,

our greatest fear was menstruation and its rituals that have been followed in Nepal for ages. Generally, people are strict when women menstruate. We must follow an arbitrary set of values that restricts our activities from the kitchen to the temples. So, we had a constant mental fear of how people would perceive the issues if it happened to us during our time in the field.

The next big question was about security. We had to travel through the forest, the paddy fields, dusty motor roads, and foot trails that were not traveled often. But the reality turned out to be so different from our fears. We had a wonderful time traveling through the villages (Figs. 9.4 and 9.5).

We collected information regarding institutions, health facilities, religious places, community centers, governmental offices, roads, foot trails, tourist attractions, agricultural farms, and more using a GPS survey. We used self-developed mobile applications for the survey. After completion of the field survey, we uploaded the data to OpenStreetMap. Then the data was preprocessed, checked for redundancies, filtered, formatted, and finally validated. The next step of the project was to prepare different kinds of thematic maps for which we used Quantum GIS (QGIS). Our final output contained various types of maps visualizing the following types of data: administrative boundaries of each of the eight

Fig. 9.4 We jump for joy when data collection is complete

Fig. 9.5 The team enjoys the scenic view from Poonhill after completing data collection

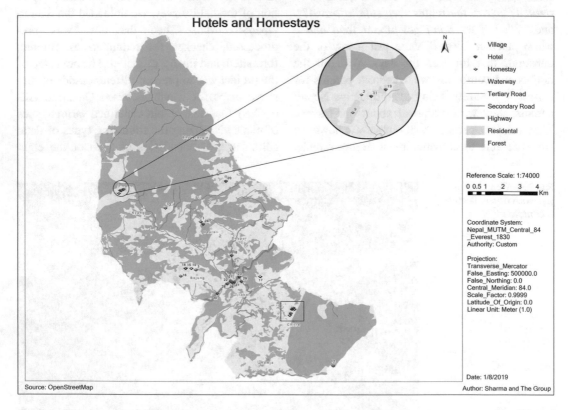

Fig. 9.6 Maps visualize hotels and homestays of the rural municipality and provide a transparent, open source of information for the whole community

Fig. 9.7 The
municipality uses the
resulting maps for the
construction and
maintenance of drivable
roads

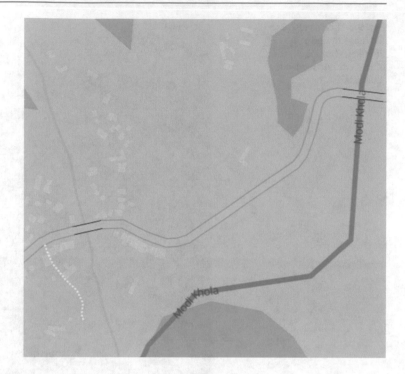

wards and demographics by ward, including population, population density, number of households, and literacy rate (Fig. 9.6).

4.1   Impact on the Community

The end products of the project formed the basis
for spatial decision-making for various stakeholders across the wards. All the data was
uploaded to OpenStreetMap and could be updated
easily. The municipality used the resulting maps
for the construction and maintenance of drivable
roads. It also became easy for them to monitor
the quality of education and the quality of services provided by various organizations and
institutions. Data was also now available to help
to promote tourism. We uploaded the contact
information of homestays and the hotels, which
helped tourists contact them more efficiently. The
locals also used the maps to find an alternative
route during landslide incidents. Because these
maps existed in an open spatial platform rather
than relying on the tabular information in the
municipality profile documents, there was
accountability and transparency to the users and
community (Figs. 9.7 and 9.8).

5   Achievements

Being female and growing up in a patriarchal
society, we had heard a lot about how we were
not suitable for fieldwork in our profession.
Regardless of our capacity, we were frequently
demotivated by these harmful stereotypes. Our
safety and our strength were so-called barriers to
our succeeding in our profession of choice. We
took the courage to break the stereotype and went
to the field. Walking through the rough terrain
and dense forests and interacting with the people
made us realize how wrong we were to doubt our
strengths. We realized that regardless of gender,
race, origin, and ethnicity, all human beings are
equally capable. Not only did we change our own
perceptions of ourselves – we were also able to
change the perceptions of our parents and other
people around us.

One of our proudest moments was inspiring
the local girls to engage in professions like ours,
which required frequent field visits. They were
curious about our work and asked if they could
pursue careers like ours. We could see ourselves
in their positions, as we too had similar doubts
before we set on this journey. We encouraged
them to pursue their dreams and tackle the chal-

Fig. 9.8 Locals also use the maps to find alternative routes as options during landslide incidents

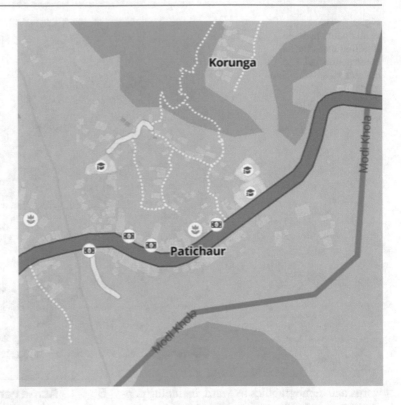

lenges that hinder women's involvement in field and technology-based careers.

6 Women in the Spotlight: Hearing from the Ones Leading the Chapters

Shraddha interviewed several women leaders of the Geomatics Engineering Students' Association of Nepal (GESAN) YouthMappers chapter about their experiences and perspectives on inclusivity and mapping. Their stories illustrate the powerful impact that YouthMappers' participation and leadership can have on a student's life and are reminders of the importance of ensuring women and nonbinary people have equitable access to the YouthMappers network. Sandhya Dhakal, vice president (3rd Executive Committee), shared the following:

I served as the vice-president of Geomatics Engineering Students' Association of Nepal, an inaugural chapter of YouthMappers for a year. The tenureship was full of passion, motivation which helped me enhance my communication and presentation skills as well as build up my confidence. We, the GESAN family conducted various programmes like mapathon, ArcGIS Training, Open Data Day Celebration, OSM training etc., where the whole team united wholeheartedly for better results. I participated in the programme 'Empowering Women in Geospatial Information Technology' organized by the International Centre for Integrated Mountain Development (ICIMOD) as a GESAN representative. There, I learned various programming and spatial analysis skills which I shared enthusiastically with my juniors through the chapter. In addition, we also published the yearly magazine 'Geoworld-II' overcoming numerous obstacles. As beginners, we lacked behind in many aspects such as proper strategies and abundant links. But the unity, dedication and zeal of the team finally led to the successful publication of the magazine which refined us in interpersonal, managerial, and professional capabilities. The chapter helped enhance the geospatial knowledge in Geomatics Engineering students, provided a platform for exposure of different issues as well as bonded all students as a single family.

Being a vice-president, I learned to conquer my insecurities and thrive in difficulties. It left me with a remark that the bond among the diverse people is what really matters in the prosperity of an organization.

Saugat Nepal, secretary (6th Executive Committee), told us:

I am currently working as a secretary for our YouthMappers Chapter, Geomatics Engineering Students' Association. Initially, I had very little experience about leadership and conducting a program. In November of 2020, I got to be a part of the 5 days Innovation Bootcamp program held at our campus. After the participation, I was very much inspired and wanted to implement the learnings. GESAN was the perfect opportunity for me to execute those. There were some slight reservations about if I was fit for the role, but my passion and hunger to serve for the chapter were the positive assets by my side.

After becoming the secretary of GESAN, our team has worked collaboratively and has organized various programs from which we have got positive responses. The responses really motivated us and pushed us harder to thrive for higher pinnacles. At the end of our every program, we used to take feedback as well as suggestions from participants which helped us to improve the program strategy as well as to fix our next move. We had to face a pandemic during our tenure: lockdown all over the world. Despite the challenge to be physically present, we were all motivated to follow in the footsteps of our previous committees and thus conducted programs like OSM training, Web Mapping, Drone data processing and many more virtually. We faced some issues in our very first virtual program, but those were the learning outcomes for us to improve as a committee. We had organised almost 10 programs during lockdown collaborating with many organisations like Kathmandu Living Lab (KLL), IT Maps, Naxa, and Geo 3D Modelling. Continuing the trend of our organisation, we also published the fourth volume of our annual GeoSpatial Magazine, "GeoWorld" which contains articles related to the Geomatics Field, interviews of the reputed personalities related to our field as well as suggestions from alumni.

While being a secretary, I got the opportunity to learn, teach, implement as well as sharpen my leadership skills. I also learned about time management, communication, and public speaking skills. But most importantly, the opportunity has taught me not to lose hope, stay motivated in the lowest situations and see my mistakes as the opportunities to improve.

7 Engaging Future Generations of Women Mappers

OpenStreetMap provides a platform for everyone to contribute to the global map, where their contributions are visible and saved. YouthMappers provides a global stage to amplify our voice of what we have been doing as students. Our YouthMappers chapter, Geomatics Engineering Students' Association of Nepal (GESAN), has made every possible effort to engage women and girls at different levels.

One of our highly successful programs is Map Literacy, an annual program organized by our female members who are motivated to educate high school students, especially girls, about the basics of map science and the use of OpenStreetMap. Women chapter members lead planning, coordination, and implementation of the program. After the program, young girls are very much motivated to see these women as role models leading a technical campaign and are aspired to expand their ambitions and reach for higher summits. The program provides an opportunity for YouthMappers students to share their knowledge with future generations, who will benefit from map literacy whether they pursue formal careers in geography or not.

Maps are integral to development and the face of a society. The data within should be something that every individual can rely upon, and it is possible to achieve this with youth leading the way and including underrepresented groups in the process. The inclusion of female leaders helps to increase the participation of females in YouthMappers, and such inclusion can help themselves and other females get more comfortable in the spotlight. The involvement of female mappers in field surveys helps break the stereotype toward female participation in fieldwork. Hence, organizations seeking better results and high-quality data should ensure the inclusion of women and youth on their teams.

Likewise, the achievement of true advances to any and all of the SDGs requires this level of attention – to gender-sensitive data about gender-relevant topics and an inclusive set of data creators – which at the same time will advance SDG 5, gender equality within and across nations, to the benefit of everyone, SDG 10 generally reducing inequalities in our societies and our world.

References

Everywhere She Maps (2020) From YouthMappers: https://www.youthmappers.org/everywhereshemaps. Cited 21 Jan 2022

Gates B (2019) Our 2019 annual letter. Available via GatesNote. https://www.gatesnotes.com/2019-Annual-Letter?WT.mc_id=02_12_2019_05_AL2019_GF-TW_&WT.tsrc=GFTW. Cited 21 Jan 2022

Moloney A (2020) "Visible women": feminist mappers bridge data gap in urban design. Available from Reuters, 7 March. https://reut.rs/2YFM3wH. Available from Medium, 11 March. https://medium.com/up42/visible-women-bridging-the-data-gap-in-urban-design-69abb84718e0. Cited 21 Jan 2022

Namitala G (2017) News. Available via Humanitarian OpenStreetMap Team, 17 February. https://www.hotosm.org/updates/2017-02-28_why_women_in_mapping_matters_-_mapping_for_women_and_girls. Cited 21 Jan 2022

Youth Engagement and the Water–Energy–Land Nexus in Costa Rica

10

Jasson Mora-Mussio

Abstract

The water–energy–land nexus methodology proposes land management treatment based on watershed areas, considering the Sustainable Development Goals (SDG) objectives of SDG 6 Clean Water and Sanitation and SDG 7 Affordable and Clean Energy, and the evolution of governance. This framework was applied in two volcanic watersheds using open-source geodata methods. The results explain the governance evolution by stakeholder's types and their activities around the water, land, and energy distribution and land evolution.

Keywords

Water · Energy · Costa Rica · Stakeholders · Land · Geodata

J. Mora-Mussio (✉)
Universidad de Costa Rica, San José, Costa Rica

1 Linking SDGs with a Nexus Framework

I propose to use a nexus methodology with YouthMappers techniques to evaluate the governance within a natural watershed location in my home country, Costa Rica. This idea links the water–energy–land nexus tropical model through the lens of open mapping, led by youth engagement in support of the Sustainable Development Goals (SDGs) with the local community's participation in Cartago, Costa Rica, including that of YouthMappers.

The objectives of my work were identifying the principal stakeholders and defining their relationship with the water, energy, and land SDGs and the position of local governance. However, geodata is not available for the watersheds. The principal challenges I faced included that the research and fieldwork needed to explain the hydrological and the governance

P. Solís, M. Zeballos (eds.), *Open Mapping towards Sustainable Development Goals*, Sustainable Development Goals Series, https://doi.org/10.1007/978-3-031-05182-1_10

relationship. In the end, this work shows the obligation and need to improve the communication among the multiple stakeholders who interact within the watershed and use the catchment resources – as well as with youth who can provide framework ideas, mapping, and input as to how to sustainably develop our water, our energy, and our lands, for our future. All figures and maps have been made using free and open-source software.

1.1   Water, Energy, and Land

To begin with, we used the nexus framework for evaluating the SDGs 6 and 7. The nexus or "the water, energy, and food security resource platform," as Mekonnen et al. (2020: 9) calls it, is an analytical framework that understands, in a methodical way, the monitoring of water resources in different socioeconomic sectors (Fig. 10.2). He uses a triple support sector: (1) water, (2) energy, and (3) land. Those pillars are the supporting index where all the socioecological practices and policies are implemented into the system. Nexus can organize all the trade-off of the socioecological system to demonstrate the current situation, allowing numerous methodical interpretations in the present, and use them to estimate the future SDG principles (Transforming our world: the 2030 agenda for sustainable development, 2021). My work will apply this to a site local to me and incorporate a vision imagined with YouthMappers' experience to assess the development.

1.2   The Birris and Paez Basins

The study area is located in two hydrological units called Birris and Paez in Costa Rica, Central America. Both are described as watershed units (geo-administration scale) and catchments units (natural hydrographical area). These water catchments have a geometric characteristic of 52.64 km² (Birris) and 33.36 km² (Paez). In addition, the perimeter of those watersheds is around 41.39 km in Birris and 49.88 km for Paez, and both are considered small basins by their morpho-dynamics and physical parameters. The Birris and Paez rivers are located on the northeast side in the province of Cartago, Costa Rica, exactly in the Alvarado and Oreamuno canton. Both originate from the Irazú volcano, the tallest volcano in the country. In general, the landscape is folded and faulted by volcanic activity since the Pleistocene age (Rojas et al., 2006). The Birris and Paez rivers cross through eight different rural villages in which are living around 500,000 and 600,000 inhabitants (INEC 2011). They are focused on agricultural activities, public transport, and local ecotourism (INDER 2016: 54). These rivers drain into the Cachí water basin, which is an important hydroelectric reservoir for all of the country.

1.3   Land-Use Distribution and Resource Management Context

The principal land use distribution includes agriculture activities and conservation areas, which include national parks and wildlife corridors. This area can produce more than 60 percent of all the vegetables for all the country, and some products are for international exportation (INDER 2016: 57). The agricultural products that are cultivated are potatoes, onions, pepper, cavendish, cilantro, and other kinds of green legumes (INDER 2016: 23; Solórzano 2004: 3). This area contains three ecological corridors that connect with national parks, Turrialba and Chírripo National Parks, and protected areas such as Carpintera and Cerro de La Muerte. On the national scale, these areas are connected to the Amistad International Park (Costa Rica–Panamá) and the Central American corridor. The principal objective of these corridors is the ecological connection between the local and international migration of species in all the American continents (Plan General de Manejo Zona Protectora Cerros de La Carpintera Gestión Participativa Para La Conservación 2012). All these watershed characteristics have a common point of comparison: the water resources related with the socio-ecological effects.

The resource management process is the responsibility of the Costa Rican State by the National Conservation System or SINAC (Brenes and Soto n.d.) linked to the Environment and Energy Ministry (MINAE), protected by the laws as the Canon (legal rate of water use) of water and the future water law (currently in discussion at the time of writing) can provide a legal basement of the resource conservation. However, the local monitors are responsible for the environment with the SINAC. Each city hall are partners in the resource management (Plan Gen. Manejo Del Parq. Nac. Volcán Irazú, 2013, p. 8). However, for these catchments, the local organism is the COBIRRIS-Paez (Watershed Commission of Birris and Paez) who are responsible for the management and planning of activities inside the basins. The most important thing in this commission is the local governance promoted by stakeholders as academic, government, private, and community leaders.

1.4   A Stakeholder Governance System Re-imagined by Youth

The COBIRRIS has an open governance composed of multiple stakeholders (Fig. 10.1), and everyone has the same level of relevance in the decision-making. The governance and policy relationship can be described as linear scale (Koontz and Newig 2014: 416–417). This framework could be explained to understand the current roles of stakeholders and their position in a geopolitical system. Using this framework exposes dependencies or independencies of the sociopolitical system, governance and government actions, impact and opportunities, and vulnerabilities and weaknesses in the management of resources.

The objective of the COBIRRIS is to discuss all projects and situations related to the water resources for human consumption, such as the state of the pipelines or maintenance of the water system, and the water evolution through the dry and precipitation seasons. The COBIRRIS can share tools, gear, and replacement parts when necessary to provide fresh and potable water for all villages within the watershed. Additionally, the COBIRRIS teaches about the environmental impacts and the watershed status, for example, the agricultural impacts and their evolution, the research of any situation that impacts the basin, and the funding of transdisciplinary studies. All the budget for each project, reparation, or any other logistic expenses are covered by institutional and community stakeholders. In the communication and logistic field, the COBIRRIS-Paez has developed an assembly each month and a WhatsApp group in which the group has open discussion and publicity/reports of present situations or future actions.

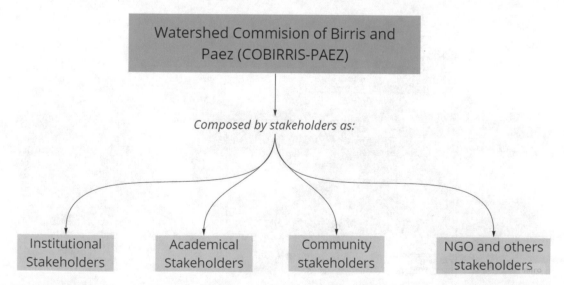

Fig. 10.1   Stakeholder relationships are framed for the Watershed Commission of Birris and Paez, Costa Rica

The COBIRRIS-Paez represents a non-linear governance framework where the stakeholders can present, help, improve, share, and grow the watershed management. However, the geographical system implementation and the open-source methods and technologies have not been used or implemented, opening a new line of research and applications. The current composition and status of the governance have not been evaluated in order to detect weaknesses nor then to propose opportunities for developing a better framework. This is where re-imagination from this project and the YouthMappers methodology could contribute.

1.5 The Role of Local Water Associations

These organizations are conformed to local leaders with the objective of the administration of the water for human consumption (AyA 2020: 12). These groups are coordinated by the National Administration of Aqueducts and Sewage (AyA). The ASADAS are responsible for the maintenance, operation, decontamination, and water supply for all villages. All the operations of the ASADAS are government sponsored by the AyA. Inside the COBIRRIS-Paez, ASADAS play a relevant role in the monitoring and detection of the water issues during the season, reporting of damage in any pipelines sectors, claims of the

replacements parts, and presentation of future projects where the funding and the capital will be aimed. Members of ASADAS living inside the Birris and Paez basins can understand the upcoming principal vulnerabilities and dangers for months and even years.

2 Methodology

2.1 Local Governance

Governance is a challenge, and any discipline can understand it in multiple ways. However, the governance is important to measure the government actions' effects and community resilience at any scale and space. The governance of COBIRRIS-Paez is related to volcanic space (folded with high slope rates) with the presence of a dendritic hydrographic distribution located in government policies and socioeconomic activities inside a watershed-scale and outside a regional-scale process.

One scientific method that can show the principal variation and many interpretations of the governance is the nexus framework. This methodology was created to measure the hydrologic governance status, and the nexus will expose more accurate governance in a hydrological administration area (Fig. 10.2).

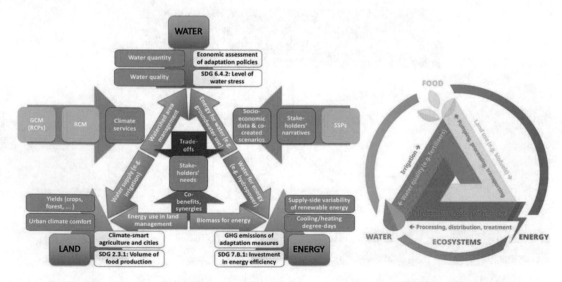

Fig. 10.2 A Land-Water-Energy nexus framework exemplifies complexity. (Adapted from Nexus.org, 2021)

In my own research, I adapted it to the local context for the case of COBIRRIS-Paez including (1) the local stakeholder's participation as a new fourth pillar and (2) using a timescale variation of the activities and actions of the administration. This last point took the communicative abilities of the commission for communicating their projects, actions, and achievements to the communities, institutions, or any other watershed commission. This communicative point is useful to see the divulgation and sharing of good practices to others. This measures the influence of the actions taken by COBIRRIS-Paez. My methodology required fieldwork to understand the system variations in the social, economic, and environmental fields.

2.2   Open-Source and Geodata Applications

Kobo Toolbox v1.29, an open-source application, was used in an android phone, version Android 9. The principal objective of the use of this application is the open and online survey that can be easily filled in the field. This application can create a form and be completed remotely making it easy to use and share with community leaders. This is a common tool for many YouthMappers to adapt to local projects.

Two forms were developed: (1) biophysical variables and (2) socioecological variables. The first one exposes all the geophysical and ecological variations as lithology, slope, soil color, and hand texture and photos with their geolocation as .gpx format file. The second one was made for the people, to understand their perception of the system and the acceptance of the COBIRRIS-Paez. This form contains all the observations, opinions, and commentaries about the performance or nonperformance of the commission.

Next, the National Space Agency (NASA) dataset called "Vertex ASF" was used to download the ALOS PALSAR product. ALOS is a digital elevation model (DEM) with a spatial resolution of 12.5 meters made by JAXA (Japan Aerospace Exploration Agency). This DEM is open and free to use and is one of the best products because of its spatial and temporal resolution for geosciences studies. The model will be used for delimitating each basin using the open software's SAGA (System for Automated Geoscientific Analyses), GRASS GIS, and QGIS platform. The version of SAGA used is 2.3, GRASS 7.8, and QGIS 3.1.5.

The delimitation of the basin is necessary to define the principal ecological area where the socioecological activities and the most relevant stakeholders can interact. The first action before managing any resources is the delimitation of the monitoring area (Shannak et al. 2018: 213). SAGA is a good idea for preprocessing any DEM and then returning some interesting hydrography products using algorithms. One of them is the Wang–Liu filling, which can fill and repair any "No data" from the DEM in order to ensure the preprocessing phase.

In Fig. 10.3, one can compare the pink shape, where SAGA basin was returned after the algorithm, to the light blue shape, depicting a geomorphological correction by topography isolines interpretation using QGIS 3.1.5 . The last one exposes the depuration shape.

Another method used is called "basic terrain analysis," which can return the automatic delimitation of the basin, the principal hydrography, the mean slope, the hypsometry, the analytical hillshade, and the topographic wetness index. These are necessary for the physical hydrological interpretation. The last algorithm can return more automatized products; however, these were not necessary for this study. After the detection of the basin, QGIS can help to improve and validate the data using a hand control between the SAGA shapefile and the local topography isolines, and in this case, the National Geographic Institute has corrected the topography isolines. This information of the IGN is open and free to use. The objective is return the best geomorphological interpretation to get the most accurate precision in the basin delimitation and then get located the

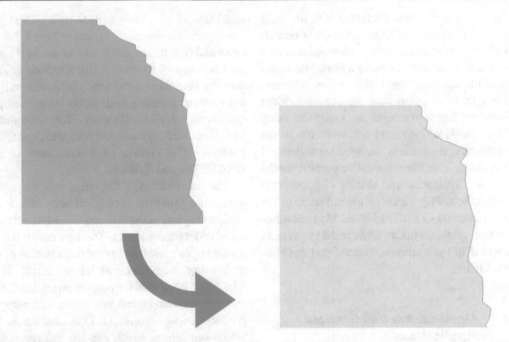

Fig. 10.3 The pink shape depicts the SAGA basin returned by the algorithm. The light blue shape includes geomorphological correction after use topography isolines using QGIS. This last one exposes the depuration shape

human and environmental components inside each watershed.

The location of each geographical information as enterprises, villages, agricultural crops, local tourism, and type and length of street was taken by OpenStreetMap (OSM) website free geodata information. QGIS plugin called "QuickOSM" version 0.19 (also free to use) was used. The IGN could provide access to all the geodata dataset for the basins. The government actions and the current policies bring a legal support of each action inside the watersheds. The Costa Rican System of Juristic Information (SCIJ) provided all the updated law and any statement bibliography. Similarly, the SINAC and MINAE have made available the access to all the documentation written for the basin and any other digital information.

Using all these international, national, and local datasets, the objective is the application of nexus methodology inside both watersheds using a new version of adaptive nexus for tropical and volcanic basins.

3 Putting the Methodology in Place

3.1 Geophysical and Spatial Modeling Interpretation

The application of Vertex (from NASA) DEM provided a spatial resolution of 12.5 ms. The SAGA preprocessing returned a list of algorithms, which each product has been analyzed for understanding the energy transfer in the hydrological system. The scale of the basins to subbasins can be described using the GRASS software, which can return the basin ≤100 ms (Fig. 10.4) registering 1,000,075 subbasins. This result could expose the high variation in the geographical scale that shows the socioecological system as the multiple fusions of little catchments and their little effect in all the watershed systems. Another contribution is the index list that SAGA can give. That list has appropriate importance for the local distribution between the socioecology spaces and their trade-off.

Fig. 10.4 Results are returned from the SAGA Basic Terrain algorithm products. Sub basin analyses used GRASS. Each unit colored represents each sub basin of 500 m² using the watershed module. The red lines correspond to each watershed of Paez (left) and Birris (right)

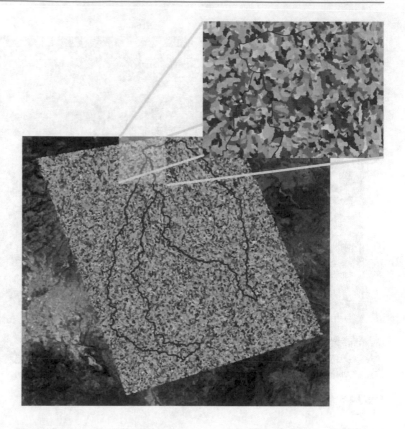

Figure 10.4 shows the resulting subbasin analyses using GRASS. Each unit colored represents each subbasin of 500 m² using the watershed module. The red lines correspond for each watershed of Paez (left) and Birris (right).

As an example, the topographic wetness index (TWI) explains the principal spatial humidity distribution inside a basin under the condition and orientation of the topography (Fig. 10.5). The distribution of water can be interpreted as potential sources of high ecology connectivity or a potential site of crop installation. The location of the water can be interpreted as the next water supply for the ASADAS in the dry season. The high distribution of water supply is commencing to be located downstream, a normal process in a volcanic basin by the water recollection of the upstream. Figure 10.5 shows the orange color as dry (lowest water concentration), whereas the light blue and blue colors represent moderately high water concentration levels.

Another example is the cross-sectional curvature (CSC) value, which is one principle of the landslide modeling. This index can expose the geomorphology variation of the topography units (mountains or any other orogenic structure) based on the geometry of their facies. That means the principle between concave and convex in the relief can be interpreted as a nonstructural tightness index (Fig. 10.6). The potential threat detection, especially of volcanic landslides (the vulnerability), could be detected inside the socioecological system. This volcanic landscape exposes the physical and potential hazards, which are the lava and pyroclastic flow or any volcanic eruptions and, more frequently, the earthquakes and volcanic tremors. The detection of this spatial variation locates the vulnerability and physical threats inside the watershed.

Figure 10.6 exposes the heterogeneous spreading of the CSC in both basins. The values ≤0 correspond to the convex geomorphology and ≥0 as concave geomorphology. Here we can find the rivers as maximum convex features and the principal mountain pikes as concave units. This kind

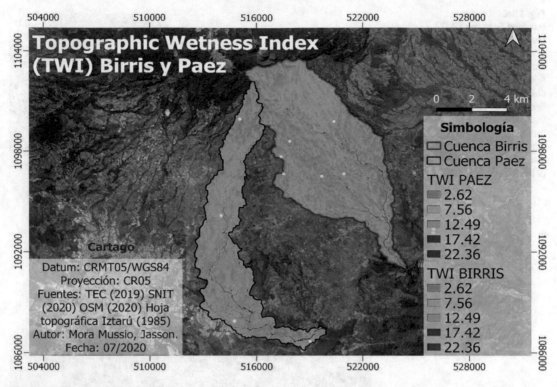

Fig. 10.5 The map shows TWI location for Birris and Paez catchments

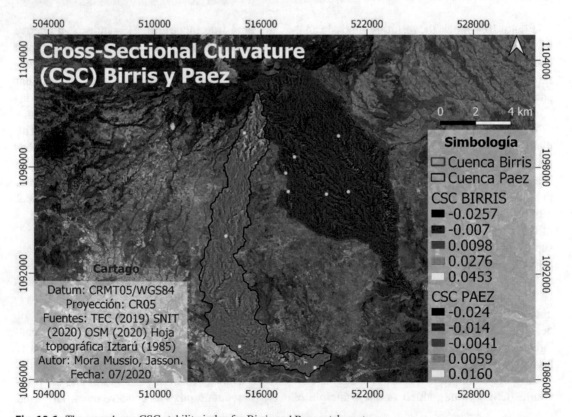

Fig. 10.6 The map shows CSC stability index for Birris and Paez catchments

of analyses can express and qualify the relief and how they could differ in these basins.

The morphodynamics and hydrographic variation can define the energy, amount of water, and water reservoir that each basin can store and the variation of each one with their area, perimeter, mean length, and width of the streamflow or using the Horton (low value = high water energy and vice versa) and capacity or Gravelius index where if the number is close to 1 that means high volume of water charge (for more, read Cardona n.d: 3–10).

3.2   Nexus Application

Based on the Nexus application (Fig. 10.2) in the COBIRRIS-Paez and including the biophysical variables described in adobe, the result is in Fig. 10.7.

The timescale and climate variables are included to understand better the governance relationship inside and outside the hydrological system.

Figure 10.7 was made similar to Fig. 10.2 in order to compare both results. In the center of the diagram, we find water, energy, and land (SDGs), which express the relationship with the COBIRRIS-Paez stakeholders using a line. Each line connects these three pillars and the COBIRRIS-Paez influence. Light blue lines are the trade-off between the SDGs, yellow squares are the negative effects in the socioecological system, and green squares are the action taken by the COBIRRIS-Paez to reduce or improve better impacts. The brown lines expose the indirect causes that are outside the basin system. The policy action that promotes open and updates information is one of these indirect causes and a local climate variability.

4   Discussion of Strengths and Challenges

I presented a nexus framework application in a tropical volcano landscape in Costa Rica, using open geospatial data and tools and YouthMappers

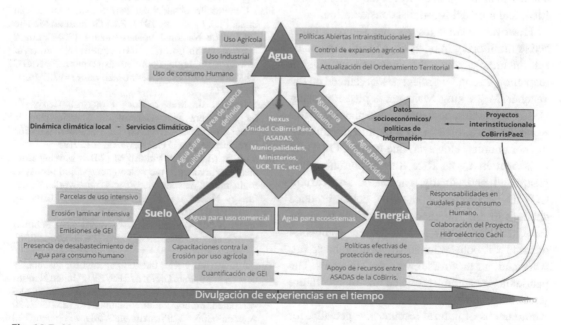

Fig. 10.7 Nexus approximation results from this analysis using the COBIRRIS-PAEZ context. The time scale and climate variables are included to better understand the governance relationship inside and outside the hydrological system

experiences. The phases to obtain all the data were mixed, including fieldwork and virtual meetings. Platforms such as Android and Windows have the most accurate results using free and open software. The data and geodata information were open and free to use. The geophysical analysis shows two heterogenous basins due to the closer distance (around 15 km). The COBIRRIS-Paez nexus exposes an adaptive framework that recalls, by a competitive way, the race between the negative impacts and the local management actions for the restoration of the system. The nexus exposes the land management inside both basins and the quick detection of negative impacts.

Access to open and up-to-date information is a right of the people because misinformation produces vulnerability (UNDP 2021). This study could get updated information and free data (as necessary for the scale of research) to understand the governance variation and biophysical effects. The conversation with the members and leaders of the community showed the pride in being part of the COBIRRIS-Paez and then the acceptance of this commission. This, in turn, advances SDGs 6 and 7 in an integrated way for the nation, Costa Rica, and a model for global consideration.

However, challenges for the future include active monitoring in open ways for all stakeholders, including youth. My research included a mapathon through a project task organized on the TeachOSM Tasking Manager, a platform using OSM for learning how to map. We planned to use Kobo Toolbox software as a monitoring tool for the community. However, due to the COVID-19 lockdown in Costa Rica, this opportunity was postponed and, finally, cancelled. The current COVID-19 pandemic situation could produce changes in the governance and actions taken by the COBIRRIS-Paez and in the environmental demand of both watersheds, setting back our work and our progress toward the UN goals. The pandemic must be accounted for in the future governance and governance studies around a whole nexus of natural resources, especially for the role of water, energy, and land.

References

AyA (2020) Aspectos básicos para la gestión de las nuevas Juntas Directivas de las ASADAS

Brenes C, Soto, V. (n.d.) Sistematización del Proceso de Creación y Desarrollo del Sinac. MINAE. Retrieved June 25, 2018, from http://www.sinac.go.cr/ES/partciudygober/LibrosSistematizacion/Creacion y Desarrollo del SINAC.pdf

Cardona B (n.d.) Conceptos básicos de Morfometría de Cuencas Hidrográficas. Retrieved July 19, 2021, from http://www.repositorio.usac.edu.gt/4482/1/Conceptos%20b%C3%A1sicos%20de%20Morfometr%C3%ADa%20de%20Cuencas%20Hidrogr%C3%A1ficas.pdf

INDER (2016) Caracterización del Territorio: Cartago-Oreamuno-El Guarco-La Unión. https://www.inder.go.cr/territorios_inder/region_central/caracterizaciones/Caracterizacion-territorio-Cartago-Oreamuno-El-Guarco-La-Union.pdf

INEC, I. N. de C. y E (2011) X censo nacional de población y de vivienda 2011: características sociales y demográficas (INEC (ed.); 1a ed, pp 1–120). INEC

Koontz T, Newig J (2014) From planning to implementation: top down and bottom up approaches for collaborative watershed management. Policy Stud J 42:416–442. https://doi.org/10.1111/psj.12067

Mekonnen Y, Sarwat A, Bhansali S (2020) Food, energy and water (FEW) nexus modeling framework. Adv Intell Syst Comput 1069:346–364. https://doi.org/10.1007/978-3-030-32520-6_28

Plan General de Manejo del Parque Nacional Volcán Irazú, Pub. L. No. N° 1917, Plan General de Manejo del Parque Nacional Volcán Irazú 1 (2013) http://www.pgrweb.go.cr/scij/Busqueda/Normativa/Normas/nrm_texto_completo.aspx?param1=NRTC&nValor1=1&nValor2=74664&nValor3=92294&strTipM=TC

Plan General de Manejo Zona Protectora Cerros de La Carpintera Plan General de Manejo para la Zona Protectora Cerros de La Carpintera Gestión Participativa para la Conservación, 1 (2012)

Shannak S, Mabrey D, Vittorio M (2018) Moving from theory to practice in the water–energy–food nexus: an evaluation of existing models and frameworks. Water-Energy Nexus 1(1):17–25. https://doi.org/10.1016/j.wen.2018.04.001

Solórzano N (2004) Sistemas Silvopastoriles No. 3. http://www.mag.go.cr/bibioteca_virtual_ciencia/brochure_silvopast.pdf

Transforming our world: the 2030 agenda for sustainable development, Pub. L. No. A/RES/70/1, United Nations 1. Retrieved July 15, 2021 from https://sustainabledevelopment.un.org/content/documents/21252030%20Agenda%20for%20Sustainable%20Development%20web.pdf

Rojas MA, Vargas DD, Corrales KF, Araya LDH, Sánchez RV (2006) Amenazas y vulnerabilidad: El caso de los ríos Reventado y Toyogres, Cartago. Revista Reflexiones, 85(1–2):331–349. https://doi.org/ ISSN:1021-1209/2006

UNDP (2021) The Sustainable Development Goals Report 2021 https://www.undp.org/sustainable-development-goals

Power Grid Mapping in West Africa

11

Tommy G. D. Charles

Abstract

YouthMappers helps to map fundamental features of rural communities across the countryside of Sierra Leone as an innovative model for how to bring power to villages across West Africa. Charting location of buildings, tracing streets, and pinpointing where utility poles dot the landscape inform efforts to design and install mini-grids in places without power or with insufficient service. By understanding settlement patterns, road connectivity, and the layout of current low-voltage distribution networks, the team speeds up and scales up design for rural electrification, contributing directly to SDG 7 to bring affordable and clean energy to communities and to SDG 9 to build resilient infrastructure in ways that foster innovation for mapping and beyond.

Keywords

Energy · Sierra Leone · Power grid · Innovation · Infrastructure · Fieldwork

1 The Status of Power Access in West Africa

Limited access to information and electricity are related challenges affecting West Africans. In this chapter, we look at an innovative example from the YouthMappers chapters in Sierra Leone regarding both limitations. The lack of information, especially geospatial information, has made decision-making difficult and sometimes wrong or unreliable when they are made available. Similarly, access to electricity is also a huge challenge that affects the livelihood of West Africans, especially those in rural communities, where there is limited access to electricity when compared to urban communities and advanced communities around the world. Approximately, only 15% of the Sierra Leone's population have access to electricity that is derived from fossil fuel. The low level of access has had crippling effects on the economic growth and human capital development of the country. Women, girls, and children living in rural communities are the most affected. Economic development can be constrained in general when this industry is not

T. G. D. Charles (✉)
OpenStreetMap Sierra Leone, Freetown, Sierra Leone

P. Solís, M. Zeballos (eds.), *Open Mapping towards Sustainable Development Goals*, Sustainable Development Goals Series, https://doi.org/10.1007/978-3-031-05182-1_11

ubiquitous, and innovation is stifled. These issues directly speak to the need to address SDG 7 for clean reliable energy for all, as well as the larger development potential of West Africa represented by SDG 9, regarding industry and innovation.

2 Geospatial Solutions for Power Access

The growing need to access and use geospatial information at a more cost-effective rate and the provision of electricity across the country have led to the development of methodologies to create and utilize geospatial information that can be used in making decisions relating to the generation, supply, and distribution of electricity.

2.1 Trajectory of YouthMappers in Sierra Leone

Sierra Leone is experiencing significant growth in the use of renewable energy. There has been erection of utility poles, wiring, and construction of solar-powered stations across the country. To make the provision of electricity affordable and efficient, there must be reliable data and information such as the number of poles to be erected in the community, the approximate number of households, and the roads within the community to make rational decisions. Development partners like USAID and electricity companies realize the need to have reliable data about existing power grids in rural communities in Sierra Leone.

We were contacted by the YouthMappers director at Arizona State University (ASU) to coordinate a pilot phase of a power grid mapping project that would directly provide data for ongoing needs for information with research and development projects in the country. However, a methodology that smoothly links remote and field data had yet to be fully developed, so we worked collaboratively giving input for innovating new approaches to help speed up and improve data that is necessary for the grid system in rural communities, which goes beyond current approaches for the main national grid. This approach seeks to connect all the way to households, using artificial intelligence (AI) tools and remote and local engagement together from the start.

Before the intervention of YouthMappers in Sierra Leone, there was little or no information about electricity infrastructures in open source geospatial platforms. With the intervention of YouthMappers, utility poles are being mapped, especially in rural and suburban communities where solar grids have been erected and proposed sites across the country.

It is on this backdrop that YouthMappers embarked on extensive collection and processing of images that can be used for decision-making relating to the provision of electricity across the country. YouthMappers utilize open source geospatial tools to do the work and are receiving internal and external support.

Embarking on geospatial data collection in any form requires the appropriate skills, resources, and interests or enthusiasm. Over the years, since 2016 to be precise, the OpenStreetMap (OSM) community in Sierra Leone has been giving training to students in tertiary institutions and facilitating the establishment of YouthMappers chapters. These include five YouthMappers chapters in Sierra Leone, namely, the Students Geographical Association YouthMappers, Njala Mokonde YouthMappers, Canadian YouthMappers, and UNIMAK YouthMappers and the Eastern University YouthMappers. The first one was inaugurated at Fourah Bay College, University of Sierra Leone, where student pioneers began an influential presence for OSM Sierra Leone today. Because of being motivated to engage students in other tertiary institutions to benefit from the YouthMappers network, we made contacts with students at Njala University and the INGENAES network, and we organized our first mapping parties in 2017 where we introduced students and extension workers to OpenStreetMap and data collection tools.

2.2 Disruption of a Pandemic Drives Innovation

In 2020, in the heat of the COVID-19 pandemic, we still facilitated the establishment of the Canadian College YouthMappers, the University of Makeni YouthMappers, and the Eastern University YouthMappers chapters. These chapters include students from different fields of studies ranging from law, geography, computer science, and geology. One thing they have in common despite their diverse disciplines is the enthusiasm to create and use data for social good. Members of our chapters have been receiving and also giving data collection and visualization training and techniques over the years with the use of several open source geospatial tools. But we had to innovate given the shutdowns and disruptions of the pandemic.

To ensure that the required information for the power mapping project is acquired, YouthMappers that have the required skills were given refresher training to collect aerial and ground-level images in selected communities across the country. The captured images are processed to detect utility poles that are instantly mapped or added to the OpenStreetMap platform.

In order to get the whole methodology right, a pilot phase was implemented in five (5) communities to learn the challenges and the necessary steps or techniques we can apply to overcome them. Thereafter, we went on to map 20 more communities in rural and urban areas across the country. However, we still encountered significant challenges such as gaining authorization, liaising with community stakeholders, and logistical and technological issues. But with the right mindset and perseverance, the YouthMappers were able to overcome these challenges and implemented a successful mapping campaign.

2.3 Stitching Together Open Tools to Innovate Fieldwork

From the very beginning of the process of establishing the YouthMappers chapters, students were capacitated with mapping skills using open source geospatial tools such as OpenStreetMap, Mapillary, Pic4Review, MapRoulette, and other essential tools. This included augmented mapping using artificial intelligence (AI or GeoAI), which gave us opportunities to be a part of cutting-edge applications of science. Most students used these skills for their coursework and continuously improved on them, but they had not yet worked with GeoAI nor had we applied them in solving real-life problems within society. Therefore, the mapping of utility poles using open source geospatial and GeoAI tools was a perfect opportunity to showcase and grow our skills in addressing a national issue.

The information that was produced is already being utilized by government agencies and electricity companies to know the extent and status of electricity infrastructure. It helps them to do feasibility studies and other essential analysis that boost efficiency and service delivery and hasten and streamline their workflow. The partners at Mapillary and at Arizona State University and the YouthMappers Validation team at George Washington University (GWU) served as guides and support for the methods, validation, and linking to the partners who are implementing the data.

3 Local Fieldwork Implementation with YouthMappers

The on-site part of the power grid mapping methodology was implemented in two phases in general, a pilot phase wherein 5 communities were mapped with the help of two YouthMappers chapters and the larger implementation phase wherein 20 communities were mapped with the help of 5 YouthMappers chapters. We learned some lessons from the pilot phase and rectified them in the actual implementation phase. Together with all of the partners, we developed this portion of a robust methodology that represents innovation for youth engagement and project application in local sites.

3.1   Capacity Assessment and First-Pass Remote Mapping

At the start of the project, the capacity level of students was ascertained to better understand training needs and ensure the production of quality and reliable data on every open source geospatial platform that was used. The procedure that was developed included the strategic selection of locations where data was to be collected, most of which are rural and suburban communities across the country. The selection indirectly engaged the end-user (energy industry) needs, essentially interfacing with the national ministries and private entities involved in rural electricity access.

Thereafter, remote mapping tasks were created using the Humanitarian OpenStreetMap Team's Tasking Manager through the TeachOSM instance of the platform. Students had to remotely map and validate buildings and residential roads within the selected communities where utility poles were bound to be found or erected. This was done by YouthMappers chapters in Sierra Leone and supported by others at ASU. The validation was done by GWU students remotely.

3.2   Field Mapping and Ground Truthing

With most of the communities remotely mapped, local students went to the field to capture ground-level and aerial imageries of utility poles and other electricity infrastructures with the use of an action camera, a drone, and their mobile phones. This gave the students the opportunity to ground truth and update features on the existing maps of the locations by validating them and adding new features where necessary. Some attributes were also added. The captured images were uploaded to Mapillary, which is an open source street-level imagery platform that is available worldwide (Figs. 11.1 and 11.2).

Fig. 11.1 UNIMAK YouthMappers Ibrahim Kalokoh and Ibrahim Yusuf Jalloh capture street level images in Makali, Northern Sierra Leone

3.3   Setting Up and Deploying Mapping Teams per Location

Before going up to map each location, we set up teams consisting of one (1) supervisor and between 2 and 4 YouthMappers. These teams connected with district youth leaders through their supervisors. The supervisors are mostly graduates that have extensive mapping experiences, which made them knowledgeable enough to give mapping tutorials to beginning mappers and give them the necessary guidelines when in the field. All the supervisors are active members of the OpenStreetMap Sierra Leone community, most of which are YouthMappers alumni.

Since the YouthMappers chapters are autonomously led by local students, it was given their responsibility to select mappers that were to be part of each team for every location. This was

Fig. 11.2 Utility poles cannot be seen on satellite imagery, so street level images are important to identify the location of utility poles, and in turn, where the power grid reaches, and where are communities with and without access to electricity

mainly done through their chapter officers. They took a protocol to make the selections and submitted lists of their members with their contacts and locations they were to map. These lists were given to the supervisors who established connections with the mappers, where they usually created WhatsApp groups to liaise and determine the field mapping timelines and logistical arrangements. The lists were also sent to the USAID team for the names of students to be added to a generic letter that stated the purpose of the field mapping activities. These letters were given to each mapper as a backup for authorization purposes if the need arises to explain the activity.

After the teams were constituted and assigned departure dates, the necessary financial and technical support was given to each supervisor before they met with the YouthMappers in person. The support included but was not limited to stipends for lodging, Internet, transportation, YouthMappers styled T-shirts, selfie sticks, action cameras, power banks, and refresher training as needed. This support ensured that the teams were ready and well capacitated to map.

The number of mappers per location was determined by the sizes of the locations and their surrounding communities. Based on this, each team, including supervisors, usually consisted of 3–5 members. With the teams established with mapping timelines and a point of liaison with district youth leaders, we considered a location ready for mapping.

3.4 Authorization and Identification

The purpose or intent of fieldwork can be misinterpreted by local community members if they are not properly and thoroughly informed about the project. Our teams, therefore, had to make sure that we got in touch with community stakeholders before going to the field to capture images. The district youth leaders were focal persons of contact since we could relate well with them and they generally had the local connections. They were briefed about the project, and they willingly connected us with other community stakeholders.

Unfortunately, on one occasion during the pilot stage, a team of mappers were arrested by community stakeholders because they were not properly briefed ahead of time about the project.

Fig. 11.3 A drone image of Lunsar, Sierra Leone provides a glimpse of how the power grid does not reach everyone

Our mappers were misunderstood to be surveyors for land grabbers. However, the whole organizing group, including the network and agencies abroad, immediately responded and made the connections so that shortly the team was set free after a thorough description of the project and proper liaison with community stakeholders.

To prevent any further arrest or detention of our fieldworkers, we then got all the contacts of the elected district youth leaders through the Ministry of Youth Affairs. We introduced the Sierra Leone project leadership and the team and gave them insights about the work we do and how it would benefit their communities. In that light, supervisors had to alert the youth leaders at least a week ahead before they travelled to locations that were to be mapped.

When the mapping teams reached their designated locations, their supervisor ensured that their first point of contact was the community youth leaders who then took them to other community stakeholders for meetings to discuss the project and introduce them to locals of the community. At the end of the meeting, the teams were always assured of their safety when mapping within the communities (Fig. 11.3).

As a further backup plan, each member was provided with individually tailored letters from the USAID bearing the names and purpose of their mapping campaigns. This made the whole process safer for our mappers.

We also had to make sure that our mappers looked distinct and unique; therefore, we gave each of them YouthMappers styled T-shirts with the logo and name on the back of the shirt so they were clearly visible when the youth were riding around on the motorcycles. This made them identifiable and protected from negative community interferences.

While we anticipated a few of these measures during the original pilot work, knowing the importance of community connections and following the ethical guidelines that YouthMappers promotes, we had not fully considered the potential reactions given the innovative technology and the local context of the fear of land grabbing. But our solutions not only helped us to implement smoothly, they also ended up giving us an innovative way to connect to the elected local youth leaders and teach them about the importance of mapping for the SDGs.

3.5 Field Mapping and Monitoring

Most mappers made arrangements for accommodations before going to the mapping locations, while others do so upon arrival. It was the duty of supervisors to make sure that the YouthMappers had conducive accommodations during the whole mapping process, and once that was ensured, then they started mapping. The range of options for the open source tools was important so that the final choices supported local configurations best.

Mappers were sometimes given printed maps of the communities that they were to map. Sometimes they used applications such as Maps. me and OsmAnd. The supervisors then ensured that arrangements were made with local motorcyclists to ride the YouthMappers within the communities. The actual capturing of images commenced with each mapper in the YouthMappers regalia, seated behind their respective motorcyclist with selfie sticks and phones attached to them. The mappers gave directions to the motorcyclists to ride through the communities (Fig. 11.4).

The mappers were usually encouraged to upload the captured images as they mapped in order to create space in their phones or cameras for more images to be captured. Some mappers that had large enough storage spaces on their phones uploaded images at the end of the mapping exercise each day. The supervisors ensured that all the images were uploaded and each mapper had sufficient cellular data to do so from the resource allocation support provided.

At first, it generally took a minimum of 3 days for images to be processed and published on the Mapillary platform; however, it sometimes took over a month for them to be processed and published because the platform happened to be undergoing an important hosting transition at that very same time unrelated to our project. We directly engaged with the Mapillary team to help keep tabs on the process and receive advice and support. The latter made it a bit difficult to evaluate the extent of the mapping after the 3 days of mapping exercises. However, about 70% of the images were then normally processed and published within a week; therefore, we could tell that mapping had happened and in a correct way. At the end of each 3-day mapping exercise, our

Fig. 11.4 YouthMappers member Fatmata Kabia checks that the phone camera remains in place, snapping street view images every few seconds

teams got in touch with community stakeholders to express gratitude and say goodbyes. Finally, the supervisors facilitated the return of students to their various campuses as needed.

3.6 Second Remote Mapping

After uploading the images, the data were to be verified for the presence of utility poles and train the artificial intelligence tool to automatically detect utility poles on the images. To achieve this, Mapillary verification tasks were created, and then YouthMappers in Sierra Leone, ASU and GWU and across the world verified all the images. The utility poles verified in the images needed to be added to the OpenStreetMap platform, and the easiest way was to use a platform called Pic4Review wherein Pic4Review missions or tasks were created for all the locations and the utility poles were added to OpenStreetMap in a standard structured way, conforming in the end to the needs for downloading that data collectively for the project end-users (power industry). It was now possible for map makers and Geographic Information System experts to extract high-quality, validated, comprehensive raw data, analyze, visualize, and make use of them for the original purpose, but also from there on, for any different purposes.

3.7 Summary of Output Metrics and Mapping Outcomes

The parameters used to measure success are as follows:

Mapping of Buildings and Roads All the fifteen (15) locations across the country were fully mapped, in terms of adding buildings and roads within them. At the start of the project, some communities were not mapped with no buildings or roads on OSM. At the end of the project, 69,988 buildings were mapped and 831 km of roads were traced. This was done

collectively with YouthMappers within and outside of Sierra Leone.

Streetview Imagery Collection In terms of images within the community, none of them had Mapillary images that would give remote mappers pictorial insights about the locations they were mapping remotely. However, with the power grid project, images were collected and uploaded within 7 locations across the country. These images were used to map utility poles and train the AI Platform (Figs. 11.5, 11.6, 11.7, and 11.8).

Mapping of Utility Poles Within all the locations, none had utility poles marked on OSM. This was a huge milestone as it helped enhanced electricity distribution and generation agencies to have access to information about the existing electricity infrastructures across the country (Figs. 11.9, 11.10, and 11.11).

Capacity Building and Skills Development OpenStreetMap Sierra Leone is a fairly small mapping and geospatial community that consists mostly of YouthMappers and recent graduates. Therefore, any project that enhances the capacity and skills of the membership is considered to be laudable. During the course of the power grid mapping, YouthMappers and recent graduates were introduced to new mapping tools and techniques, and the skills of others were improved. The community also had the opportunity to purchase and own a drone, camera, and other gadgets for the purpose of mapping. Members were taught how to use the equipment, which became added skills that were sought after and became a plus to the graduates. This helped to boost the community's drive to become sustainable.

Broader Inclusion The work of YouthMappers is reflective of how we want our societies to be, in terms of inclusion and equality. Students from all genders and backgrounds were encouraged to participate. Although we have a limited

Fig. 11.5 Imagery overlaid with OSM data before mapping buildings and roads in Lunsar is shown

Fig. 11.6 Imagery overlaid with OSM data after mapping buildings and roads in Lunsar is shown

number of women that have the opportunity to participate in geospatial activities in Sierra Leone, our team made sure that women were involved throughout the project. Women were involved in the planning stage and the remote mapping, and each team that captured ground-level and aerial images had at least one female member. Women benefit disproportionately as a result of receiving power access (YouthMappers 2021).

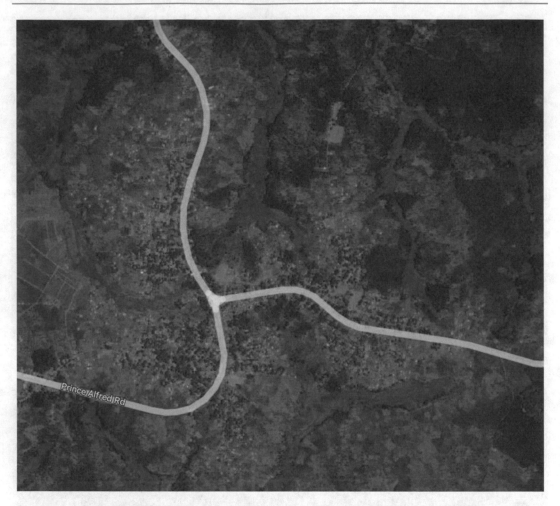

Fig. 11.7 Imagery overlaid with OSM data before mapping buildings and roads in Masiaka is shown

Exposure to Opportunities After the successful implementation of the pilot project, the community was given the assignment to map utility poles in up to 23 new locations across the country. This was considered to be a huge honor and a major boost to the YouthMappers and the OpenStreetMap community in Sierra Leone. There were interests from companies out of Sierra Leone to know about the power grid mapping project and the following blog posts, Mapillary blog (Solís et al. 2020), were published on their websites. A presentation about the project was made by the YouthMappers director (Solís 2020) during the 2020 UN Data Forum. The approach is being transferred to other countries across West Africa.

These successes of the power grid mapping are considered to be great milestones as the project was the second to be implemented by a community of mainly YouthMappers and recent graduates that were in the quest of expansion and utilizing the skills of its members for a common social good under these productivity metrics and outcomes but also for the SDGs.

4   Beyond the Map: Powerful Mapping for the SDGs

As youth, it is incumbent upon us to contribute toward achieving the Sustainable Development Goals within our countries or communities and

Fig. 11.8 Imagery overlaid with OSM data after mapping buildings and roads in Masiaka is shown

that is what YouthMappers in Sierra Leone have been doing over the years with support from both internal and external agencies. Because the power grid entailed a longer series of activities and outcomes, it is contributing to addressing some of the SDGs during the course of its implementation.

4.1   Affordable and Clean Energy

Creating affordable and clean energy is the principal goal that is being addressed by the power grid mapping, which directly relates to SDG 7. Most of the rural and suburban communities we have worked with either have installed or plans in the pipeline for solar-powered mini-grids that are in construction phases or functional. According to SDG 7, the expansion of electricity infrastructure and technology will increase the efficiency of electricity generation, distribution, and supply

and improve the livelihood of communities and the environment.

One of the things that can make this goal achievable is access to reliable information from credible sources. This is what the power grid mapping project has been doing. Electricity agencies use the data that we produce to examine and get an overview of existing electricity infrastructure within our selected communities. They also use the data to study the topography, road infrastructure, and other physical features of the locations that they use as variables to assess these communities before they start constructing electricity infrastructure.

The quick and easy access to reliable information reduces the cost of implementing electricity projects. With the power grid mapping, these agencies get firsthand information about the places of interest before physically visiting them. It makes their feasibility studies much easier, efficient, and cost-effective. This reduces the

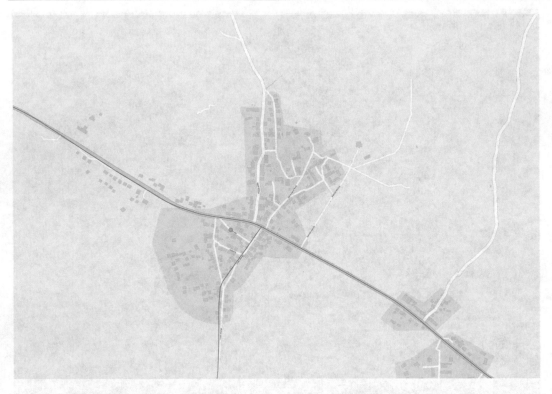

Fig. 11.9 Building and road data on OSM provides the basemap for street level imagery intake

overall cost of implementation, making access to electricity more affordable.

To be clear, the power grid mapping project was not directly involved in the construction of electricity infrastructure, but the work it did expedited the process and gave hope or increased the hopes of community members getting electricity in the near future. During the initial meetings with community stakeholders, they expressed appreciation for the initiative. They stated their ongoing plans of using electricity once generation started and, in some cases, the areas within their communities where they would want expansion. Interacting with community members gave our teams a better understanding of the latent use of electricity as they narrated their socioeconomic use of electricity when generation started. In some communities that already had partial electricity generation, they explained the socioeconomic benefits that they have been enjoying.

These socioeconomic benefits were common throughout the communities we worked. They

included the use of electricity to preserve perishable goods that are sold for economic purposes and the charging of mobile phones and electronic gadgets. The most common use among children, teenagers, and youth was for studying. They stated how the presence of electricity boosts their chances of learning and how essential it is to them and the hopes of uplifting their livelihoods.

Our teams gave hope to every community we visited and gained inspiration and motivation in return. They believed that our work could improve their chances of getting electricity in communities without and renewed the hope of expansion and resolving some of the issues they faced with electricity infrastructure.

4.2 Build Resilient Infrastructure and Foster Innovation

By mapping to provide data for efforts to bring electricity to rural communities, this project contributed to SDG 9, especially in terms of contri-

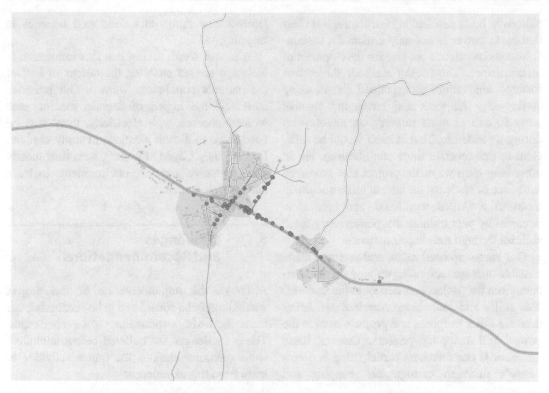

Fig. 11.10   Green dots depict where street level imagery has been added by YouthMappers at a local scale

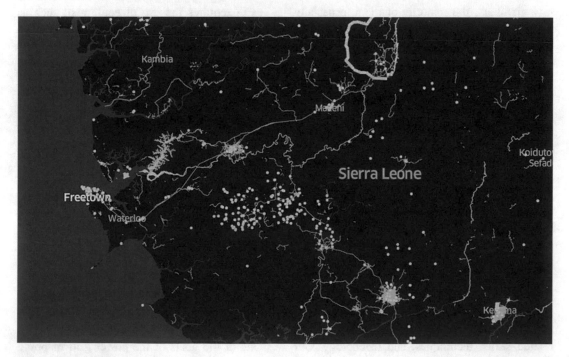

Fig. 11.11   Available open street level imagery for all of Sierra Leone has been enhanced significantly by YouthMappers in the power mapping efforts

butions to build resilient infrastructure. It is clear that clean power is not only critical for sustainability and resilience but also for development of infrastructure. The project methods themselves fostered innovation, as explained above, using cutting-edge AI tools and leveraging creative ways to collect mass imagery on motorcycles during a pandemic. That in itself would be sufficient to demonstrate these contributions, but in other long-term ways, the project also promoted resilience of students, the human infrastructure in terms of a skilled workforce, and innovation potential by peer training that passes along these abilities through the chapter network.

Our teams updated maps, and this was made possible through facilitating training for all members, from the project coordinator to the students. The skills and knowledge acquired are being used for other purposes and projects outside the power grid mapping projects. One of these instances is our members participating in Sierra Leone's midterm cartographic mapping and census.

Based on our interactions with some community members, the majority of them, especially school-going teenagers and youth, stated that their primary use of electricity was for studying. Although most of the communities had limited electricity supplies at night, students made use of the short period, usually between 6 pm and 8 pm to study and do their assignments. Very few communities had digital infrastructure ready like Internet cafes and resource centers owned and controlled by community entrepreneurs and non-governmental organizations. These cafes utilize electricity and are learning centers for students and everyone within the communities. The power mapping work is building simultaneously this electricity infrastructure, the digital infrastructure, and the human infrastructure that will enable many kinds of economic development in the future.

Another interesting thing was that community members wanted to learn about mapping and its related skills, techniques, and technology. Our team members gave them basic yet sufficient information about mapping and how they could participate. Through this, community members learned new things and developed interests in mapping.

It is also worth noting that this innovation is inclusive toward enabling the talents of half of the nation's population: women. Our presence increased the hopes of female students and women because with electricity, there was no need to go to distant locations to study at night. Instead, they stayed at home where they mostly had conducive and safe environments to learn and do their chores.

5   Challenges and Recommendations

Although the implementation of the project would largely be considered to be successful, our team, however, experienced some challenges. These challenges are bulleted below, including some recommendations for future activities to learn from this experience:

- The mapping of features on the surface of the earth requires the use of equipment such as Global Positioning Station services, smartphones, action cameras, and Internet. In this case, most of the required equipment was unavailable and had to be imported. The procurement process was bureaucratic, and this led to the delay in starting the project. However, the delay gave us more time to do more remote mapping and better understand the selected communities.
- Access to the Internet was also a challenge; because of the COVID-19 restrictions, students or mappers were unable to meet in person for training, but this was difficult as most of the training was done online and students had limited access to the Internet, which slowly altered their participation. The unreliable Internet source made it difficult to upload collected data, and this slowed down the workflow of the project.
- One of the goals of the power grid mapping was to increase women and girls' access to electricity, and the inclusion of women in such an initiative was a must. Unfortunately, having

women and girls onboard was a critical challenge. There persists a stereotype that technical activities like mapping with the use of drones were limited to men. In order to enhance inclusion, the team decided to establish the first all-female mapping club that would specifically spearhead the recruitment and inclusion of women and girls into our mapping community.

- Some miscommunication between mappers who are not community members and actual community members occurred because to deploy across such a large area, we developed a chain of communication, but YouthMappers were not always in direct contact with community stakeholders, which created a chaotic scene in one of the communities that was being mapped.

- It was evident that rural communities not only experience lack of power but also suffer a land administration crisis. Therefore, it would be a significant opportunity if we can collaborate with them and other communities faced with similar challenges by putting the skills of YouthMappers into use and collaborating with community stakeholders on this topic as well.

- It is important that we don't just build maps; we build mappers. Our students are not just mobile "sensors" collecting data, but also innovators in the field mapping process. Knowing the full context of humanitarian mapping improves the participation of students (Solís and Delucia 2019). In order to ensure that the YouthMappers in Sierra Leone are fully engaged with the entire range of geospatial data and provide deeper insights into this innovation process, the team at ASU has offered to provide additional training on the analytical framework of how spatial data is used in later stages of the project for grid feasibility and definition to all participating local mappers.

With the plan of the government and other organizations to construct solar-powered minigrids in 94 locations across the country, there is a significant need to continue mapping buildings, roads, and utility poles. This would ensure continuous creation and availability of geospatial information that relates to electricity generation and distribution. For future continuation of these and similar efforts, we are working toward building further upon this YouthMappers infrastructure put into place. To ensure the expansion and even participation of YouthMappers in yet unmapped places, there are plans to establish more chapters in tertiary institutions across the country. This would help propagate the importance of mapping and the existence of a large pool of individuals with the required skills. In terms of inclusion, to overcome some constraints to have female and mappers with disabilities on board, we have developed a plan to set up clubs or associations that would spearhead the inclusion of underrepresented people.

In summary, one can say that the power grid mapping project represents a giant leap for the OpenStreetMap community in Sierra Leone, especially for YouthMappers. It helped to unearth the potential of all those that were involved, giving everyone insights about our intellectual capabilities and emotional growth and areas we can improve on. We also discovered numerous underlying issues affecting development and community mobilization. We contributed directly to SDG 7 toward affordable and clean energy, by mapping data that will, among other things, support local solar-powered grids and also SDG 9 for the national infrastructure that will imply not only development but also resilience. This last point underscores the innovation that came from our local youth, together with the entire global team. The skills acquired during the course of the power grid mapping are already being applied to other geospatial projects or initiatives as part of the short- and long-term endeavor of YouthMappers in Sierra Leone. These include other potential infrastructure-building campaigns like the mapping of safe spaces for girls and dumpsites and street signs and then making them easily accessible to the public. This model is also being applied in other locations in West Africa. This work would help to make communities in Sierra Leone, West Africa, and become beyond more sustainable, thereby addressing several Sustainable Development Goals and our own ambitions.

References

Solís P (2020) Power Mapping: youth and GeoAI for augmented humanitarian action. United Nations Data Forum, October 19, Online Forum from New York, New York

Solís P, DeLucia P (2019) Exploring the impact of contextual information on student performance and interest in Open Humanitarian Mapping. Prof Geogr 71(3):523–535. https://doi.org/10.1080/00330124.2018.1559655

Solís P, Charles T, van Hove E, Beddow C (2020) Power mapping brings rapid reliable energy to rural communities. Mapillary Blog. 20 October. Available from Mapillary. http://bit.ly/2XM2qqf. Cited 2 Jan 2022

YouthMappers (2021) Everywhere She Maps, Power is Generated. Available from YouthMappers website. https://www.youthmappers.org/everywhereshemap-spowerisgenerated. Cited 2 Jan 2022

Mapping Access to Electricity in Urban and Rural Nigeria

Emmanuel Jolaiya, Mercy Akintola, and Opeyemi Nafiu

Abstract

YouthMappers use OpenStreetMap and IoT tools to help map where rural communities have good, poor, or no access to electricity throughout Nigeria, contributing to efforts that bring stable power across West Africa. An ongoing energy crisis is hampering other efforts to make sustainable cities and communities, so assessing and locating the reality of power access are critical not only to address SDGs related to affordable and clean energy but also as a fundamental element that serves as an engine for sustainable development. Through these efforts, we students have built our technology capacity and community networks, while also advancing practices about how to build inclusive, sustainable cities and communities in both urban and rural places where we live and study.

Keywords

Electricity · Sustainable energy · Urbanization · Rural development · Nigeria

E. Jolaiya (✉) · M. Akintola · O. Nafiu
Federal University of Technology, Akure, Nigeria
e-mail: jolaiyarsg152097@futa.edu.ng;
akintolarsg172802@futa.edu.ng;
nafiursg152103@futa.edu.ng

12.1 Interrupted, Unstable Electricity Access Across Nigeria

Electricity is unarguably one of the crucial basic amenities needed by humanity. As the population of the world grows, the need for sustainable and reliable energy significantly increases proportionally. This is true in both cities and villages.

Access to stable, affordable electricity is a major problem in developing countries like Nigeria. Many rural communities are not even connected to the national grid, while cities connected to the grid frequently experience epileptic power supply, outages, or blackouts.

The Nigerian energy supply crisis is a huge problem, evident in continuous failure of the Nigerian power sector to effectively supply and provide electricity to domestic and industrial users of electricity. Even though the nation is one of the world's largest oil-producing countries,

P. Solís, M. Zeballos (eds.), *Open Mapping towards Sustainable Development Goals*, Sustainable Development Goals Series, https://doi.org/10.1007/978-3-031-05182-1_12

with some of the world's largest deposits of coal, oil, and gas, electricity supply is still not ubiquitous, stable, and affordable. This is a very huge and massive problem for sustainability and for development.

As of 2020, only 45% of Nigeria's population is actively connected to the Nigeria energy grid. Epileptic power supply is experienced 85% of the time, where sometimes some regions have no supply of electricity for years (Aliyu et al. 2013). The power supply situation in Nigeria is so poor that several communities, even though connected to the grid, have no power supply for days, running into months and years. Coupled with flawed legislation and practices, unplanned and unscheduled power cuts are experienced regularly. This reflects the unpredictable nature of power supply in the country as a whole (Aliyu et al. 2013).

The resultant effect is a lack of access to a dependable and affordable power supply situation in Nigeria, a context that cripples various parts of the economy of the country, ranging from industrial, agricultural, educational, and more. The problem spans all sizes of communities, from urban places to rural villages. Needless to say, the current development situation impedes Nigeria's economic growth significantly.

The causes of this poor power supply situation are very complex and multifaceted, with deep roots in administrative issues, corruption, and many others (Aliyu et al. 2013). What is clear is that the energy crisis has been going on for decades. Although several power reforms and projects have been carried out by the country leadership, no tangible long-lasting solution has been proffered. The centerline of the major power reform undertaken by the country and the most promising so far is the relatively recent privatization of the power sector. This involved the sale of primary and most important assets of the Nigeria power sector to private sectors. There has, however, not been a full conversion to total privatization, as the federal government still continues to control and manage major transmission assets (Ogunleye 2017).

12.2 Geospatial Solutions for Electricity Provision and Management

Although the electricity problem is multifaceted, it is a well-understood fact that the present power facilities are under intense pressure, as a result of increasing population density. This pressure, in turn, has impacted the quality and health of such facilities resulting in the breakdown of electricity access even in places where it exists. Unfortunately, there is still little implementation of technologies such as a geospatial framework in the planning and structuring of electrical facilities in Nigeria so far. We fear that with such little to no regard concerning the implementation of geospatial solutions in planning the locations and repair of electrical facilities in this latest reform, so many communities and cities will continue to face this poor power supply problem for decades to come.

12.2.1 Trajectory of YouthMappers in Nigeria

YouthMappers is very strong in West Africa and impressively strong in Nigeria. There are no less than 23 chapters at universities in this nation (at the end of 2021). The very first chapter established in Nigeria was at the Federal University of Technology in Akure (FUTA), being named one of only 2 inaugural chapters in the country. This designation indicates that we started in the first year of the global network (2016). Our chapter has had a storied history of fellows and ambassadors serving the network and a rich set of experiences in mapping with OpenStreetMap and other geospatial tools for a large set of applied solutions.

YouthMappers at FUTA is like the others, a university-led chapter that is connected to the global YouthMappers community. YouthMappers at FUTA is made up of members with a desire to learn, evolve, and adapt to necessary technologies and skill sets needed to solve vari-

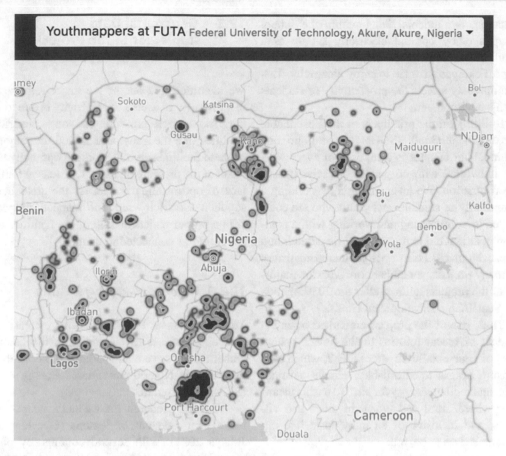

Fig. 12.1 YouthMappers at FUTA have contributed to OSM across Nigeria since establishment. (Credit: J. Anderson, 2022)

ous problems around them. The chapter is dedicated to humanitarian action and community development through open mapping, which is evident in a series of activities and projects executed in the chapter, in which members work together with open data to solve various problems around them.

The series of activities and projects carried out in the chapter has contributed to the achievement of the Sustainable Development Goals, by mapping vulnerable areas of the world to enhance effective humanitarian activities. The results of the series of activities undertaken in this chapter have significantly improved the global map data quality as members of this chapter assiduously and dexterously mapped different parts of the world, and this has also indirectly and directly

improved the mapping skills of participating members and has strengthened community resilience (Fig. 12.1).

12.2.2 Disruption of Power Sparks Action Toward SDGs

YouthMappers at FUTA is domiciled in the Federal University of Technology, Akure, which faces a similar electricity supply problem ravaging the nation. Although the school campus has optimal electricity access, the communities surrounding the school campus still struggle with access to electricity. A significant proportion of students in the school live in these surrounding communities, and this creates an uneven power situation, as students living within the school campus enjoy optimal electricity supply as a result of the unaltered access, while students living off campus struggle to enjoy electricity. This continuously affects the productivity of students in this environment.

Building on the principles of the Sustainable Development Goals (SDGs) "leaving no one behind," our chapter is composed of every willing individual with no gender or racial (tribal) discrimination. We leveraged this by building a community of students and youth who are committed to volunteering and working with a common goal in a bid to make the world a better and improved place. This principle, therefore, drives so many projects we embark on, with the majority of the projects tailored after the 2030 Agenda for Sustainable Development Goals.

Thus, one of the projects embarked on in the context of recent reforms to the power system was, of course, tailored after SDG 7, which says "ensure access to affordable, reliable, sustainable, and modern energy for all." To us this means everywhere, rural and urban, in cities and villages, and inclusive of all individuals like our chapter strives to be, too.

12.2.3 Geospatial Knowledge Needed to Understand Electricity Access

The project to map access to electricity problems was identified by YouthMappers at FUTA; however, we partnered with Energy Detectors Technologies Limited, a fast-growing startup focused on remote data collection, analysis, and visualization in the energy sectors with the use of IoT technology and advanced algorithms, to map the electricity transformers in the communities surrounding FUTA for better electricity planning, so as to bridge the gap and create a sustainable environment for all. We are also informed by other efforts in other chapters in the region and all over Africa and supported by the ambassador network and core team and by the Humanitarian OpenStreetMap Team through data internships.

12.3 Geospatial Data Implementation Projects with YouthMappers

We identified that one of the major reasons for this disparity in access to electricity is due to the huge population pressure on some electricity facilities like the transformers, and the majority of these facilities are not working optimally as a result of the pressure on them. One reason for this lack of optimization is because the areas being supplied electricity are not mapped properly, with a proper understanding of the facilities and the buildings connected to them.

12.3.1 Remote Mapping Results

YouthMappers at FUTA aimed to increase the data about access to electricity in the country embarked on several mapathons. One of those programs is Nigeria Sustainable Energy 4 All (SEA4LL) program.

When the pandemic hit, we had to adapt – and make lemonade out of lemons (Opeyemi and Jolaiya 2021). With several contingency plans from the government, as an effort to reduce the spread and impact of the virus, so many activities turned out a new leaf, as we were required to hold several meetings online rather than the normal in-person meetings we were all used to. However, we were able to adapt and also evolve in a very deciduous manner. We held different meetings to train members of the chapter as a way of refining the skill set of individual members of the group, using several teleconferencing mediums like Zoom, Microsoft Teams, and Jitsi. Workshops were also held in a virtual mode, periodically to keep members abreast about what is needed to be an effective member of YouthMappers (Figs. 12.2, 12.3, and 12.4).

In the end, the chapter massively supported the Nigeria Sustainable Energy 4 All (SE4ALL) program in remote mapathons. The progress in number (so far) is 60,183 km of electricity grid tracked. We also reached 3937 settlements remotely mapped and 3,244,605 buildings mapped.

Fig. 12.2 Mapathons by YouthMappers at FUTA typically include remote mapping training and tasks

Fig. 12.3 Officers of YouthMappers at FUTA plan the SEA4LL collaboration

12.3.2 Field Mapping Results

We also embarked on a field mapping exercise to capture locations of electrical facilities in the area, we also noted the number of buildings connected to each electrical facility, with an aim of determining the amount of load on them. Our partner – the Energy Detector Technologies Limited – provided us with IoT devices, which were installed to track and monitor if a transformer was powered on or off and the duration when it was on or off.

We collaborated with this private entity, Energy Detectors Technologies Limited, by sharing the spatial data obtained from the field. The company shared this data with the electricity distribution company responsible for managing electrical facilities and distributing electricity in the region – Benin Electricity Distribution Company (Figs. 12.5, 12.6, and 12.7).

The data obtained from the IoT devices was used in a mobile application named UPNEPA. The application notifies students/users of the current power situation within the school campus and also off campus, and this solved a great decision challenge as it prompted students automatically when electricity is supplied.

The project end goal was to improve electricity availability alerts in the mobile app developed by the E-Detectors Energy Limited.

Fig. 12.4 Students validate remote mapping for each other during the YouthMappers at FUTA events

Fig. 12.5 YouthMappers at FUTA undertake field mapping of electricity assets

Fig. 12.6 Transmission lines are important features that require field mapping or validation

12.3.3 Mapping Challenges

The project was exciting and impactful as we saw the result of our exercise in the notable increase and improved access to power supply around the school campus. However, it is noteworthy that we faced some challenges during the course of the project, one of which is the Internet connec-

Fig. 12.7 IOT devices enable mapping of attributes in electricity asset maps

tivity problem, as some of the student mappers did not have Internet data, which hindered them from participating during the course of the project.

Also, during the course of the field mapping, access to some rugged terrains was difficult, as the mappers had to walk on field visiting several locations in unsuitable weather conditions. We worked through these challenges.

Finally, there was a problem we faced, where the inhabitants of the communities being mapped were unaware of what we were doing. They were reluctant at first when they were approached about information about their residence and power supply situation where they live. We addressed these by providing the information needed for voluntary responses.

Despite the challenges faced during the course of the project, the experience cannot be compared to the tremendous impact created (Fig. 12.8).

Fig. 12.8 Skills developed through YouthMappers at FUTA projects can impact the future of student leaders

12.4 Beyond Making a Map: Mappers for the SDGs

In the course of making these projects happen, our chapter periodically embarks on exploring, teaching, and educating other members through mapping, to develop their mapping skills, contribute more to the digital world, and engage them in global funding and scholarships to attend conferences and other activities.

YouthMappers at FUTA achieved quite a lot through these experiences. Our community members Ademoyero Victor Ayomide, Olanrewaju Michael, Oke Mathew, Akintola Mercy Onaopemipo, Okikiri Favour, and Babalola Bukola participated in a data labeling contest as part of the Cloud Native Geospatial Outreach Day sponsored by Planet, Microsoft, and Azavea. Out of 231 participants from across the world, Ademoyero Victor won the 3rd Place Labeler prize award, Babalola Oluwabukola won the Top Woman Labeler Prize Award, and Akintola Mercy Onaopemipo and Olanrewaju Michael won the next 5 Top Labeler Prize Award. These outstanding participants were issued certificates for their participation at the end of the event.

Furthermore, the chapter massively supported the Nigeria Sustainable Energy 4 All (SE4ALL) program in remote mapathons, which were held five times from July through November of 2020. Two of our chapter members were awarded mapping champions thrice during the events, and we made in total over 53,000 edits on OSM. The erstwhile chapter presidents, Dennis Irorere and Emmanuel Jolaiya, were part of the program coordinators and mapathon instructors.

Moreover, we learned to adapt. Mapathons and mapping parties at first looked difficult trans-ferring all of that to a virtual space. However, because our group consisted of youth with an adaptable mindset, we were able to quickly evolve and adjust to the virtual reality of the world made necessary by the pandemic. Remote mapping wasn't a new reality, as we as a group have been involved in this for a long time, and this was done regularly by members. But the chapter engagement had to learn to be resilient, despite obstacles like access to Internet, and ironically electricity.

In a bid to contribute to the achievement and actualization of the UN Sustainable Development Goals, the YouthMappers at FUTA decidedly contributed to SDG 7, which is to "ensure access to affordable, reliable, sustainable, and modern energy for all." The goal comprises five targets, which are to be achieved before the year 2030 and can be measured by the use of six indicators. But it goes further than this. Because electricity is a foundation for sustainable cities and villages, the contribution to this SDG also implies progress for SDG 11 made as a result of the very pertinent and persistent electricity supply problem around us.

References

Aliyu A, Ramli A, Saleh M (2013) Nigeria electricity crisis: power generation capacity expansion and environmental ramifications. Energy 61:354–367

Ogunleye EK (2017) Political economy of Nigerian power sector reform. In: Arent D, Arndt C, Miller M, Tarp F, Zinaman O (eds) The political economy of clean energy transitions, UNU-wider studies in development economics. Oxford University Press, Oxford, pp 391–409

Opeyemi N, Jolaiya E (2021) 2020: a year we made lemonade out of lemons. YouthMappers Blog, 28 January. Available from YouthMappers. https://bit.ly/3frl7am. Cited 8 Jan 2022

Youth Action on Work, Leadership, Innovation, Inequality, Cities, Production and Land

Stories from Students Building Sustainability Through Transfer of Leadership

13

Saurav Gautam, K. C. Aman, Rabin Ojha, and Gaurav Parajuli

Abstract

Dedicated to designing and organizing student-centered work, we highlight the Geomatics Engineering Students' Association of Nepal (GESAN)'s collaborative effort to create training, internship, and job opportunities for the student members and alumni and with local institutions. Through putting into play a cycle of leadership, chapter activities find a sustainable way to continue to support SDG 8 Decent Work and Economic Growth and SDG 9 Industry, Innovation, and Infrastructure by preparing students to navigate newer applications of geospatial technology and tools. This work extends beyond campus and into the local community where we have trained secondary school students in map literacy – planting seeds of future leadership.

Keywords

Economic development · Youth leadership · Local collaboration · Nepal · Industry

1 Where the Cycle Began

Even though we were a small group of students in Nepal and engaging a subject that we didn't know the full scope of, in reality, our affiliation to YouthMappers gave us a valuable connection that we never thought we would ever have. We found ourselves together with the students of the world who were thinking just as us and were like-minded in terms of the importance, use, and value of open geospatial information. The different challenges and activities organized by YouthMappers encouraged us to organize mapathons, conduct map literacy programs and educate female students of our locality about geographic data and led us to design and conduct geocaching on our campus and many more activities.

S. Gautam (✉) · K. C. Aman · G. Parajuli
Pashchimanchal Campus, Institute of Engineering, Tribhuvan University, Pokhara, Nepal

R. Ojha
Kathmandu Living Labs, Kathmandu, Nepal

© The Author(s) 2023
P. Solís, M. Zeballos (eds.), *Open Mapping towards Sustainable Development Goals*, Sustainable Development Goals Series, https://doi.org/10.1007/978-3-031-05182-1_13

2   YouthMappers in Nepal

The Geomatics Engineering Students' Association of Nepal (GESAN 2017) is a student association that was founded in November 2014, and I was appointed as the president of the 2nd executive committee on August 16, 2015. Our student-led leadership committee organized a grand program on March 5, 2016, to celebrate Open Data Day (Kathmandu Living Labs). This event started my journey, as well as our chapter's journey, of being affiliated with YouthMappers. Through this program, we were introduced to YouthMappers by Dr. Nama Raj Budhathoki of Kathmandu Living Labs. Our journey of learning and collaborating with the YouthMappers international university consortium was just getting started. It was a moment of jubilant joy when we officially affiliated GESAN with YouthMappers and we were listed as a chapter that year, which was the first of its kind on our campus – and among the very first in the region as a whole, too.

> Before, I joined GESAN as a volunteer, I didn't know what geomatics really was, I even didn't know the real meaning and applications of the field. I used to take it only as a part of surveying. As I started involving in the activities of GESAN as a volunteer, I came to know it as one of the multidisciplinary subjects with a wide range of scopes. As GESAN used to organize different programs, webinars, workshops, I got many platforms and opportunities to play, enjoy and learn many things about this field. During this, I became able to judge and define geomatics as not only surveying (Napi) but also the sector of mapping, remote sensing, Geographic Information System (GIS), GNSS and many more. During this volunteering experience I can never forget about the love, teamwork and helpful environment that I got from the GESAN family (Sangeeta BC, volunteer of 6th executive committee).

2.1   Designing Student-Centered Activities

Through our years of actively participating with YouthMappers, regular mapathons are the place for everyone to connect and get to know each other better in person. The environment for teaching and learning mapping skills among the members is mutual and feels like everyone is on the same page with an intention to contribute to the world and help people in need by providing missing data to the map. Our collaborations with humanitarian mapping agencies that provide critical support through mapping campaigns to capture vulnerable, and often missing, parts of the earth are something we strongly advocate for (O'Hara 2018).

During the execution of our mapping activities, there have been times when we were met with challenges. For instance, while we were in search of potential sponsors for the publication of our annual magazine, we experienced many setbacks, receiving some very direct "No's" to "We'll inform you later." Our hopes to launch a magazine in a predetermined time had us feeling crestfallen until our situation changed. With every positive response, we were inspired much more than before. The positive rays helped us keep our hopes held high, and after finding sponsorship, we were able to successfully launch the magazine right on time along with a grand geospatial meet-up. The setbacks were something we struggled with when we encountered them. But in the end, they made us stronger and better prepared for future problems. The challenges our team has overcome have certainly helped in developing our communication skills, multitasking abilities, and time management.

> Working as a committee member in GESAN, helped me a lot in my career. It is a great platform to learn different aspects of Geomatics. As Geomatics was a new faculty at that time, GESAN let me know the scope of our field, what we can learn and how we can serve our society. I got a

chance to meet respected personnel while conducting and attending programs of GESAN, which is providing additional support in my career. During one of the trainings, I came to know about Kathmandu Living Labs (KLL), a non-profit organization and later this organization became my working place for about one year and I learnt so many things. I don't think this would be possible without GESAN. Now I have been working as Survey Officer for the last 3 years and the management capacity, public dealing capacity, teamwork, I learnt from GESAN is helping me a lot (Nisha Adhikari, executive member of 1st executive committee).

One of our most successful chapter activities resulted in more than 40 members from GESAN mapping for more than 40 days to digitize an area covering two districts with more than 50,000 buildings and 1,500 km of road networks on OpenStreetMap (OSM; GESAN). Our members were highly involved in trial and community infrastructure mapping of Bajhang and Bajura District. Top mappers were awarded with gift hampers and were provided an opportunity to get involved in field data collection in Bajhang District.

2.2 Adaptability During the COVID-19 Global Pandemic

The year 2020 began with the COVID-19 global pandemic, and we all were locked inside the house due to lockdown. In this pandemic too, the concept of online programs, especially webinars and online training, helped students gain some extra knowledge and spent their free time learning something new and useful. Our first attempts to conduct OSM training and a mapathon to map Pyuthan District of Nepal were successful with 20 participants mapping for a week. To bring different tasks to our members, we conducted a vari-

ety of programs that involved webinars on light detection and ranging (LiDAR), which is a new topic in Nepal. Similarly, we conducted trainings on web mapping with Leaflet and GeoDjango and processing drone data. We continued the publication of our annual magazine *Geo World*, which was inaugurated online and then distributed in a digital format so that it could reach a larger audience throughout the whole world.

I never thought working virtual from everything physical would be that easy and simple, though I had never heard or used online platforms like Zoom/Teams/ Google meet. Going up virtual had its own privilege. We could take training from seniors and teachers all over the world. We could conduct programs like OSM mapping, GIS training, web-mapping virtually with participants from college as well as from outside college. I think online platforms gave us the chance to broaden our horizons (Kiran Bhusal, treasurer of 6th executive committee).

This challenging year provided us all the opportunity to learn. It definitely enriched us with new paths to explore. Every member of the organization worked as a group to learn new things, implement them, and, more importantly, pass it on to the new generation of students who were joining the chapter.

My experience with GESAN has helped me in many ways. Before working for the committee, I had very little experience on leadership strategies, time management and decision making. I remember when I got a chance to organize a training program under my coordination, I was made more aware of the need for these important skills. Simply from writing a mail to writing the agendas of the meeting, the opportunity definitely helped me flourish my critical skills. My engagement with GESAN made me more prepared for my professional career and, I can surely say, it has had a significant impact on the execution of the works after I landed my first job as a research associate at Asian Institute of Technology, Thailand (Tek Bahadur Kshetri, executive member of 3rd executive committee).

2.3 Valuable Collaborations Happen Locally

Even during the difficult times of COVID, we adapted with the new normal and partnered with other institutions for different mapathons and trainings. Collaboration is something we strongly advocate for and have designed successful collaborations with different national and international organizations like NAXA, the World Food Program (WFP) Nepal, and the Youth Innovation Lab that have delivered fruitful results. Locally many student members of GESAN have participated in activities, trainings, and projects organized by Kathmandu Living Labs (Khanal et al. 2019). Most of our current members and alumni are still active contributors to OSM and are active members of the OSM community in Nepal.

All year, GESAN organizes different events that foster personal and professional growth opportunities in the students. These events include introductory OSM workshops, the map literacy program, mapathons, talk sessions, and trainings focused on using the different tools that integrate with OSM, to name a few. The activities mainly focus on student empowerment, female participation, and creating dialogue and partnership with different organizations. We are continuously working to fill the geospatial data gaps in Nepal and other countries. Involvement in all these activities either through organizing them or participating in them has developed leadership, communication, and other different soft skills in us. It is to be noted that all the efforts done by the members are voluntary rather than paid works. So far, as a YouthMappers chapter, GESAN has been able to set an example to the other chapters, and we still have a long way to go (Fig. 13.1).

The collaborative effort of GESAN with other institutions has resulted in several internships and job opportunities for the members and alumni. This is one strong indication of our contribution to SDG 8, decent work and economic growth through youth employment. According to SDG 8.6, the goal is to reduce the proportion of youth not in employment, education, or training. In a similar line, we have trained secondary school students with map literacy. Likewise, we have encouraged female students to take part in our training with a motive to instill in them the idea of geography-related subjects and reduce the gender bias in the geospatial industry.

Working as GESAN's treasurer was a fantastic experience that boosted my confidence and helped me develop my networking skills. GESAN served as a platform for us to gain geoinformation and build our networks. It gave me the opportunity to participate in a number of conferences and seminars, which introduced me to the world's developing geoinformation community. Being a part of this community influenced my decision to begin my career as a research associate at the Asian Institute of Technology. During my time at AIT, the leadership I gained at GESAN aided me in completing my task and delivering it well to the end-user. This GESAN-developed quality in me has also been extremely beneficial in obtaining a highly regarded Erasmus scholarship. Currently, I am an Erasmus student pursuing a master's degree in Copernicus Digital Earth, and I believe that GESAN has been a good influence throughout my academic career (Pratichhya Sharma, treasurer of 2nd executive committee).

Fig. 13.1 The GESAN Chapter of YouthMappers pivots to a robust virtual program of activity in the wake of the COVID-19 pandemic shutdown of in-person meetings

3   Prioritizing Chapter Sustainability

We have prioritized setting a proper tone in the chapter. We, as a chapter, have always focused on maintaining a culture of well-being within the members of the chapter, fostering education and training, promoting inclusivity, and building a resilient, student-led team. *Map Literacy* is an annual program organized and led by our female members that target high school girls and teach them the basics of maps, while encouraging and inspiring female leadership. These activities have helped our own GESAN members in multiple ways, and through continued practice of transferring skills and leadership to a new generation of students, we are preparing and inviting future students to join us.

As confirmation of our priorities, every year GESAN has been awarded for the contributions our members have been continuously making to fill the data gap through mapping. In 2021 our effort to bring about change has been recognized with the *YouthMappers Top Editors Award* and *Chapter Inspiration Award*. One of our members and alumna received the *2021 Best YouthMappers Blog Award* and the *2021 Best Alumni Blog Award*, respectively. By cultivating this tradition of supporting and mentoring chapter members, GESAN has been represented in multiple YouthMappers fellowship cohorts and has presented at international conferences organized by the OSM community. All these awards have encouraged us to continue putting effort in the sector.All these warm and beautiful experiences shared by each member gives a kind of happiness of what you have done so far for GESAN and its members. It also gives confidence that the upcoming generation holding the backbone of GESAN will lead to the greatest height. As time passes, the opportunity for the upcoming members will increase along with more and more challenges to overcome.

The work that we did as a committee, that I am very proud of, is advancing GESAN from a campus-based club to TU chapter of YouthMappers. That decision has helped us and future generations to develop skills and learn more about open data. Being a member of GESAN helped me throughout my career, I could embrace the new opportunities and the new world of scientific and academic works. In my career, while working with the Asian Institute of Technology and also during my MSc at ITC, I could use many things that I learned at GESAN such as holding meetings, preparing minutes, and listening to ideas from others. I had also built my network bigger through the conferences that I attended when I was a member of GESAN. I have always had a deep interest and am eager to learn more about geomatics and using geospatial data to solve real-world problems. My current Ph.D. research is on geospatial modeling and monitoring of earthquake-induced landslide hazards which is mostly related to the use of geospatial data and modeling using the geospatial data (Ashok Dahal, secretary of 2nd executive committee).

4   Future Outlook for Us and for the SDGs

GESAN will focus on collaboration with more diverse organizations, because even though we have been able to engage with many local actors, we have not been able to work with the national mapping agency and other government bodies. A partnership with a government institution would expand OSM to a wider audience and create more opportunities for youth in the field. Contemplating the fact that there is always room for improvement, we have been trying to reduce the gender disparity in executive and general members of the committee. One of the reasons hindering this effort is the unequal ratio of male and female students in the engineering colleges of Nepal. Inclusion of women mappers in various mapathons, map literacy programs, and other activities can boost their confidence, and promoting their work is a good start.

Recalling the journey from being a fledgling university chapter to having matured into a globally recognized chapter, each and every step on the way has been a learning experience for everyone engaged. The problems we faced and the success we achieved are lessons we can treasure for life. Ending on a positive note, we see this sustainable leadership of GESAN in the future as well, with a greater focus on collective leadership rather than individual. And all the while, all of this adds up to our own incremental contributions to advancing SDGs, especially when it comes to SDG 8 Decent Work and Economic Growth, for our careers, and SDG 9 Industry, Innovation, and Infrastructure, for our institutions that we transfer forward.

As a student who was just going into the sophomore year at university, I didn't know how to make a CV, as well as to write answers which were 800 words long. Being one of the 20 participants selected for the YouthMappers Leadership Fellowship from chapters around the world was obviously a dream come true. Meeting many people for the first time from different countries gave me international experience. The leadership workshop taught me key skill sets in an easy and interesting way which has been an asset to me till date and served my chapter as well (Saurav Gautam, YouthMappers Regional Ambassador and former Leadership Fellow, President of 2nd executive committee).

References

GESAN (2017) Geomatics engineering students' association of Nepal. "Gesan." GESAN – OpenStreetMap Wiki, 10 July. https://wiki.openstreetmap.org/wiki/GESAN. Cited 29 Jan 2022

Kathmandu Living Labs (2016) KLL Holds Open Data Day in Pokhara. Kathmandu Living Labs, 5 March. https://www.kathmandulivinglabs.org/news/2016-understanding-risk-forum-in-venice-ignite-speech-by-kll. Cited 29 Jan 2022

Khanal K, Budhathoki NR, Erbstein N (2019) Filling OpenStreetMap data gaps in rural Nepal: a digital youth internship and leadership Programme. Open Geospatial Data Softw Stand 4(12). https://doi.org/10.1186/s40965-019-0071-1

O'Hara M (2018) Digitising Kathmandu from above. Humanitarian OpenStreetMap Team, 13 December. https://www.hotosm.org/updates/kathmandu-from-above/. Cited 29 Jan 2022

Drones for Good: Mapping Out the SDGs Using Innovative Technology in Malawi

14

Ndapile Mkuwu, Alexander D. C. Mtambo, and Zola Manyungwa

Abstract

Drones are being used in various industries, focusing on youth involvement and mapping activities as a key component of the multiple initiatives. YouthMappers are creating open-source geospatial data using geospatial technology to address local developmental and humanitarian issues in Malawi and beyond. As the world works to meet the Sustainable Development Goals, this new technology, combined with significant youth involvement, is being used to its full potential because of its sophistication, which allows for more efficient operations in support of industry, innovation and infrastructure (SDG 9) and responsible consumption and production (SDG 12) in the health sector for delivery, precision agriculture and the education sector, to name a few.

Keywords

Innovation · Drone technology · Malaria · Malawi · Geospatial industry · Youth

1 Evolution of an Innovation

The use of unmanned aerial vehicles (UAVs), colloquially referred to as drones, has grown tremendously following the decision by aviation authorities to grant commercial permits to non-military drones in 2006 (Cain Kitonsa and Kruglikov 2018). Since then, the technology has progressed from its traditional applications in security and surveillance to becoming the first line of defence during emergency response, land surveying, construction site inspection and healthcare delivery. Drones are also increasingly being used to aid in promoting development and humanitarian assistance (UNICEF 2018). As a result of the coronavirus (Covid-19) pandemic, drones have gained popularity in the healthcare

N. Mkuwu (✉)
Chancellor College, University of Malawi, Zomba, Malawi

A. D. C. Mtambo
The Polytechnic, University of Malawi, Zomba, Malawi

Z. Manyungwa
Chancellor College, University of Malawi, Zomba, Malawi

West Virginia University, Morgantown, WV, USA

© The Author(s) 2023
P. Solís, M. Zeballos (eds.), *Open Mapping towards Sustainable Development Goals*, Sustainable Development Goals Series, https://doi.org/10.1007/978-3-031-05182-1_14

industry. UNICEF is the leading organisation in implementing drone projects in various countries.

Developing countries such as Malawi have benefitted from UNICEF's deployment of drones in the transportation sector, especially for delivering medical samples, which is hindered by the country's inadequate road infrastructure. There have been cases in which drones have been able to transport blood to a remote hospital in a timely manner when a woman was in a life-threatening condition during childbirth. A drone testing corridor has been established in Kasungu, located in Malawi's central region. The purpose of the corridor is to facilitate tests for interested drone companies in the following areas: (i) imagery, generating and analysing aerial images; (ii) connectivity, exploring the possibility for drones to extend telecommunication connectivity across difficult terrain; and (iii) transport, delivery of low-weight supplies such as medicines. Children and marginalised communities will benefit from drones in this corridor because it provides a controlled environment for testing their use. Information gathered from the flights is used to inform the government's plans for the future use of unmanned aerial vehicles/drones (UNICEF 2018).

As a result of this initiative, the idea of establishing an academy dedicated to imparting twenty-first-century skills to the African youth in drone and data technology was conceived, and the concept evolved into what is now known as the African Drone and Data Academy (ADDA).

2 Drones to Promote Inclusion, Empower Youth and Foster Innovation

The goal is to build resilient infrastructure, promote inclusive and sustainable industrialisation and foster innovation. Why? Economic growth and social development are dependent on investments in infrastructure, sustainable industrial development and technological progress. Drones

are a sustainable and energy-efficient alternative for tasks involving other methods of transport or for developing and inspecting infrastructure. In this way, this technology contributes to Sustainable Development Goal 9, sustainable infrastructure.

UNICEF saw the potential for drones in Malawi, with infrastructures such as the humanitarian drone corridor being accessible for interested drone companies, academic institutions (Virginia Tech University) and local Malawian universities willing to spearhead the drone industry in Malawi. With every structure in place, one key component was missing from the mix; the local human resource was lacking. This was when capacity building for drone pilots and data analysts was started through the African Drone and Data Academy.

The African Drone and Data Academy aspires to establish an inclusive and creative environment through its one-of-a-kind curriculum covering drone technology and data analysis. Students from across Africa are given a chance to stay ahead of the curve in the technology sector. It was launched in 2020 and has already made a significant difference in the lives of the youth who participate in the program.

2.1 Youth and Drones

As the only Trusted Operator Program (TOP) training provider outside the USA, the African Drone and Data Academy (ADDA) has contributed to the empowerment of African youth. More than 500 youth across Africa have been in online and in-person drone and data technology courses. Graduates from the academy have gone on to work for a variety of drone companies, each feeling empowered and eager to make a difference in their communities after completing the program. The majority of graduates have been employed in three primary industries, including agriculture, health, and education.

Ndapile Mkuwu and Alexander DC Mtambo are graduates from the first cohort of the African

Drone and Data Academy (ADDA). The two are currently working as senior instructors at the academy. Sharing her experience, Ndapile Mkuwu stated that the program alters one's perspective and significantly increases one's self-worth. Initially, she was interested in the data analysis aspect of the program. However, as time went on, Ndapile developed a growing interest in learning how to build a drone and design fundamentals. It became even more exciting when she discovered that drones can be constructed from locally sourced materials and still perform the same functions as a $10,000 drone. "You get to learn how to think outside the box, while still thinking of local solutions to local problems", she shared. One of the activities involved building a fixed-wing drone out of foam board and 3D-printed components. It was fascinating to see how what appeared to be poster boards transformed into a drone that could save lives or collect imagery, depending on the type of payload integrated into the craft. Some students took up the idea of utilising low-cost materials and constructed a quadcopter entirely out of plywood, a material abundant in Malawi.

After their graduation, Ndapile and Alexander participated in a mapping activity at Dzaleka Refugee Camp in Malawi which was headed by their instructors at that point. The goal of this project was to acquire high-resolution imagery in order to create a flood model of the area. This was the beginning of their journal as young drone pilots in Malawi. Having joined the academy's task force as national instructors, they are at the forefront of teaching twenty-first-century skills. It has been an exciting experience. However, there are also challenges with difficulties ranging from increasing the number of women engaged in an academy to trying to complete a jam-packed curriculum in a short period of time (5 weeks). Speaking in this, Ndapile said that the only way to withstand the intense nature of both providing the program and receiving the program is to have a strong sense of enthusiasm and passion. In order to boost participation in the tech industry, strategies such as outreach programs have been used.

2.2 An Innovation Region

YouthMappers currently has four chapters based at universities located in Malawi's southern and central regions. The ADDA has so far trained 6 female YouthMappers alumni from these different universities. Having undergone the program and appreciated the technology and its applicability in OpenStreetMap (OSM), we advocated for the training of student members from the LUANAR YouthMappers chapter at Lilongwe University of Agriculture and Natural Resources. Ndapile conducted this training over two days, and it familiarised a group of people who knew nothing about drones and drone flights with the fundamentals of the technology. It was designed so that individuals who participated in the training would have an advantage in any future programs that may need participants to have a basic understanding of drones and other related technologies.

Being a YouthMappers volunteer affiliated with the Chanco chapter is the starting point of the narrative. I began as a volunteer mapper in 2018 and was chosen to be a YouthMappers leadership fellow not long after. As a leadership fellow, I received training in the soft skills that are necessary for managing a chapter and was given the opportunity to network with other youth from across the world. The next year, after the conclusion of the fellowship, I was introduced to a person in Malawi who was involved in drone research, which was relatively fresh knowledge at the time. We connected and were introduced to the academy and what it does; since then, I have progressed from being a student to later being a member of the task force (Fig. 14.1).

3 Drones to Promote Good Health and Well-Being

The introduction of drone technology to the youth in Malawi is already showing positive impacts. Various youth-led initiatives that address multiple issues in different communities have used drone technology.

Fig. 14.1 YouthMappers alumni serving as instructor for the African Drone & Data Academy works with a student during a drone flight

3.1   Dzaleka Mapping Project

MapMalawi is an organisation that was founded by two Malawian females, Ndapile Mkuwu and Zola Manyungwa. The two are former YouthMappers leadership fellows and graduates of the African Drone and Data Academy (ADDA).

It was established to create, engage and work with youth-based communities to create open-source geospatial data using geospatial technology as one of the means of contributing to addressing local developmental and humanitarian issues in Malawi.

The Dzaleka Mapping Project was the first project under the organisation, and it was funded by the Humanitarian OpenStreetMap Team (HOT) community impact grant. Dzaleka Refugee Camp is the largest camp in Malawi, located 50 km just outside the capital city, Lilongwe. It was established in 1994 by UNHCR in response to a surge of forcibly displaced people fleeing genocide, violence and wars in Burundi, Rwanda and the Democratic Republic of Congo. It was meant to house 10,000 people but has reached its maximum absorption and now has 52,000 registered refugees and asylum seekers (There Is Hope 2018). This increase means that the resources in the camp barely cater to the needs of the already vulnerable population.

The Dzaleka Mapping Project generated geospatial data that will show the provision of basic needs at the refugee camp by integrating drones, OpenStreetMap and other geospatial technologies. Centres within or around the refugee camp that allow the refugees to access the different basic needs were mapped out. These include education, healthcare, water/sanitation and buildings/housing. The relevant statistics of the mapped amenities present a picture of how people living in the camp have access to the different amenities.

3.2   UAV Imagery Giving an Essential Perspective

Drones were an essential part of the project. The Dzaleka Refugee Camp is a high-density settlement with clustering buildings. Therefore, there was a need for imagery with high spatial resolution. MapMalawi volunteers, in collaboration with CAGE, used drones to capture images of the refugee camp. The imagery was captured using a Mavic 2 enterprise dual; the mission was flown at the height of 100 m to achieve an image resolution of 3 cm/pixel. Individual images were combined to form an orthomosaic, and the final image was published on openaerialmap.org.

The HOT Tasking Manager was used for organised and coordinated OpenStreetMap mapping. The *Malawi – Mapping for People Living in Protracted Crisis – Dzaleka Refugee Camp* task was created to manage the project. The high-resolution drone imagery was used as a background image for the project. Volunteers expressed that the clarity of the drone imagery made feature identification and mapping easy. It is with no doubt that the presence of the drones contributed to the success of the mapping activities.

3.3   Collaborations, Youth and Community Involvement

In the words of Hellen Keller, "alone we can do so little; together we can do so much". Collaborations, youth and community involvement were critical to the project success. Youth that are graduates of the drone academy were recruited as volunteers under MapMalawi. Ndapile and Zola believe that community members should take part in mapping activities. Apart from the fact that their local knowledge is beneficial to the project activities, involving the community empowers them and lets them see themselves in the created maps. Therefore, MapMalawi also collaborated with a local tech lab, TakenoLab, located at Dzaleka Refugee Camp, and works with people living within and around the refugee camp. The MapMalawi volunteers and youth affiliated with TakenoLab were trained to make edits on OpenStreetMap using iD editor. Remote mapping activities using iD editor focused on mapping buildings within and around the refugee camp. The youth at the camp were also trained on how to collect point data using KoBoCollect, and Ndapile also facilitated a class on drone technology at the refugee camp.

Collaborations went beyond Malawian boarders. The Malawi OSM community hosted OSM Africa's monthly mapathon in September 2021. The Dzaleka Mapping Project was the project mapped on this day, with OpenStreetMap users from different parts of Africa contributing towards the generation of the OpenStreetMap-based geospatial data for the refugee camp. The Humanitarian OpenStreetMap Team (HOT) data quality interns validated the mapping. With the

Fig. 14.2 Drone imagery provides an excellent source of high-resolution imagery that enables remote feature mapping onto OpenStreetMap. (Credit: N. Mkuwu)

efforts of the groups mentioned, a total of 16,715 buildings, 23 schools, 11 toilets, 16 waste disposal sites, five health centres, one police station and 47 water points were mapped out (Fig. 14.2).

4   Drones for Innovation in Health Prevention Efforts

4.1   Novel Technology to Control Malaria

According to the World Malaria Report 2018, Malawi is one of the top 20 countries globally in terms of malaria prevalence and fatality rate (2% of global cases and deaths). Malawi accounts for approximately 7.4% of all malaria cases in Eastern and Southern Africa (World Health Organization 2018).

The Malawi Liverpool Wellcome Trust carried out a pilot study in Kasungu, Malawi. According to the preliminary studies, malaria transmission occurs all year round in this area, with parasite prevalence in children between 2 and 10 years old estimated at 19% in 2017. Its transmission is potentially driven by a number of reservoirs (artificial lakes) that provide permanent water sources within which female *Anopheles* can lay their eggs (Stanton et al. 2021). The project's

goal was to perform field research inside the UNICEF humanitarian drone testing corridor in Kasungu district, Central Malawi, to see whether mosquito breeding areas could be detected pragmatically using drones and to provide a framework for the National Malaria Control Program to adopt this technology (MESA 2021).

In addition to the production of orthomosaics, the Malawi Liverpool Wellcome Trust team uses additional geospatial methods, such as machine learning, to quickly identify tiny ponds of water that may be a possible breeding location for the mosquito. Because of the geotagging feature of drones, identifying locations where breeding sites are likely to be found is very simple, this reduces the amount of effort required for humans to find the potential sites for water sample collection to have some lab work done to determine the presence of mosquito larvae that causes malaria.

4.2   Mosquito Breeding Sites in Kasungu

Mwase is a village in Kasungu with a dam that goes by the same name. Mwase village is surrounded by two other villages (Chiponde and Chinkhobwe) with high malaria cases (Fig. 14.3).

Fig. 14.3 Mwase village in Kasungu has a dam and reservoir which needs monitoring to control malaria

While on site, Ndapile Mkuwu and Alexander Mtambo were struck by how important the initiative is since they were able to see first-hand how bad malaria is in the area. They randomly went to two different houses to introduce themselves, and the first house had two people suffering from malaria, and the other had four people. Sharing her experience, Ndapile said, "I must say that this gave us more motivation as we knew that the work we are doing is something that will actually impact real people with real problems".

Although it is clear that the dam is a mosquito breeding site, there are other smaller water bodies inside the villages that have the potential to be mosquito breeding sites. However, the size of these smaller bodies makes them relatively harder to detect. This is where drone imagery becomes quite useful. After creating an orthomosaic, areas with water that were difficult to identify through manual surveying by simply strolling about the towns on the ground became easier to identify (Fig. 14.4).

4.3   Pre-flight Preparation

Malawi Liverpool Wellcome Trust contracted Alexander Mtambo and Ndapile Mkuwu to conduct mapping activities in the areas of interest and produce maps through GLOBHE, a global crowd droning platform. As per the Malawi regulations, for an individual to pilot a drone, you must have the remote pilot license (RPL). Ndapile and Alexander are certified drone pilots by the Malawi Department of Civil Aviation (DCA) and holders of the Trusted Operator Program Level 2 (TOP 2) issued by Aerial Unmanned Vehicle Systems International (AUVSI). Nonetheless, they were required to obtain authorisation to fly or at the very least notify the DCA of the times and heights at which the drones would be flying to facilitate coordination with other airspace users and to ensure the safety of all airspace users during all operations.

Ndapile and Alexander conduct community sensitisation activities before flying the drones as per standard practice. Drones are very useful and

Fig. 14.4 Small ponds of water that may serve as breeding grounds for mosquitos that cause malaria are difficult to identify or locate through manual surveying but easily identifiable via drone imagery

make work simpler in many situations, but they also have some negative effects, such as personal privacy and safety concerns. The community is educated on what the drone is and how it is likely to behave once it takes off. Once individuals are aware that a camera is hovering above their heads, they begin to wonder what the purpose of all the pictures being recorded could be. This is one of the main reasons why community sensitisation is very crucial. The drone pilot duo always strives to make sure that relevant authorities in the community are aware that they are flying and why drones are flying. The case in Kasungu was slightly different for us despite the location being remote; Malawi Liverpool Wellcome Trust had already done much sensitisation, and people had already seen a drone fly. Nonetheless, Ndapile and Alexander avoid working on the assumption that everyone knows what a drone is and that they have seen it before (Fig. 14.5).

When sharing how they handle flying drones with people around, Alexander said:

To avoid drawing negative attention to ourselves or our activities, we constantly emphasise safety while using drones, particularly when working around young children. We also make sure that all flight crew members are properly attired, which makes us far more recognisable in the event that someone has a question or requires our assistance with anything related to our operations. We make sure that we have a minimum of three people: one who serves as the chief pilot, another who assists with crowd management, and another who serves as a visual observer.

4.4 Data Collection

After evaluating the terrain, which was mainly flat, Ndapile and Alexander decided to fly at the height of 100 m. The Department of Civil Aviation specifies a limit of 120 m as the maximum height allowed. It should be noted that the resolution that the project managers request also defines the altitude to which the mission will be set. The lower the height at which the drone flies,

Fig. 14.5 Each member of a drone team has specific responsibilities to pilot and manage the operations

the higher the image resolution. Finally, the drone pilot duo makes sure that they are flying higher than the highest building or natural landmark in the region. This is to prevent the drone from crashing, despite the drone having an inbuilt obstacle avoidance system. The overlap percentage for the mission flight was set on 70% frontal and side, which produced approximately 2000+ images for the 600-ha area that was being mapped.

4.5   Working Through Challenges

Batteries are a constant source of frustration during flight activities. This includes batteries for the remote controller and the batteries for the drone itself. Usually, a battery would provide about 20–25 min of flying time for a mission. With the study area being approximately 600 ha in size and to avoid spending a significant amount of time on site, Ndapile and Alexander are obliged to accelerate the procedures while not compromising on quality. Despite all efforts, batteries still prove to be a challenge. Car battery chargers are used to tackle this, and sometimes an uninterrupted power supply (UPS) is utilised. In addition to this, the duo makes sure that the take-off

home point is as close to the trading centres as possible to charge the batteries in the nearby houses and shops within the community.

5   Reflecting on the Potential of Drones for SDGs in the Hands of Youth

With the two activities that we covered in this chapter, it is evident that drone technology is dependable and can be used for a broad range of applications. Drones have the potential to contribute to the achievement of the Sustainable Development Goals more efficiently and sustainably; the fast development of the drone industry is an indicator of the promise that they possess in this regard. Various youth engagement programs have been launched to promote active participation in ticking the boxes of the different Sustainable Development Goals (SDGs). The African Drone and Data Academy is one of the key players in this regard; it trained more than 500 youth across Africa in online and in-person drone and data technology courses. A good number of its graduates are now employed by drone companies that specialise in either health or agriculture. Several others have start-up companies, with the core activity being the provision of mapping services.

We believe that with the appropriate amount of empowerment, most of the problems that plague our communities would be resolved, and lives would be spared. As of 2019, drones delivered over 27,361 vaccines and 95,011 doses of antibiotics in Malawi; drones are bringing us a step closer to excellent health and well-being. Who better head these initiatives than our African youth?

References

Cain Kitonsa H, Kruglikov SV (2018) Significance of drone technology for the achievement of the United Nations sustainable development goals. R-Economy 4(3):115–120. https://doi.org/10.15826/recon.2018.4.3.016

MESA (2021) Assessing the operational usage of drone imagery for malaria mosquito breeding site mapping in Malawi (Maladrone). Available via Mesa Malaria. http://mesamalaria.org/mesa-track/assessing-operational-usage-drone-imagery-malaria-mosquito-breeding-site-mapping-malawi. Cited 29 Jan 2022

Stanton MC, Kalonde P, Zembere K, Spaans RH, Jones CM (2021) The application of drones for mosquito larval habitat identification in rural environments: a practical approach for malaria control? Malaria J 20(1). https://doi.org/10.1186/s12936-021-03759-2

There Is Hope (2018) The situation in Dzaleka Refugee Camp. Available via Dzaleka Refugee Camp – There Is Hope Malawi. http://thereishopemalawi.org/dzaleka-refugee-camp/. Cited 29 Jan 2022

UNICEF Malawi (2018) Drones In Malawi – Factsheet. Available via UNICEF. https://www.unicef.org/malawi/media/651/file/The%20Drones%20Factsheet%202018.pdf. Cited 29 Jan 2022

World Health Organization (2018) World Malaria report 2018. Available via WHO. http://apps.who.int/iris/bitstream/handle/10665/275867/9789241565653-eng.pdf. Cited 29 Jan 2022

Assessing YouthMappers Contributions to the Generation of Open Geospatial Data in Africa

15

Ebenezer N. K. Boateng, Zola Manyungwa, and Jennings Anderson

Abstract

As leaders of tomorrow, we, the African YouthMappers, are taking the initiative of contributing toward the attainment of the Sustainable Development Goals that are a blueprint for a better and more sustainable future for all. Our contributions center on the generation of open geospatial data, which is critical when making decisions on achieving desired development across all seventeen goals. In response to the African continent's alleged data inadequacies, YouthMappers, as vital members of the OpenStreetMap community, have made significant contributions to open geospatial data in Africa. This chapter highlights the contribution of YouthMappers *to not just build maps* – which especially supports reduced inequalities around data access in Africa (SDG 10) – but also *to build mappers* – by advancing the geospatial capacity of young people across the continent, addressing (SDG 8) decent work and economic growth.

Keywords

YouthMappers in Africa · Mapping contributions · Digital divide · Geospatial skills · Ghana · Malawi · Youth

E. N. K. Boateng (✉)
University of Cape Coast, Cape Coast, Ghana
e-mail: ebenezer.boateng@stu.ucc.edu.gh

Z. Manyungwa
Chancellor College, University of Malawi, Zomba, Malawi

J. Anderson
West Virginia University, Morgantown, WV, USA

YetiGeoLabs, LLC, and YouthMappers, Helena, MT, USA

1 The Landscape of Youth and Open Data in Africa

Let us imagine an Africa where there is enough data.

Enough data to meet all our development objectives, from economic to social to environmental. Enough data to know the location of every house in Mozambique and every road leading to it. Enough data for the easy delivery of life-saving malaria prevention services. Where we can see all the rural villages that are without electricity, pointing the path to electrification initiatives that will bring light and energy to them.

Imagine an Africa where young people see themselves in the maps they create. An Africa where students can build a social infrastructure

P. Solís, M. Zeballos (eds.), *Open Mapping towards Sustainable Development Goals*, Sustainable Development Goals Series, https://doi.org/10.1007/978-3-031-05182-1_15

around knowledge and technology that empowers the whole continent to achieve all of the Sustainable Development Goals.

While we are not there yet, map contributions by YouthMappers residing in Africa and beyond are charting aspirational paths. Aside from creating spatial data, our core mandate is to raise the youth to engage in sustainable economic growth and reduce inequality by building our workforce for the future. This fundamental mandate is aligned with SDG 10 (Reduced Inequality) in that we work with open data and 8 (Decent Work and Economic Growth) in that improving geospatial skills increases labor capacity in a fast-paced, high-growth industry.

The SDGs have been described as one of the noblest decisions to promote prosperity to millions of people by addressing social, economic, and environmental challenges (UN Communications Group [UNCG] in Ghana and CSO Platform on SDGs 2017). These goals target the sustainable development of all forms of life across all spheres of the earth (biosphere, lithosphere, atmosphere, and hydrosphere). SDGs have largely been globally accepted, and conscious efforts seek to successfully implement these goals to ensure national, regional, and global peace, prosperity, and stability. The Global Partnership for Sustainable Development Data, a coalition of 280 organizations, states that "it will require unprecedented amounts of data to ensure we leave a better planet for future generations" (Global Partnership for Sustainable Development Data 2015). A conscious effort by governments, the private sector, civil society, citizens, and youth movements such as YouthMappers is thereby required to contribute to this unprecedented amount of data needed for sustainable development. In our data-driven age, the lack of readily available data for decision-making motivated us to create enough data for the achievement of the SDGs.

One of the key factors that restrained the full implementation of the precursor global targets for development, known as the Millennium Development Goals (MDGs), in some parts of the world, particularly Africa, was limited spatial data (UN-Habitat 2019). Thus, the absence of spatial data to aid in informed decision-making was not achieved, which hindered the MDGs' implementation. Even after the launch of the SDGs, a report from the United Nation-HLPF (2018) showed that there is limited data available to track the performance of the SDGs, especially spatial data. The forebears of MDGs suggested the generation and utilization of spatial data since it offers much more data visualization opportunities that aid in informed decisions resulting in target-oriented policies and programs within a specific area. This led to several initiatives and interventions through global and local movements and programs to help bridge the spatial data gap, particularly in Africa.

African regions have been identified as some of the poorest in the world, with limited data for informed decision-making (Keeton and Nijhuis 2019). According to the International Labor Organization (ILO) (2021), the youth unemployment rate in Africa is 10.6%. Although this represents a rate below the global youth unemployment rate (13.8%), a majority of the youth in Africa cannot afford not to work. Simply put, the youth suffer from underemployment and lack of decent working conditions (ILO 2021). The majority of young people lack the requisite skills and abilities to find suitable work or start their own business. Based on the ILO (2021) statistics, the rate of unemployment among young women (11%) is relatively higher compared to the number of young men (10.4%). Interventions by organizations and groups such as YouthMappers, the United Nations, the World Bank, the Global Partnership for Sustainable Development Data, and Humanitarian OpenStreetMap Team (HOT) channel their focus on grooming the youth in creating and using spatial data for the achievement of all 17 Sustainable Development Goals.

This chapter thus focuses on how the activities of YouthMappers help in the achievement of SDG 8 (Reduced Inequalities) and 10 (Decent Work and Economic Growth). SDG 10 seeks to reduce inequalities within and among countries (UN Communications Group [UNCG] in Ghana and CSO Platform on SDGs 2017). According to ILO (2021), there is a gender gap in youth unemployment in Africa.

Therefore, it is essential that this gap is reduced so that young women enjoy the same access to opportunities for employment as boys and men. The purpose of SDG 8 is to promote inclusive and sustainable economic growth, employment, and decent work for all. According to the UNCG (2017), decent jobs mean safe working conditions with fair wages that allow people to lift themselves out of poverty. Linking this to the challenge of youth being underemployed and/or lacking decent working conditions in Africa implies that there is a lot of work left to attain SDG 8. Efforts to achieve SDG 8 might not be complete if there is unequal access to job opportunities based on characteristics such as gender, race, income, disability, and others. This makes it essential to couple SDG 8 with SDG 10 to ensure a holistic achievement of the SDGs. To address these issues, the question is asked: How can SDG 8 and 10 be achieved aside from government interventions? Based on this question, we focus on the contributions of YouthMappers in Africa toward achieving SDGs 8 and 10.

Several studies have been conducted about YouthMappers in terms of bridging the digital divide generally and the global potential for workforce capacity building in geospatial skill sets (Coetzee et al. 2018; Solís et al. 2018, 2020a, b). However, there is less information assessing the specific regional contribution of YouthMappers in creating spatial data by African youth toward attaining these goals as they relate to the SDG 8 and 10 here. In the subsequent sections, we assess the creation of open-source spatial data by YouthMappers and examine its implications for SDG 8 and 10 among the African youth and challenges encountered, lending our own experiences as YouthMappers to this assessment.

2 Generation of Open Spatial Data in Africa

OpenStreetMap (OSM) is an openly editable map of the world to which anyone may freely contribute. Globally, OSM has seen contributions from more than 1.7 million users worldwide and is considered the largest free and open geospatial database available today. On the African continent, mapping activities in OSM have been increasing the availability of free, open geospatial data for African countries since 2012. This growth can be partly attributed to the drive toward humanitarian mapping across the continent (Herfort et al. 2021). Since 2013, just over half of all OSM changesets with a #hotosm comment or hashtag are located in Africa, indicating that the editing was related to a campaign issued by the largest organizer of humanitarian mapping efforts in the world, the Humanitarian OpenStreetMap Team (HOT; see also Chapter 30 in this volume). Conversely, Fig. 15.1 also shows that nearly half (47%) of the changesets submitted for Africa to OSM as a whole include a #hotosm-related hashtag. This pattern spikes between mid-2017 and late-2019, showing that most of the changesets in Africa during these years included a #hotosm-related hashtag.

Africa is among the latest continents to receive considerable mapping attention from OSM editors compared to other continents. Figure 15.2 presents the total edits per continent since the beginning of OSM. The rate of editing in Europe, while higher, has been relatively consistent since about 2010, appearing more linear in Fig. 15.2 of cumulative edit counts. In contrast, while other locations have steadily increased over time, edits only began to surge, as indicated by the dark blue line representing Africa that began curving upward since 2015, the year YouthMappers was formally founded.

When YouthMappers was founded, the first inaugural university to join the network of three organizing universities was the University of Cape Coast in Ghana, Africa. The addition of new university chapters has been vital in the region, to where among the nearly 300 campuses at the end of 2021, half of them are in Africa. Of

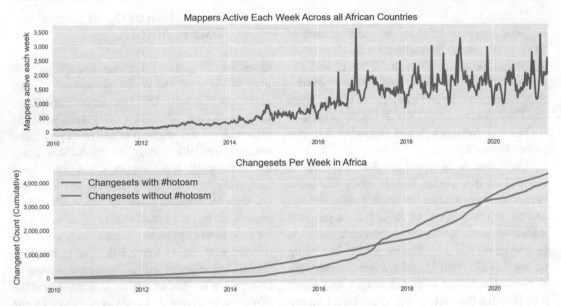

Fig. 15.1 The number of active contributors across Africa fluctuates but has grown over time

the 64 countries with chapters, half of those countries are in Africa.

It is no surprise that the YouthMappers in African universities have made considerable contributions to the generation of OpenStreetMap data, and YouthMappers everywhere have been committed to mapping in the continent. While YouthMappers have been active across all continents, Fig. 15.3 shows that YouthMappers are most prolific in Asia and Africa, with 48% and 45% of all features edited by YouthMappers located on these continents, respectively.

Figure 15.4 presents a cumulative count of features edited by YouthMappers in Africa. Among African countries, Uganda, Zimbabwe, and Mali have seen the most editing (by feature count) among all YouthMappers.

In sum, YouthMappers have performed 7.1M edits across African countries from 2015 through the end of 2021. Of these edits to OpenStreetMap, 6.2M were building objects, and 98,000 were amenities, such as schools, businesses, or points of interest. As of the start of 2022, 113,000 km of roads and paths across the continent were last edited by YouthMappers. Buildings and highways are more easily traced from satellite imagery by remote mappers. Amenities require having some level of localized knowledge. The chapters

in African universities and the network collaboration to map the continent with joint campaigns have produced these results (Fig. 15.5).

3 Implications for the Sustainable Development Goals

Our contributions have significantly increased the amount of open geospatial data accessible by all for several purposes geared toward the attainment of the SDGs. The engagement of the African youth in YouthMappers activities offers some opportunities aside from creating open geospatial data.

3.1 On SDG 10: "We Don't Just Build Maps" That Bridge the Digital Divide

In addition to the contributions to OSM outlined above, which represent an important fundamental publicly available dataset for locations that had previously been poorly mapped, our mapping activities aid in human resource development in Africa that breaks the "gender digital divide."

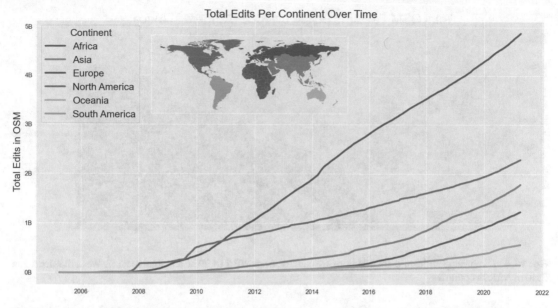

Fig. 15.2 The total OpenStreetMap edits contributed over time by continent shows significant increases since the establishment of YouthMappers for Africa

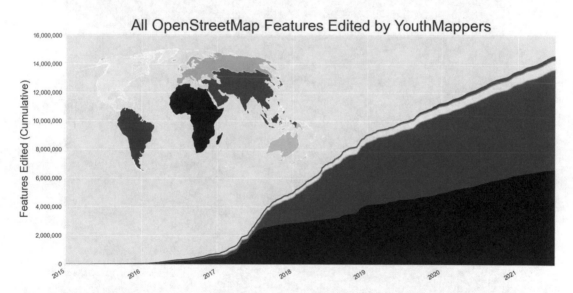

Fig. 15.3 YouthMappers contributions by region from 2016 to 2022 are significant in Africa and Asia

Regarding SGD 10, Reduced Inequalities, several programs have been put in place to bridge the gender gap in geospatial sciences. Two of these initiatives are "LetGirlsMap" and "EverywhereSheMaps." To bridge the gender gap in geospatial science, YouthMappers have awarded African women an opportunity to further their skills and knowledge in geospatial science through advanced degree study at higher education institutions with international fellowships. This contributes to the achievement of SDG 10, where inequality is reduced with access not just to data but also to the knowledge and networking circles. Aside from bridging the gender gap, these initiatives promote education, aiding in the achievement of SDG 10.

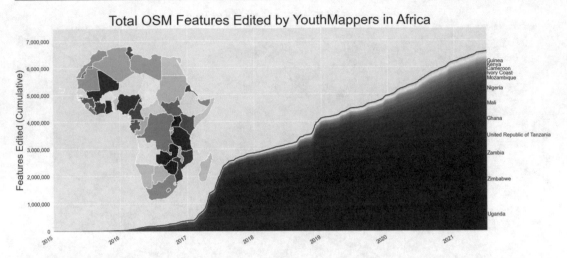

Fig. 15.4 YouthMappers feature contributions by country from 2016 to 2022 are significant in Western Africa and Eastern African countries

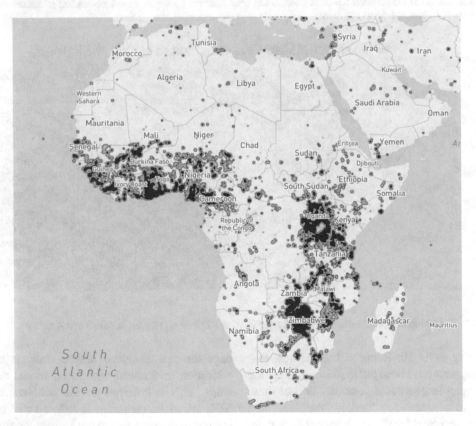

Fig. 15.5 The pattern of edits contributed by YouthMappers across the African continent have supported many projects related to the UN SDGs

3.2 On SDG 8: "We Also Build Mappers" with Geospatial Skills

The billion-dollar geospatial industry has grown exponentially in recent years. Jobs of the future will require spatial literacy and technological abilities that intersect with geospatial data and tools. A study by Solís et al. (2020a) revealed that mapping activities build the capacity of all participating youth to prepare for geospatial careers. One such activity is the introduction of internships. This is a key element to unlocking job opportunities for the African youth. Further, the study found a growing self-reported proficiency in the use of open geospatial tools that promote workforce capacity. In our current dispensation, acquiring geospatial skills offers one the opportunity to establish their firm with a little capital. This would reduce the overreliance on governments and other large firms for job opportunities. Once such a purpose is achieved, there is a higher chance that SDG 8 could be achieved in innovative ways.

4 Challenges and Opportunities

Our experiences elucidate the challenges encountered and opportunities available to YouthMappers in Africa as we contribute to the generation and use of open-source data in Africa.

The YouthMappers program has provided us with a platform to network and acquire the necessary soft and technological skills to transition into professional careers in a world rapidly adopting geospatial technologies. Beginning with the fundamentals, the YouthMappers program provides guidance and training on how to map with OpenStreetMap and how these also offer networking opportunities. Since YouthMappers chapters are student led, we gain employability skills like teamwork, resource mobilization, communication, and interpersonal skills. These soft skills prepare us for real-world scenarios that involve critical thinking, problem-solving, and working with others on a day-to-day basis.

We are also enabled to pursue opportunities outside of the university, such as internships with government and non-governmental organizations. Some YouthMappers have worked as USAID GeoCenter and YouthMappers virtual interns, interns with the Humanitarian OpenStreetMap Team (HOT), and interns with organizations in their countries of residence.

In Africa, as a testimony to this is University of Cape Coast YouthMappers alumnus Benedict T.K. Nartey, who secured an internship at the Center for Remote Sensing and Geographic Information Services at the University of Ghana. He credits his affiliation with YouthMappers as an added advantage in obtaining this internship. Laura Mugeha, a pioneer of YouthMappers in Kenya, has thrived with her geospatial career, both drawing from and building beyond her student experiences, to support the first TechFugees hackathon in Africa; to support the organization of the first Nairobi Ideation Lab with the Archimedes Project; to an appointment with the African Union Youth Envoy; to the Steering Committee of Women in Geospatial+; to service on the gender equality, disability, and social inclusion working group of Digital Earth Africa; to coordinator of OSM Kenya; and to employment with Sanergy. She often gives keynote speeches on behalf of youth, women, and technology. Temidayo Isaiah Oniosun, an inaugural YouthMappers leadership fellow from Nigeria, has launched his career to quickly rise to establish and become the manager director at Space in Africa and be named one of Forbes 30 under 30 for Africa in 2021. There are countless other examples of our students gaining excellent career opportunities, becoming entrepreneurs, and rising to new leadership positions.

This happens partly because of the YouthMappers leadership fellowship, research fellowship, and regional ambassador programs, which have greatly improved the capacity building of many university students. It works as a talent incubator and launchpad. YouthMappers ensures that all students develop a capacity for local empowerment by teaching us to be mindful of the challenges in our communities and promoting chapter initiatives that use geospatial technology to solve these challenges. The network provides an open and supportive environment that enables us to demonstrate our skills and talents. YouthMappers goes even further by acknowledging our contributions through blog posts, social media articles, and the annual YouthMappers awards. On its Award and Recognition page on the consortium's website, the YouthMappers network recognizes and appreciates award and grant winners and people who carry on different leadership positions across the network. This gesture boosts recipients' morale and makes us more recognizable to future employers and partners worldwide. As YouthMappers, we can also apply for and attend international conferences that widen our professional networks.

Women and girls are explicitly supported by the YouthMappers project in a coordinated effort to build an inclusive mapping community. The 30% participation rate of females in YouthMappers (Solís et al. 2020a) demonstrates this. Females have been supported in attending or organizing mapathons, receiving training, initiating local chapter-led projects, performing field mapping, acting as a mapping intern, and receiving a YouthMappers Leadership or Research Fellowship, according to the results of open geospatial participation in YouthMappers (Solís et al. 2020a).

Furthermore, YouthMappers has provided women with opportunities to serve in various capacities within the network to date, i.e., as regional ambassadors and mentors. Female-centered YouthMappers campaigns such as "LetGirlsMap" and "Everywhere She Maps" seek to empower women members of the mapping community while also steering mapping efforts toward addressing women-centered concerns such as gender-based violence and female genital mutilation (FGM). The various opportunities that women are exposed to as part of the program help close the gender digital gap while also empowering and preparing women for entry into the global workforce.

However, the nature of the aforementioned opportunities is not without its drawbacks. Chapters of YouthMappers, especially in Africa, face challenges due to a lack of resources. One of the most significant issues we face is a lack of a reliable and affordable internet connection, which impedes our mapping activities, such as using the in-browser iD editor on osm.org. In one case, YouthMappers from the University of Cape Coast resorted to hiring a generator for electric power and using student phones as Internet hotspots. The University of Malawi YouthMappers partnered with a local tech hub called Mhub that brought in Wi-Fi devices for mapping exercises. In addition, some university chapters still lack the requisite equipment, such as computers and GPS, for remote and field mapping activities. Furthermore, students face institutional obstacles like a lack of support from university administration or approval to use university labs, thereby limiting their ability to fully engage in the program's activities in universities where the concept of student-led humanitarian mapping is unfamiliar.

5   Imagining Africa's Tomorrow, Today

As members of the YouthMappers project who live or work in Africa, we are proud to have made significant contributions to open-source geospatial data for Africa over the last five years. These open spatial data cut across a wide range of features such as buildings, roads, and water bodies. We have shown through the data that YouthMappers have contributed millions of edits in Africa, predominantly within the sub-Saharan region, and are a large part of the major force of open data edits that have come about in that region. Through YouthMappers activities, youth in Africa can attain geospatial skills and knowledge that would increase their levels of employability. This advances achievements toward SDG 8 and 10 in particular and provides the underlying data and knowledge capacity that can ultimately serve to inform all seventeen of the SDGs in a region where data has notoriously been a challenge to meet global targets. Although the

YouthMappers community has built our capacity in diverse ways toward the generation and use of open-source geospatial data for the SDGs, we remain committed today to solving a handful of challenges for Africa's tomorrow.

References

Coetzee S, Minghini M, Solis P, Rautenbach V, Green C (2018) Towards understanding the impact of mapathons: reflecting on YouthMappers experiences. Int Arch Photogramm Remote Sens Spat Inf Sci XLII-4/W8:35–42. https://doi.org/10.5194/isprs-archives-XLII-4-W8-35-2018

Global Partnership for Sustainable Development Data (2015) Better data. Better decisions. Better lives. Available via Data7SDGs. https://www.data4sdgs.org/. Cited 28 Dec 2021

Herfort B, Lautenbach S, Porto de Albuquerque J, Anderson J, Zipf A (2021) The evolution of humanitarian mapping within the OpenStreetMap community. Sci Rep 11(1):1–16. https://doi.org/10.1038/s41598-021-82404-z

ILO (2021) Global employment trends for youth 2020: Africa. Extract 4–6

Keeton R, Nijhuis S (2019) Spatial challenges in contemporary African New Towns and potentials for alternative planning strategies. Int Plan Stud 24(3–4):218–234. https://doi.org/10.1080/13563475.2019.1660625

Solís P, Mccusker B, Menkiti N, Cowan N, Blevins C (2018) Engaging global youth in participatory spatial data creation for the UN sustainable development goals: the case of open mapping for malaria prevention. Appl Geogr 98:143–155. https://doi.org/10.1016/j.apgeog.2018.07.013

Solís P, Anderson J, Rajagopalan S (2020a) Open geospatial tools for humanitarian data creation, analysis, and learning through the global lens of YouthMappers. J Geogr Syst 23(4):599–625. https://doi.org/10.1007/s10109-020-00339-x

Solís P, Rajagopalan S, Villa L, Mohiuddin MB, Boateng E, Wavamunno Nakacwa S, Peña Valencia MF (2020b) Digital humanitarians for the sustainable development goals: YouthMappers as a hybrid movement. J Geogr High Educ 1–21. https://doi.org/10.1080/03098265.2020.1849067

UNCG (2017) The Sustainable Development Goals (SDGs) in Ghana. UN Communications Group (UNCG) in Ghana, and CSO Platform on SDGs. Available from United Nations, 30 November. https://ghana.un.org/en/19077-sdgs-ghana-why-they-matter-how-we-can-help. Cited 28 Dec 2021

UN-Habitat (2019) Sustainable development goal 11 – make cities and human settlements. Monitoring framework. In: A guide to assist national and local governments to monitor and report on SGD goal 11+ indicators. Available from United Nations. https://sustainabledevelopment.un.org/sdg11. Cited 28 Dec 2021

United Nation-HLPF (2018) 2018 review of SDGs Implementation: SDG 11 – make cities and human settlements inclusive, safe, resilient and sustainable. In: High-level political forum on sustainable development, pp 1–11

Mapping Invisible and Inaccessible Areas of Brazilian Cities to Reduce Inequalities

Elias Nasr Naim Elias, Everton Bortolini,
Jaqueline Alves Pisetta, Kauê de Moraes Vestena,
Maurielle Felix da Silva, Nathan Damas,
and Silvana Philippi Camboim

Abstract

This account of the experiences of the first YouthMappers chapter in Brazil aims to present the work done by Mapeadores Livres UFPR focusing on the themes of accessibility (SDG 10 Reduced Inequalities) and favela mapping (SDG 11 Sustainable Cities and Communities), bringing as a transversal discussion of the quality of the geospatial data obtained collaboratively. Mapeadores Livres UFPR proposes to be a formative space in collaborative mapping, and the theme of favela mapping reinforces our character as an extension and research group and as a teaching group with the inclusion of this topic in the course developed. YouthMappers functions as an outreach project at the Universidade Federal do Paraná and is an interface between the universe of social organizations and their humanitarian projects and that of universities and their research topics.

Keywords

Inequalities · Digital invisibility ·
Accessibility · Methodology ·
OpenStreetMap · Favelas · Brazil

1 A Scenario of Digital Inequality

In the context of urban and regional planning, geospatial data is a useful tool that can be used in several ways, particularly for conducting spatial analysis for evaluating and predicting expansion and growth. Unfortunately, in developing coun-

E. N. N. Elias (✉) · E. Bortolini · J. A. Pisetta ·
K. de Moraes Vestena · M. F. da Silva · N. Damas ·
S. P. Camboim
Federal University of Paraná, Curitiba, Brazil
e-mail: elias.naim@ufpr.br; evertonbortolini@ufpr.br;
jaquelinepisetta@ufpr.br; kauevestena@ufpr.br;
mfdsilva@mppr.mp.br; nathandamas@ufpr.br;
silvanacamboim@ufpr.br

tries like Brazil, where the combination of a very dynamic urban structure and the lack of investment in cartography results in a chronic dearth of information about the city, the integration of geospatial data in decision-making becomes quite challenging. This deficiency is even more severe when we think of data on informal settlements, such as favelas, or information of interest to populations chronically invisible to public authorities, such as people with disabilities. Favelas are spaces in cities characterized by the precariousness of their infrastructure, the social vulnerability of their population, and their informality. It is difficult to access geographic data from underrepresented precarious settlements on official maps. At the same time, these spaces demand information to make decisions about improvement projects on varied fronts. Accessibility is a fundamental element for guaranteeing human dignity in urban environments and a right guaranteed by law in Brazil by the statute for persons with disabilities and the statute for the elderly. However, the Brazilian reality concerning accessibility is far from ideal. The inadequacy of measures that would guarantee access to urban environments for several groups, especially wheelchair users, the elderly, and the visually impaired, is frequent.

An associated concern is the little capacity building of civil servants, professionals, and the general population on open geospatial data and tools for collecting and utilizing this vital information. This scenario increases inequalities in the country, creating a data gap in territorial management and limiting the full participation of society.

Fig. 16.1 The Mapeadores Livres UFPR chapter logo pays tribute to the flag of Brazil

Geospatial Foundation (OSGeo), International Society for Photogrammetry and Remote Sensing (ISPRS), and International Cartographic Association (ICA) to make spatial information available to everyone.

Our team brings together undergraduate and graduate students, researchers, and professionals to expand the use and understanding of open, collaborative information in Brazil. In 2019, Mapeadores Livres became the first YouthMappers chapter in the country, and we expanded our projects in spatial data quality, slum mapping, and urban mobility. We research collaborative mapping in the abovementioned themes and produce text and video resources in the Portuguese language. These materials contribute to mappers' training in Portuguese-speaking countries and have supported the expansion of the YouthMappers' presence in Brazil. This chapter also develops partnerships with groups active in the chosen themes and creates open-source software to expand the mapping capacity and data availability on OpenStreetMap (OSM). More recently, we are promoting free remote courses to the community on mapping with open tools in the OpenStreetMap ecosystem. We aim to expand the mapping community and foster those in Portuguese-speaking countries. In order to make the resources our team has created, some of the courses that Mapeadores Livres have offered can be found on the chapter's YouTube channel.

2   YouthMappers at the Federal University of Paraná

Mapeadores Livres, the YouthMappers chapter at the Federal University of Paraná (UFPR), arises from the research group *Laboratório Geoespacial Livre* that since 2012 works on open-source cartography and collaborative mapping, focusing on our reality as a developing country. The laboratory was created as part of the *Geo For All* network, a joint effort between the Open-Source

As seen in Fig. 16.1, the Mapeadores Livres UFPR chapter logo is a modification of the YouthMappers logo and features the colors of the Brazilian flag. The hemisphere of the globe is in

South America; the marker is a stylized pine nut (the pine nut is an edible seed typical of the region), which is the symbol of the city of Curitiba, where the group is based.

3   Systematized Collaborative Mapping

In these nine years of history, dozens of students, courses on free software, scientific articles, and open applications have been developed by the laboratory team. Turning our attention to the collaborative data of open platforms like OpenStreetMap was a natural passage in our teaching and research activities and an opportunity to interact with and impact communities more directly. From a research context on open spatial data, we began to dialogue with public institutions, receiving professionals such as analysts from municipalities and the National Mapping Agency (IBGE) to be part of our team to extend the idea of using collaborative information to help build up the geospatial data in the country.

We maintain a workflow for the organization of activities. The working theme of one or more members of the group or an external demand is what guides the workflow. The main stages are

(1) an interdisciplinary meeting, (2) the production of training material, and (3) organizing a mapathon (Fig. 16.2).

The process begins with contacting the community acting in that region or theme and seeking to broaden the group's knowledge of concepts related to the issues being addressed. These spaces are either closed meetings or lectures open to the public. In these meetings, we also discuss the user's perspective on applying geospatial data to the topic under analysis.

We consider the creation of tutorials, videos, lectures, or courses to disseminate knowledge on the subject and make them available in different media, such as manuals on the chapter's website, social media, and audio or video through the group's YouTube channel. Tutorials, lectures, or short courses focus on specific themes or tools. Mapathons are the final step in our workflow cycle and are a space for dissemination, training, testing, and validation of systematized collaborative mapping processes. These are also opportunities to publicize our activities and strengthen the social ties that permeate the group. Following the mapathon, a new cycle of the workflow begins.

Prospecting for new themes is open at every workflow step, allowing the group to run multiple projects and cycles in parallel. The pandemic

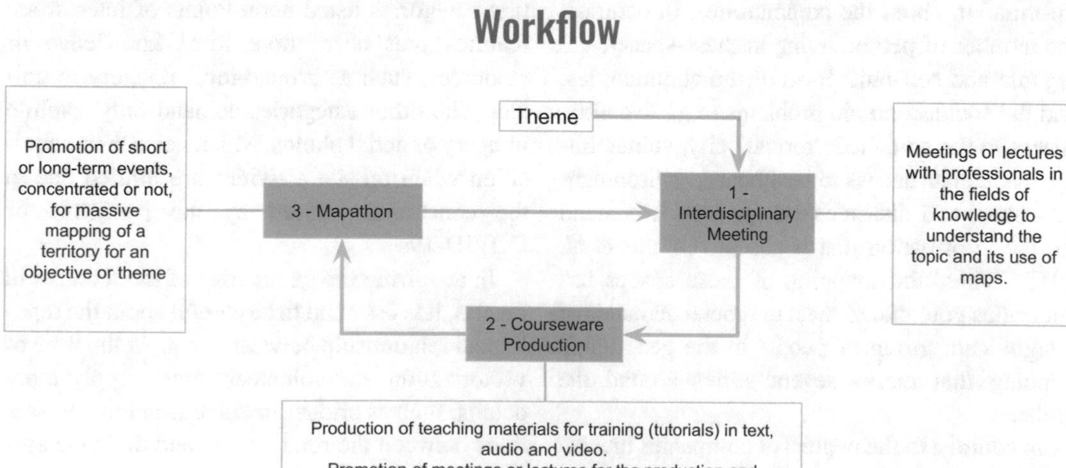

Fig. 16.2   A project workflow integrating remote and local mapping ensures quality, student learning and community engagement in YouthMappers activities

of COVID-19 had a rather significant and prolonged impact on the country. Our chapter was created in mid-2019, and within a few months, we would be under an extended regime of remote activities in Brazilian universities. This scenario was quite challenging for our group, but throughout 2020 and 2021, we managed to maintain our remote activities, even managing to expand the number of participants from other regions of Brazil. We believe that even with the return of face-to-face activities, the experience of this period will lead us to maintain, at least partially, the remote activities to minimize geographical barriers to the participation of interested participants.

4 Mapping Favelas

Favelas are spaces in cities characterized by their precarious and informal infrastructure and equipment and the social and economic vulnerability of its population (UN HABITAT 2003; Davis 2004; IBGE 2011; Cardoso 2016). In Brazil, 6%, or 11 million Brazilians, live in this condition (IBGE 2010).

Public authorities and service providers treat the favelas as informal and invisible spaces. Consequently, informality and invisibility by these institutions affect the lack of geographic information about the communities. In contrast, the number of people living in these spaces, the organic and self-built form of the communities, and the socioeconomic problems (e.g., livability, access to the so-called "formal" city, vulnerability, inadequate access to healthcare, environmental issues, and disaster risk) generate a demand for the information that is absent (Olthuis et al. 2015). Thus, the mapping of these spaces has enormous potential to meet the social demands of a significant group of people in the geographic clipping that marks several cities around the globe.

In contrast to the neglect of companies or even state entities, various social organizations map marginalized urban and rural areas. Some are the Humanitarian OpenStreetMap Team (HOTOSM);

MAPKIBERA, in Kenya; and TECHO, more specifically, in Latin America. These social entities use distinct processes, including collaborative mapping techniques and tools such as OpenStreetMap (Bortolini and Camboim 2019). Therefore, they show interest and use approaches that are not the traditional ones.

The approach to collaborative favela mapping in the project considers the presentation of best practices. These include the correct acquisition of features considering three essential points: (1) completeness of the features, including geometry, (2) topological relationship between feature geometries, and (3) thematic and attribute completeness. Thus, we describe them in some detail (Fig. 16.3).

4.1 Ensuring Quality in Data Collection

Geometric completeness deals with the acquisition of all features in a given area. If it is not possible to capture all features, it is necessary to consider prioritizing them. For favelas, one way to list information categorically, in order of priority for acquisition, is as follows: buildings, roads, hydrography and other drainage features, points of interest and land cover, and vegetation features. Humanitarian projects commonly map the first categories listed here. Points of interest are features that need more local knowledge or resources, such as ground-level imagery to collect. The other categories demand only satellite imagery or aerial photos, which helps their acquisition when on-site activities are limited due to the constraints caused by the pandemic of COVID-19.

In acquiring the geometries of the features in favelas, it is essential to be careful about the topological relationship between them. At the time of vectorization, the volunteers must supply more details, such as bridges or culverts at the intersections between the road system and drainage system elements. Other basic precautions are not to intersect buildings with other elements. In addition, all roads or hydrographs must have the

Fig. 6.3 YouthMappers in the UFPR chapter utilize the TeachOSM Tasking manager to map in favelas

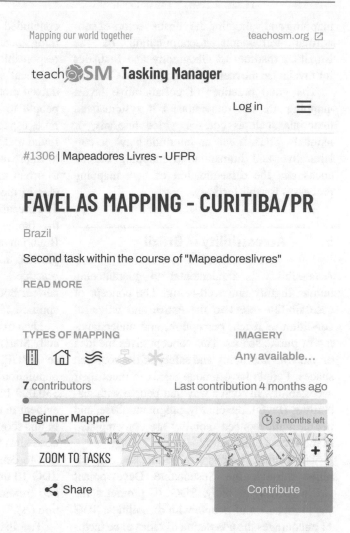

direction of flow and be intra-connected to be used in network analysis or routing. Mapping that pays attention to topological relationships allows using that favela database in products beyond a static map.

Finally, when vectorizing features with good practices in mind, we should not focus only on the geometry but also categorization and attribute completion. For example, roads in favelas can have distinct characteristics from others in the formal city, making it essential to define the correct category or fill attributes according to their trafficability. The mapper needs to consider the type of surface, the width of the road, and the existence of sidewalks, stairs, or bridges, as in some cases for vector geometry, filling in the category and attributes may require local knowledge

or ground-level imagery for a more accurate feature description.

4.2   Open Tools and Best Practices

The tools used are linked to the OpenStreetMap ecosystem and include some editors of its database. iD is the web editor with the most extensive user base and has a more accessible interface and more simplified functionality. In contrast, JOSM is a completer and more complex desktop editor, which allows for more functions, but with a higher learning curve. Therefore, the choice of the tool must be made according to the experience and interests of the audience. We rely on the TeachOSM platform, which can help us organize mapping and validating the census sectors of sub-normal settlements (denomination of the Brazilian Institute of Geography and Statistics for favelas) in the metropolitan region of Curitiba.

The good practices of collaborative favela mapping are not unanimous but systematized materials such as courses, video tutorials, or manuals. Thus, it was an interlude between collaborative and humanitarian mapping experiences and the dissemination of new mapping processes for this territory.

5 Accessibility in Brazil

Accessibility is fundamental to guaranteeing human dignity and well-being. The concept of accessibility refers to the entire and universal condition of reach, perception, and understanding of public spaces. This concept gives the idea of use with autonomy and safety of the common spaces. People have unique needs to meet their accessibility. In such a way that people with disabilities (PwD), the elderly, pregnant, obese, and people with reduced mobility are contemplated in the accessibility laws.

The United Nations (UN) addresses accessibility through the Sustainable Development Goals (SDGs) globally. SDG 10 promotes the social inclusion of people with disabilities. SDG 11 encourages the adaptation of cities to be inclusive and accessible. In Brazil, the adaptation of urban equipment with adequate conditions for accessibility still does not meet the necessary demand. As a result, it is possible to see non-compliance with the technical standards regulated by the Brazilian Association of Technical Standards (ABNT) nationally in public spaces.

The Brazilian press has reported the incongruence in urban spaces making access and mobility impossible in cities such as Jundiaí/Sorocaba-São Paulo, Brasília-Distrito Federal, and Porto Alegre-Rio Grande do Sul, showing disrespect for the disabled and for technical standards. A study by Remião (2012) highlighted the lack of accessibility in schools in Viamão-RS and has concluded that the problems met are not something localized but a general problem that the responsible authorities should address based on technical standards. Azevedo (2020) has conducted a study on the access of PwD and older people to transportation in their first and last mile, especially on rainy days in São Paulo and found that it is necessary to apply public policies to improve and guarantee equality and the right to urban mobility without having to face obstacles in the access to and between urban cities. The precarious situation and the disregard toward the theme can be due to ableism present in Brazilian society. Ableism is a discriminatory practice toward people with disabilities. This practice is based on the social construction of a perfect body and underestimates the ability and aptitude of people due to their disabilities.

Data from the Brazilian Institute of Geography and Statistics (IBGE), from the 2010 census (IBGE, 2010), show that 6.7% of the Brazilian population has some disability although only 330,000 have a signed work card. Prejudice is evident in quite common colloquial expressions. In this scope, the mapping of accessibility conditions aims to contribute to the supply of information to denote such reality. This action supports SDG 10 and 11, especially the latter, impacting two fronts: inspection (A) and access optimization (B).

The inspection of the degree of compliance with accessibility conditions throughout the geographic space can be applied in two main ways: to expose the neglect/disinterest in its implementation (A-I) and to draw up strategic plans to achieve adequacy (A-II), such as the revision of municipal master plans.

About the optimization of the use of space by people with accessibility limitations, there are two main applications: the provision of improved routes avoiding "barely" or "non-accessible" stretches (B-I) and the classification of the accessibility conditions of public facilities and spaces (B-II). In both cases, the aim is to supply proper solutions for the distinct types of limitations/disabilities.

5.1   Contributing Accessibility Data to OSM

For the collaborative mapping activity, in compliance with the SDGs, we suggest the OpenStreetMap (OSM), which has various tools and an active community that develops solutions and contributes with mapped data made available on its platform. It is possible to develop maps on the platform to meet specific needs. Many of these tools are designed for the needs of users with disabilities, highlighting the Wheelmap tool (which infers the degree of accessibility of facilities for wheelchair users, meeting the B-II). Another tool is the OpenRouteService (which meets the B-I, generating optimized routes for wheelchair users). These tools highlight the importance of open data to support accessibility.

In this context, Mapeadores Livre acts by researching, producing content and materials, and conducting training on "how to map accessibility in OpenStreetMap." In addition, the group put this theme on the discussion for their first extension course. The organizers divided the content into three parts: (1) introduction and motivation about accessibility, (2) geometries for the theme, and (3) tags for the theme.

5.2   Contributing Concepts to the OSM Data Model

Part 1 introduced the concepts presented here throughout this text, emphasizing how lack of accessibility is the product of an ableist society. In part 2, we present the importance of mapping sidewalks as separate features from streets and connecting the two using intersections, emphasizing the importance of connectivity of the elements and teaching students about topological relationships between features. Part 3 shows how to characterize the degree of accessibility of the features mapped in part 2. This part introduces the students to OSM's integration with collaborative terrestrial image databases such as Mapillary and KartaView. These services allow detailed visual inspection of features, and the students performed the mapping tasks using Teach OSM's Tasking Manager.

Thus, by addressing this topic, the Mapeadores Livre group aims to expand the cartographic community on such a topic and increase and broaden the availability of such data in OSM, in addition to various thematic activities such as mapathons and courses that are planned in the chapter's future activities.

6   Final Considerations

More than a university activity, Mapeadores Livres connects various social organizations, their humanitarian projects, and the university in its research themes. The projects undertaken within the group aim to expand the mapping community trained in favela mapping and accessibility and increase the availability of data in OSM.

Our group's mission is to continue linking teaching, research, and connection with the community. In teaching, participating in the group helps create more proactive professionals to create and develop their projects in an autonomous and participative way. Furthermore, through the various activities, we can integrate this knowledge into the training of various professionals within the university, both at the undergraduate and postgraduate levels. In terms of research, emerging themes such as data quality, the integration of diverse sources, and the dynamics of contributions in diverse contexts help us propose technological solutions and expand the frontier of cartographic knowledge in these new scenarios. Additionally, the university as part of the community, helping to solve the country's real problems, is the complement both for building citizenry and for the integration so necessary in the construction of a fairer society.

In addition to these relevant internal aspects, being part of a Brazilian network, exchanging experiences with our colleagues from such different realities only motivates us to continue expanding the work of the YouthMappers in Brazil and Latin America. Finally, expanding this vision even further, in an interconnected world, being part of an international network, especially one with a significant presence in the global south and a real connection to a humanitarian vision, contributing to the UN SDGs, is a unique experience.

References

Azevedo G A (2020) Análise da acessibilidade da primeira e última milha de pessoas com deficiência (PcD) e idosos que utilizam o transporte de ônibus público na cidade de São Paulo diante de condições climáticas com a chuva

Bortolini E, Camboim S (2019) Mapeamento Colaborativo de Favelas com o OpenStreetMap. In: Pessoa Colombo V, Bassani J, Torricelli GP, Araújo SA (eds) Mapeamento Participativo: Tecnologia e Cidadania. FAU-USP, São Paulo, pp 53–61

Cardoso AL (2016) Assentamentos Precários no Brasil: Discutindo conceitos. In: Morais MP, Krause C, Lima Neto VC (eds) Caracterização e tipologia de assentamentos precários: estudos de caso brasileiros. Ipea, Brasília

Davis M (2004) Planet of slums. Verso, London

Instituto Brasileiro de Geografia e Estatística (IBGE) (2010) Censo Demográfico de 2010. https://censo2010.ibge.gov.br/. Accessed 15 July 2021

Instituto Brasileiro de Geografia e Estatística (IBGE) (2011) Aglomerados Subnormais. https://www.ibge.gov.br/geociencias/organizacao-do-territorio/tipologias-do-territorio/15788-aglomerados-subnormais.html. Accessed 15 July 2021

Olthuis et al (2015) Slum upgrading: assesing the importance of location and a plea for a spatial approach. Habit Int 50:270–288

Remião J L (2012) Acessibilidade em ambientes escolares: dificuldades dos cadeirantes. https://www.lume.ufrgs.br/bitstream/handle/10183/63191/000863847.pdf?sequen

United Nations Human Settlements Programme (UN-Habitat) (2003) The challenge of slums: global report on human settlements. Earthscan Publications Ltd, London

Visualizing YouthMappers' Contributions to Environmental Resilience in Latin America

Nayreth Walachosky, Cristina Gómez,
Karen Martínez, Marianne Amaya,
Maritza Rodríguez, Mariela Centeno,
and Jennings Anderson

Abstract

YouthMappers throughout Latin America are working to advance the attainment of the SDGs. Our contributions here centre on open data but also the maps and visualized analysis derived from them, which supports making decisions on how to achieve desired development across all seventeen goals. This chapter highlights the contribution of YouthMappers in ways that reduce inequalities around information access in Latin America (SDG 10), meanwhile advancing environmental resilience across the hemisphere, addressing a number of projects promoting biodiversity, conservation, and tourism that improve life on land (SDG 15).

Keywords

YouthMappers in Latin America · Mapping contributions · Conservation · Biodiversity · Tourism · Panama · Ecuador

1 A Picture of Youth and Open Data in Latin America

Let us visualize a world where youth are seen.

To meet all of our development objectives, from economic to social to environmental, we need more than data and capacity to analyse it. We also need to build information and knowledge that can inform decisions. Mapping open data is an incredible way to visualize what needs to be seen in order to make the right choices, for all of

N. Walachosky (✉) · C. Gómez · K. Martínez
M. Amaya · M. Rodríguez
Jóvenes Mapeadores, University of Panama,
Panama City, Panama

M. Centeno
GeoMap-ESPE YouthMappers, Universidad de las Fuerzas Armadas, Quito, Ecuador

J. Anderson
YetiGeoLabs, LLC, and YouthMappers,
Helena, Montana, USA

P. Solís, M. Zeballos (eds.), *Open Mapping towards Sustainable Development Goals*, Sustainable Development Goals Series, https://doi.org/10.1007/978-3-031-05182-1_17

the SDGs. Mapping with youth, and led by youth, goes even further. It can not only ensure that the right data and information are visualized, but also that the youth of the country are seen, and that their interests are part of the picture.

Picture the hemisphere of the Americas, where our lands and communities can be viewed by those who make the decisions and policies that affect our lives, livelihoods, and futures. This viewpoint might influence what is decided in ways that better prepare us for the uncertain future of our planet, the one that we as youth will inherit. It could make all of us more resilient, inspiring the whole continent to make progress on any of the Sustainable Development Goals.

While we are not there yet, map contributions by YouthMappers, both residing in Latin America and beyond, are making ourselves seen. Aside from the creation of open spatial data, students are also using this data to visualize important patterns, trends, and potential resilient solutions in spaces where there are the greatest vulnerabilities. This work helps to reduce inequality through access not only to information but also to the processes where smart decisions are made based upon that information. For YouthMappers in Latin America, this vision aligns with SDGs 10 (Reduced Inequality) in the sense that we work with *open* data, and with 15 (Life on Land) in the sense that building this link affects the territories where we live, our biodiversity in protected areas, and how we value and navigate through them.

In this chapter, we first review some of the contributions that YouthMappers within and outside of Latin America have made for the region, in terms of creating open spatial data that can be visualized. We then present a few cases that we have carried out, some with a little success in being seen in the community and in decision-making circles.

Each of the cases highlights ideas about SDG 10, in the sense that unequal access constrains innovation and prevents the smartest choices (Global Partnership for Sustainable Development Data 2015). Each also addresses other dimensions of sustainable development goals where the impact ranges from social and economic aspects (SDG 11: Sustainable cities and communities,

inclusive participation, 3: Health and well-being, 5: Gender equality, 4: Quality education, and 17: Alliance to achieve the objectives), as well as environmental (12: On responsible production and consumption, 13: Climate action, 14: Underwater life), and in each case, a connection to what is happening in our environment (SDG 15: Life of Terrestrial Ecosystems).

The participation of YouthMappers and these chapters goes beyond just using OSM and other Geographic Information Systems for data capture, but also counting on the participation of citizens, in which they are motivated and involved to solve those social problems or that promote development activities in their communities (Shannon et al. 2020). This strengthens the possibilities of visibility even further.

The objectives and methods used for each of these projects include work with the community and for the community (Solís et al. 2015; Hawthorne et al. 2015). These were diverse, from university and local communities, with a focus on tourism and conservation, as well as on the ordering and resilience of the population.

Together, these experiences – and many more that are not described here – recount some of the specific regional contribution of YouthMappers in the creation of spatial data by Latin American youth towards the attainment of these goals and a more resilient environment.

2 Generation of Open Spatial Data for Resilience

The activities of YouthMappers in Latin America have offered an important contribution to the open spatial data community itself, including OSM, but also as part of the global network of YouthMappers as well (Anderson 2021). While YouthMappers have been especially active in Africa and Asia, the participation of students across many South American, Central American, and Caribbean nations (LAC) has been impressive for our relative size. While the world population of Asia (4.6 billion) and Africa (1.3 billion) far surpasses the whole Latin American region (527 million, 7% of the world's population), the size of participation in YouthMappers surpasses average expectations. Of the nearly 300 chapters globally, there are 30 chapters (or 10%) in countries of the LAC region. Of the 64 countries of the world where YouthMappers have been established as of the end of 2021, there are 12 of them in the LAC region (19%). Certainly, the small size of our population does not deter us as youth from the region to join and contribute to the global network.

Figures 17.1 and 17.2 show steady growth over time in the presence of YouthMappers at universities of Central and South America, respectively, particularly in the last two to three years since the establishment of Regional Ambassadors and the presence of YouthMappers organizers as leaders in cartographic diplomacy in Latin America. We also attribute some of this growth to linking YouthMappers activities to externally funded research partnerships in environmental and geographic themes (TTU, ASU, World Bank, OAS-PAIGH, HOT, USAID, etc.). In particular, countries like Peru, Argentina, Colombia, Ecuador, and Brazil are important places for mapping by YouthMappers from LAC and elsewhere. Campaigns around environment and resilience feature prominently, from earthquakes in Ecuador to the pandemic in Peru and to biodiversity in Brazil. Similarly, looking at changes over time,

we see in Central America (Fig. 17.3), attention spiked when a series of hurricanes hit Puerto Rico and other nations in the Caribbean. YouthMappers in the Latin America and Caribbean region understand the connection between open spatial data and environmental resilience.

In sum, throughout the entire region of Latin America, YouthMappers have performed nearly 600,000 edits from 2015 to the end of 2021. Of these edits to OpenStreetMap, most of them were to building objects (about 88%), while 3000 were amenities. South America accounts for about half of these contributions (an estimated 49%), followed by Central America (29%) and then the Caribbean region (22%). As of 2022, more than 6750 km of roads and paths across the continent were last edited by YouthMappers. Of course, buildings and highways are more easily traced from satellite imagery by remote mappers. Amenities, however, require having some level of localized knowledge. The chapters in LAC universities and the network collaboration to map on the continent with joint campaigns have produced these contributions so far, with more to come in the future, especially amenities, once the pandemic restrictions may permit more fieldwork.

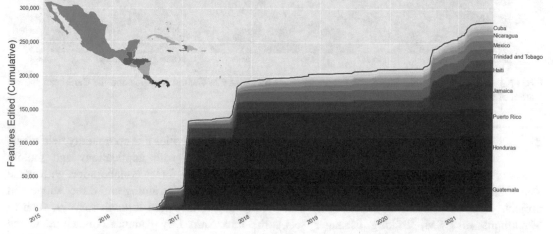

Fig. 17.1 Accumulated numbers of features edited by YouthMappers in Central America and the Caribbean show a surge in growth around 2017 to 2018

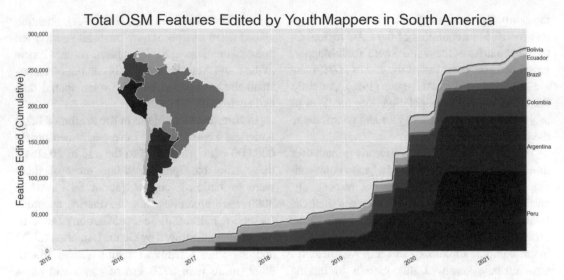

Fig. 17.2 Accumulated numbers of features edited by YouthMappers in South America show steady increases from 2017 to 2022

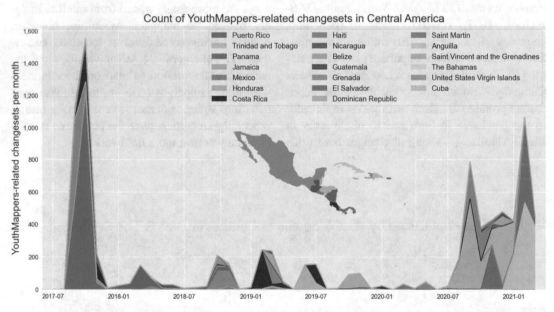

Fig. 17.3 The counts of YouthMappers related changesets by country in Central America and the Caribbean show a pattern of response to campaigns related to UN SDGs

3   Cases for Environmental Resilience in Latin America

The strategies used for the development of these projects include virtual methodologies like Mapathons in OSM, Tasking Manager sets, Webinars, ArcGIS training, Free Software distribution, and more (Coetzee et al. 2018); as well as face-to-face activities like community field visits, data collection with applications and mobile devices or from direct collaboration with some NGOs conducting training and direct support in collecting information through surveys, such as the Pink Shirt Day Panama Foundation in the children's sensations project (Rees et al. 2020). We present visualizations that marked many of

Fig. 17.4 Officers of the chapter of YouthMappers at the University of Panama point to locations on the satellite map during a presentation given at a meeting of a technical advisory group to the Organization of American States

the different stages of this work, to reinforce the idea of open spatial data being viewable and viewed in ways that also make the mapmakers and collaborating communities visible (Fig. 17.4).

3.1 Mapping a Hidden Paradise During a Pandemic

The first project of the Ecuadoran chapter GeoMap-ESPE shows how the community can go from 'hidden' to 'visible' by the geospatial approaches we have advocated for through YouthMappers. This work was a direct approach to the concerns of the community (Fig. 17.5).

The project was developed in Lloa, the most extensive rural parish of the Metropolitan District of Quito-Ecuador, as a field survey to help the economic, tourist, and social reactivation of the sector which had suffered from the necessary shutdowns due to public health concerns in the wake of the global COVID-19 pandemic. In this area you can enjoy beautiful landscapes with natural and ecological attractions which are optimal for attracting national and foreign tourists, an extra bonus of the place being a focus on agricultural and livestock production that allows its potential economic development.

Through a field visit in full observance of the biosecurity measures due to COVID-19, YouthMappers surveyed the residents of Lloa, obtaining vital information that was referenced to OSM base map data. The team was helped with field sheets and tours in tourist areas generating points of control (via the open platform), formally putting into public evidence the resources that the community possesses in terms of tourism and basic services (Fig. 17.6).

More than 50 land-use points characterizing tourist attributes were visualized. More than 180 surveys were carried out with residents of the parish, and georeferenced. YouthMappers tabulated results and produced cartography aimed to inform decision-making in the community about the strengths of tourist areas, and where to recover post-pandemic. This visualization also helped with marking points of interest such as fields, churches, health centres, and places of land use productivity, which will help inform policies to renew the growth and reactivation of the sector after facing the pandemic (Fig. 17.7).

3.2 Kuna Nega, a Community Living Amidst Pollution

YouthMappers at the University of Panama were the first in Latin America and only the second

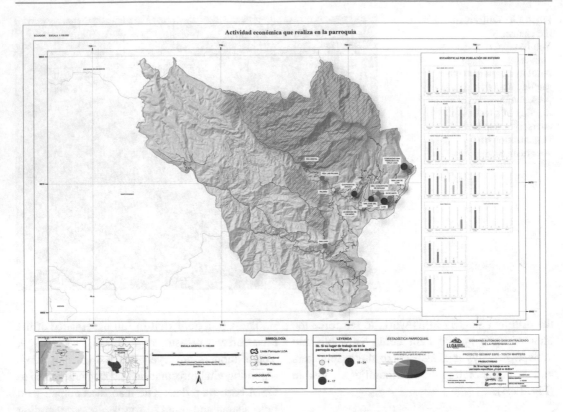

Fig. 17.5 A formal government map summarizes the results of an economic touristic survey conducted with YouthMappers in Lloa, Quito, Ecuador

Fig. 17.6 The details of feature edits on OpenStreetMap contain attribute data supporting the project

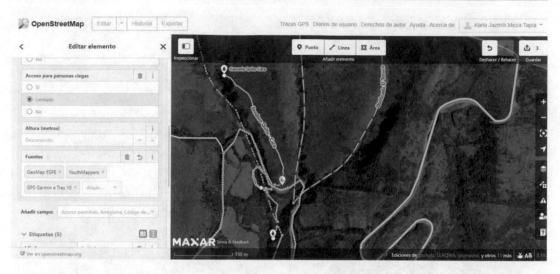

Fig. 17.7 The details of feature edits on OpenStreetMap contain hashtags identifying the project and YouthMappers

inaugural campus in the world to join the network (after UCC in Ghana). This legacy showcases a strong, long history of utilizing mapping by, for, and with communities to make visible problems that are otherwise unseen, hidden, or ignored (Fig. 17.8).

One exemplar of this character is the work with our Kuna Nega community, near the Cerro Patacón landfill that has contact with the Mocambo and Cárdenas rivers. Every day, more than 3000 tons of garbage are deposited here from all parts of Panama City, which has generated large outbreaks of diseases in the surrounding communities – largely marginalized indigenous peoples.

The mapping project was developed by the YouthMappers chapter of the University of Panama with the help of technological tools such as the OSM platform. A launch mapathon was held that intended to support this cause, through the voluntary participation of students who mapped (10 women and 5 men from the chapter and 5 volunteers at-large). This first pass at fundamental data helped to collect geographic information of features in the place, mainly of their buildings, in this case the houses, and also the roads (Fig. 17.9).

For us, it has been quite a challenge to carry out the fieldwork, since due to the pandemic the project has been stopped before reaching its second phase, but even so, we continue to advance in methodology to finish this project that will benefit many people. We will learning from chapters in different places around the world (see also Chap. 20) and begin making the connections and relationships in the meantime with entities responsible for taking decisions about solid waste management through our mentor and networks supporting YouthMappers.

3.3 Mapping 'Where Life Is Born'

A project that is considered one of the most significant and characteristic for the YouthMappers in Ecuador was a mapping effort to visualize tourist-ecological information in the Ecuadoran Amazon (Fig. 17.10).

The province of Pastaza, Canton Mera, is a beautiful place where there is a whole world to discover which with the correct land use could meet the needs of all residents, who are often marginalized. Through face-to-face collection of information in the field, YouthMappers sought to make visible endemic forests, rivers, caverns, and tourist attractions. This asset inventory helps make the community aware of the potential of its resources, aiming to contribute to the care of bio-

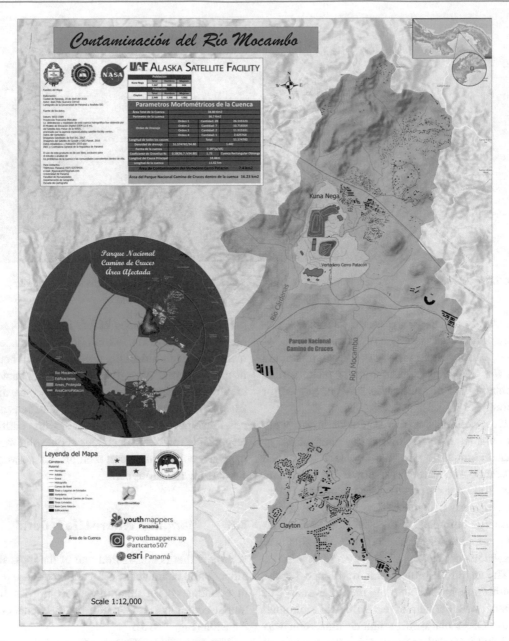

Fig. 17.8 Final products such as this map showing the results of a project monitoring river pollution acknowledge the contributions of YouthMappers at the University of Panama

diversity and showing an extra bonus for benefitting from conservation via eco-tourism (Fig. 17.11).

The community of this sector preserves and takes great care of the environment that surrounds them. However, the production of wood is destructive and damages the ecosystem. To avoid this exploitation of the territory, the residents with the team proposed to take tours to places with eco-tourism potential, representing a source of income, sustaining their communities and the habitat that surrounds them.

Thanks to the visual analysis and the tour, more than 35 eco-tourism information points

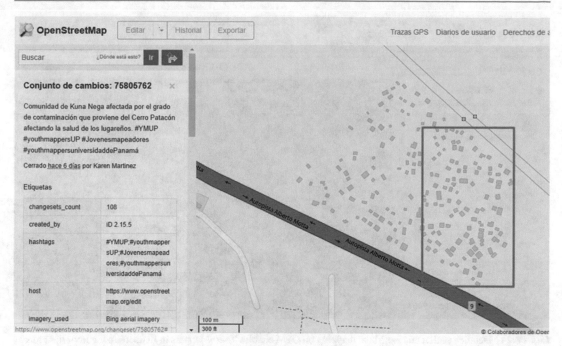

Fig. 17.9 Indigenous communities like Kuna Nega are affected by pollution, so building features are mapped to analyze their location relative to illegal solid waste disposal

Fig. 17.10 Detailed points of interest attribute mapping on OSM enables YouthMappers in Ecuador to tag important tourist features with information such as ticket costs

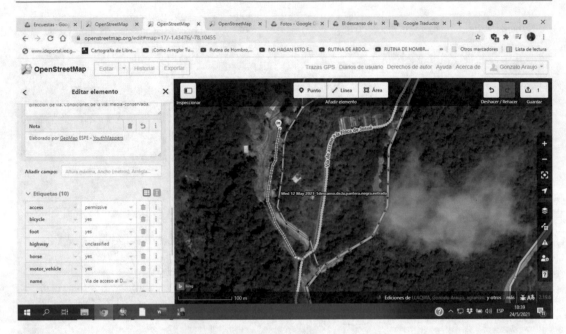

Fig. 17.11 Detailed path feature attribute mapping on OSM enables YouthMappers in Ecuador to tag important tourist routes with information such as bicycle access

were obtained that were published in OSM and are evidenced in a thematic map. The team generated an information matrix with characteristics, activities, and costs of the places that help local residents to be more organized. Finally, training was provided so that the community can learn about its resources, conserve its species and help care for them, preventing them from disappearing (Figs. 17.12, 17.13 and 17.14).

3.4 Ecological Zoning for Conservation and Sustainability

In the eastern area of the province of Panama, in the district of Chepo, a need arises to safeguard biological and hydrological wealth in the face of the negative impact caused by different phenomena such as deforestation, loss of biodiversity, pollution, among others, which from a future perspective could be accompanied by considerable environmental and socioeconomic consequences. YouthMappers propose ecological zoning as a planning and development instrument for the sustainable management of natural resources, specifi-

cally considering the relationship between water resources, protected areas, and communities located in the study area. Watersheds are considered the natural units of analysis and planning that in some cases coincide with the political-administrative divisions of the country (Fig. 17.15).

The basin of interest for this research was Number 148 of the Bayano River, represented by the Chichebre and Tranca rivers in the study area. Here there are 21 communities with mainly agricultural and service economic activities.

Panama has approximately 105 protected areas, of which two share an ecosystem relationship in the area of our study: the Tapagra Hydrological Reserve and the Panama Bay Wetlands. The forests of the Tapagra Hydrological Reserve are located on slopes, where the natural forest cover provides an effective barrier against soil erosion and reduces water sedimentation. Therefore, they are vital in maintaining high water quality. Wetlands are highly dependent on freshwater inputs and the healthy nutrients and sediments provided by rivers. In this case, a significant area of the Panama Bay Wetland requires the waters that drain from the Tapagra Hydrological Reserve.

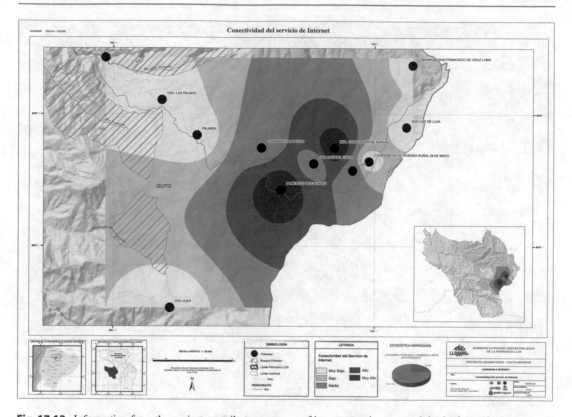

Fig. 17.12 Information from the project contributes to a map of internet service connectivity in the project area

Fig. 17.13 The project informs a master database of tourist attractions in Mera

Fig. 17.14 YouthMappers in Ecuador create data on OSM to develop a touristic map of Mera

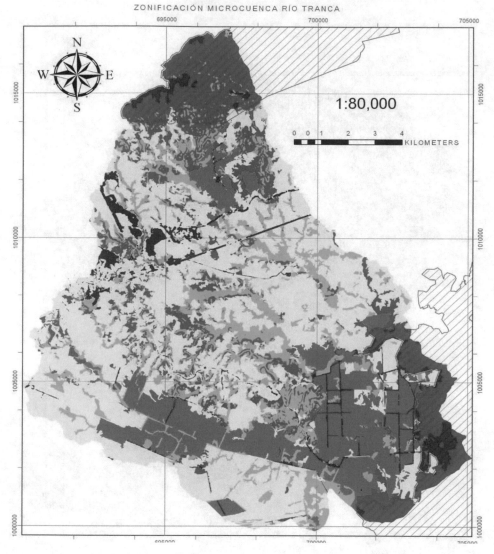

ÁREA DE INFLUENCIA DE LA RESERVA HIDROLÓGICA
TAPAGRA Y HUMEDAL DE LA BAHÍA DE PANAMÁ,
MICROCUENCAS DE LOS RÍOS CHICHEBRE Y TRANCA
ZONIFICACIÓN MICROCUENCA RÍO TRANCA

Fig. 17.15 YouthMappers in Panama create data on OSM to inform research in protected areas in the Trancha River watershed

The natural subsystem diagnosis phase of our work required an elevation model to generate morphometric data through geographic information systems tools. In addition to fundamental data from OSM, data on wild flora and fauna, ecosystem services, and socioeconomic and environmental problems were obtained from previous studies, field trips, and statistics. With the diagnostic information obtained, the information was integrated for a holistic analysis, which determined the aspects of the proposed solutions (Fig. 17.16).

Zoning recommendations took into account three factors: slope, land use, and protected areas. The result of the spatial analysis showed seven areas suitable for activities and land use in a compatible way, namely: protection zone, forest restoration, ecological connectivity, live-

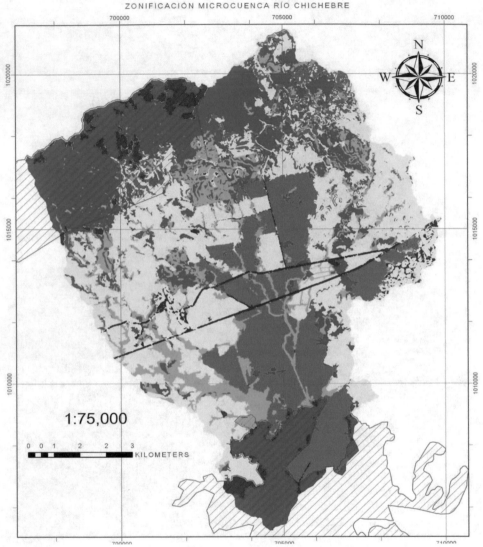

Fig. 17.16 YouthMappers in Panama create data on OSM to inform research in protected areas in the Chichebre River watershed

stock production, agricultural production, forestry production, and urban development and infrastructures.

Our project also frames strategies for food production areas, with the adaptation of agricultural, livestock, and forestry practices, promoting the adequate use of water, soil conservation, and use of trees and crops with the ability to adapt, in order to mitigate the environmental effects. This

would benefit rural development and food security alike.

Guidelines and strategies for the conservation of biodiversity are also proposed, and the importance of the environmental services offered by wetlands. In this case, a vital area of the Bay of Panama is highlighted. From visualized patterns, a framework was generated to manage and protect in a sustainable way the coastal ecosystems that are affected, for the most part, from terrestrial sources, as a result of anthropogenic activities, such as agriculture, livestock, deforestation, and poor disposal of solid waste.

Finally, we offered recommendations for policy makers to sustainably manage the forests, both in the protected areas and in the gallery forests that extend along the rivers, and reduce desertification, soil degradation, and stop the loss of wild flora and fauna, through the guidelines and strategies proposed in the Protection and Restoration Zones.

3.5 Mapping Trees and Palms Around Old Panama Historic Monument

The intersection of environmental and cultural assets is a ripe place for the kind of work we promote. In the historic old sector of Panama City, in an area administered by the Board of Trustees, is an important monument to the national heritage, a park surrounded by the modern city of Panama, off the coast of the Pacific Ocean. This site aims not only to tell the history of Panama, showing the ruins of the first European settlement on the American Pacific coast, as well as a series of vestiges of the first inhabitants of the Isthmus, but also, through the aesthetics of the trees, it shows us the natural history in a well-known, visible site, interacting the human legacy with animals and plants.

We took an inventory of the entire extent of arboreal vegetation which allowed us to know the health status of each of the trees and palms within the study area and to serve as the basis to provide necessary recommendations for their maintenance. Each one of the uses of these trees, their state of conservation, was shown to the public and alliances were achieved with the aim of conservation, among the Board, the University of Panama through YouthMappers, and Esri Panama for the continuation of related activities that highlight the importance of our national monuments. historical sites and the surrounding environment. We produced web resources such as location maps, web applications, and Story Maps of the project process itself (Fig. 17.17).

3.6 Smart Campus as a Model of Resilient Inclusive Innovation

The Smart Campus University of Panama project centres on the creation of an app that contains

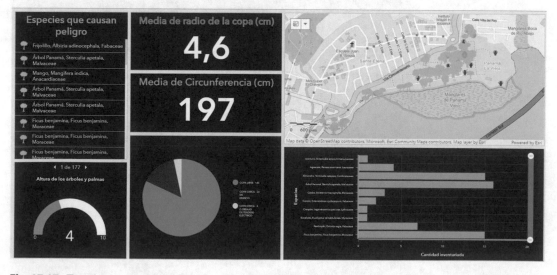

Fig. 17.17 Tree data mapped onto OpenStreetMap near a historic monument in Old Panama City combines with a dashboard to monitor species health

interactive maps in 2D and 3D. The app shows the facilities and infrastructure found within the campus – from multi-story details of the built environment to the environmental features of tree-lined paths. The project aims to create a more friendly and inclusive university from the point of view of giving the public open access to everyone on the location of its infrastructures, maintenance, and making the administrative processes more efficient (Fernández 2017; Nedwich 2018). For instance, a good understanding of where the broken sidewalks can help students

and faculty with mobility challenges, location of poor lighting to improve the safety of women, and general knowledge about how to find collaborators on campus can all make it that much easier for the community to collaborate on research. The incorporation of this technology promises to improve the quality of life not only of those who work or study there but also of the members of society who visit the University of Panama (Fig. 17.18).

The project has been developed and implemented in phases starting with training of techno-

Fig. 17.18 YouthMappers at the University of Panama launch a smart campus mapping program

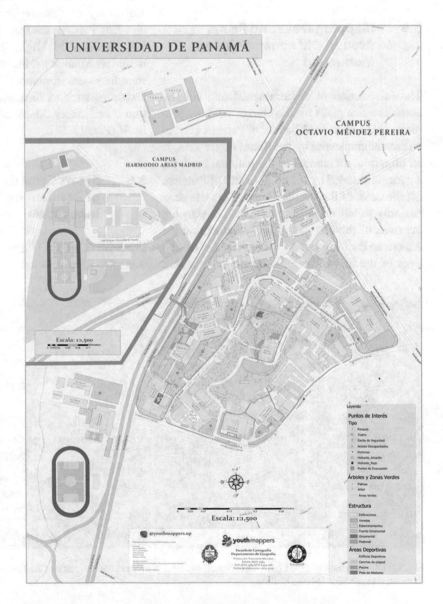

Fig. 17.19 Smart campus mapping at the University of Panama is visualized in 3D

logical tools for the collection of geographic information of the infrastructures, the elements around the campus, and inventories. We thus proceed to the creation of digital maps, achieving the survey of the campus editions in 3D. We next worked on developing web applications for use by the general public. Then a drone flight was carried out on the campus for a new methodology that supported the analysis and updating of information. Finally, a database portal was created for the University of Panama that is open and used broadly (Fig. 17.19).

The development of this project includes projects by students and professors from various campus faculties and with the support of public and private companies.

This project also makes visible the main challenge of the University of Panama, particularly the need to continue to move towards an intelligent and sustainable campus model, achieving the integration of data and the university community for decision-making and actions without losing sight of the integral way the university is part of the context of the city.

4 Towards an Equitable Resilient Future

At present, collaborative work and the use of geographic information are vitally important skills for obtaining information that benefits the Latin American community. Truly, community geography is an important framework for this action towards a more equitable future (Shannon et al. 2020), but also the spatial component that we obtain from geography within other aspects can be considered as a whole if we also seek resilience. For this, it needs components that cross boundaries that are social, economic, cultural, ecological, biodiverse, and inclusive, among others, and that comply with the sustainable development objectives, in order for these efforts to generate open data, embedded solutions, and helpful decision-making.

In this way, YouthMappers in and from Latin America made a path to different approaches that today allow us to tell our experiences lived within each project, gain great learning, and, above all, grow as human beings by serving. Each challenge and each success make us *modelers within the communities*, working with them to be resilient, entrepreneurial, and successful, having the sole impulse to help the community that surrounds us and precisely wanting to make the change; daring to do things differently and loving our world. The final proof will be in the adaptation we have made by handling digital tools for the creation of each project and its dissemination.

This work teaches us that small changes can lead to great achievements. Working hand in hand across communities as university students, neighbours, and friends, it has been possible to create a series of projects in our part of this little handkerchief called the world.

References

Anderson J (2021) YouthMappers activity dashboard. https://activity.youthmappers.org/#2.02/12.62/-68.3. Cited 29 Dec 2021

Coetzee S, Minghini M, Solis P, Rautenbach V, Green C (2018) Towards understanding the impact of mapathons: reflecting on YouthMappers experiences. Int Arch Photogramm Remote Sens Spat Inf Sci ISPRS Arch XLII-4/W8:35–42. https://doi.org/10.5194/isprs-archives-XLII-4-W8-35-2018

Fernández, M. (2017). Ciudades a escala Humana. El surgimiento de las ciudades inteligentes como nueva utopía urbana. Available from CEH, 27 November. https://www.ciudadesaescalahumana.org/2017/11/el-surgimiento-de-la-ciudad-inteligente.html. Cited 16 Jan 2022

Global Partnership for Sustainable Development Data (2015) Better data. Better decisions. Better lives. Available via Data7SDGs. https://www.data4sdgs.org/. Cited 28 Dec 2021

Hawthorne T, Solís P, Terry B, Price M, Atchison CL (2015) Critical reflection mapping as a hybrid methodology for examining socio-spatial perceptions of new research sites. Ann Assoc Am Geogr 105(1):22–47. https://doi.org/10.1080/00045608.2014.960041

Nedwich R (2018) What is a "Smart Campus" and how does it differ from a traditional learning environment? Available from DotMagazine, February. https://www.dotmagazine.online/new-work-and-digital-education/ICT4D/smart-campus-merging-smart-city-and-smart-home-in-education-for-digital-natives. Cited 16 Jan 2022

Rees A, Hawthorne T, Scott D, Spears E, Solís P (2020) Toward a community geography pedagogy: a focus on reciprocal relationships and reflection. J Geogr. https://doi.org/10.1080/00221341.2020.1841820

Shannon J, Hankins K, Shelton T, Bosse A, Scott D, Block D, Fischer H, Eaves L, Jung JK, Robinson J, Solís P, Pearsall H, Rees A, Nicolas A (2020) Community geography: toward a disciplinary framework. Prog Hum Geogr. https://doi.org/10.1177/0309132520961468

Solís P, Price M, Adames de Newbill M (2015) Building collaborative research opportunities into study abroad programs: a case study from Panama. J Geogr High Educ 39(1):51–64. https://doi.org/10.1080/03098265.2014.996849

Marking a Path to Goals on Sustainable Communities, Consumption, Climate, Oceans, Land, and Justice

Youth Engagement and Participation in Mitigating Perennial Flooding in Kampala, Uganda Using Open Geospatial Data

18

Ingrid M. Kintu and Henry N. N. Bulley

Abstract

Ingraining spatial thinking for problem-solving is critical for future decision makers and leaders. We argue that the use of open geospatial data and technology makes it easier to understand the interconnections between places and many socioecological issues facing communities. This facilitates openness to adopt the methods and strategies needed to make our communities and the world at large a better place as envisaged by UN-SDG 11. This case of two informal human settlements Uganda features low-lying areas with mostly slum conditions and urban poor migrants who settled there from rural communities in search of better livelihoods. YouthMappers documented conditions of drainage systems that impact flood vulnerability. We highlight important lessons in collaborating with local humanitarian organizations to spatially conceptualize development-related activities for underprivileged communities in a context that resonates with local people.

I. M. Kintu (✉)
Department of Geography, Geo-Informatics and Climatic Sciences, Makerere University, Kampala, Uganda

H. N. N. Bulley
Borough of Manhattan Community College, The City University of New York, New York, NY, USA
e-mail: hbulley@bmcc.cuny.edu

Keywords

Sustainable cities · Flood resilience · Participatory mapping · Urban informal settlements · Uganda

1 Rural to Urban Migration in Africa

Many African cities have experienced rapid population growth and by 2050, the African urban areas are projected to have an additional 950 million people to the current population of about 567 million people (OECD/SWAC 2020). This increasing trend of Africa's urban population growth presents many challenges including socioecological and sanitation problems related to exponential growth of informal human settlements or slums. Kampala is one such major African urban center that has experienced significant growth over the past few decades.

A primary driver of urban settlements is rural-urban migration, mostly youth and educated pro-

© The Author(s) 2023
P. Solís, M. Zeballos (eds.), *Open Mapping towards Sustainable Development Goals*, Sustainable Development Goals Series, https://doi.org/10.1007/978-3-031-05182-1_18

fessionals seeking employment in commercial hubs, in search of better employment opportunities and improved living conditions. However, an increasing number of rural migrants end up in informal settlements or slums that sprung up in different parts of Kampala district including Bwaise, Kalerwe, Kireka, Lufuka, and Najjanankumbi. The slums in these areas are not very different from those in other major cities in Africa and are generally distinguished by the poor quality of housing, the poverty of the inhabitants, the lack of public and private services, and the poor integration of the inhabitants into the broader community and its opportunities (UN-Habitat 2003).

In the Kampala area, Bwaise and Kalerwe slums are the most affected by widespread flooding during heavy rainstorms because they are low lying. This is due to increased demand for land and housing which causes the urban poor to seek shelter, "homes" or "rooms," in low-lying areas close to canals. The canals often lack the capacity to handle the increasing volume of stormwater runoff from other parts of Kampala district that are on a relatively higher elevation. Such factors coupled with the poor drainage structures and poor waste collection in Kampala increase the vulnerability of the informal settlements of Bwaise and Kalerwe.

Another factor that has greatly contributed to the floods is the land cover change in these areas, especially the sprouting of new buildings which has led to the reduction of infiltration of rainfall thus increasing the amount of runoff. Whereas the increase in flood events can be attributed to the aforementioned factors, climate change has also greatly affected local communities, due to the increased intensity and frequency of unpredictable rainfall cycles (Douglas et al. 2008).

It is important to note that amidst all these challenges, residents in informal settlements rank far lower on human development indicators than other urban residents; they have more health problems, less access to education, social services, and employment, and most have very low incomes. Residents lack access to a clean water supply, sanitation, stormwater drainage, solid waste disposal, and many essential services.

Moreover, there is very little forward planning to address even the current problems, let alone the expected future doubling of demand. The urban poor are trapped in an informal and "illegal" world – in slums where waste is not collected, where taxes are not paid, where public services are not provided and are not well reflected on maps (UN Habitat 2003). The use of updated, timely, and reliable open-source geographic data can provide the critical information for flood resilience policy and mitigation planning (Hachmann et al. 2018).

2 Urban Flood Mapping of Informal Settlements

The main aim of our community flood mapping exercise was to showcase the potential of using open data tools to identify and assess flood vulnerability factors in informal settlements of Bwaise and Kalerwe. The outcome of such activities will support planning and development strategies to make informal settlements in African cities more inclusive, safe, resilient, and sustainable in alignment with United Nations Sustainable Development Goal 11.

2.1 Bwaise and Kalerwe Communities

Bwaise and Kalerwe are parishes in Kampala Capital City, Uganda's capital. Both parishes are located North of Kampala city's Central Business District. Bwaise comprises Bwaise I, Bwaise II and Bwaise III. It is located in the Kawempe division sub-county and bordered by Kawempe to the north, Kyebando to the east, Mulago to the southeast, Makerere to the south, and Kasubi to the southwest. Kalerwe on the other hand is bordered by Kawempe to the north, Kyebando to the east, Makerere to the south, Bwaise to the southwest, and Kazo to the west (Fig. 18.1).

Both communities are characterized as slum dwellings and represent some of the poorest areas in the city. They are also located in low-lying areas, close to major drainage channels and wetlands which makes them prone to floods and vulnerable to waterborne diseases such as cholera and typhoid.

2.2 Pre-fieldwork Preparations

The community flood mapping and data collection exercises were undertaken by YouthMappers from the GeoYouthMappers Chapter at Makerere University. A number of preparations were carried out before the data collection process ensued. Before any data collection exercise is carried out in any part of Uganda, it is important to inform local authorities about the purpose of the activity and how the data collected will be used, which in this case was the Local Council (LC) I.

To aid the data collection process, we approached Uganda Red Cross about the possibility of partnering in this endeavor. They gave a positive response. This partnership was important because most of the YouthMappers from Makerere University were not familiar with the people and the living conditions in the slum areas, Bwaise and Kalerwe. Uganda Red Cross volunteers facilitated a briefing session on the morning of the fieldwork. They briefed the YouthMappers about the seven core principles of Red Cross humanitarian work, and the symbolism that people attach to anyone wearing the Red Cross jacket (Figs. 18.2 and 18.3). Working with the Ugandan Red Cross facilitated interactions with residents in the target communities of Bwaise and Kalerwe. Community members are already familiar with the humanitarian activities of the Uganda Red Cross in their communities. As a result, the authorities and the residents were more receptive and willing to participate in the data collection exercise led by YouthMappers.

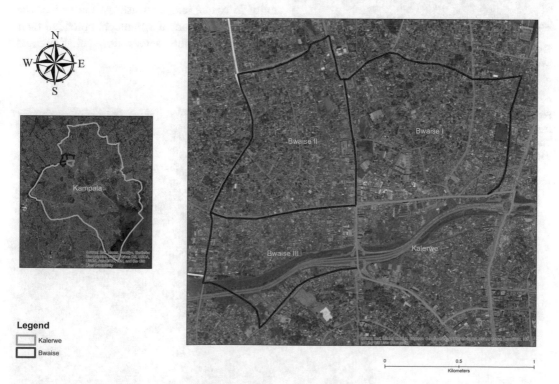

Fig. 18.1 A map of the study area shows Bwaise and Kalerwe informal settlements

2.3   Data Collection

The data collection exercise was executed by a group of YouthMappers volunteers from the GeoYouthMappers Chapter and three volunteers

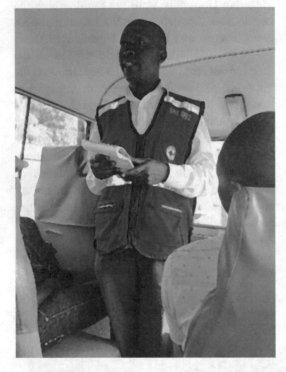

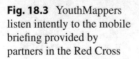

Fig. 18.2 The mapping team is briefed by a Uganda Red Cross Volunteer

from Uganda Red Cross. In all, there were 27 participants, and they were divided into 9 teams. The teams started by collecting data in Bwaise and later, in Kalerwe (Fig. 18.4).

In both places, the teams spread out to different parts of the settlement, and they also interacted with the community members who provided invaluable contextual information about the state of the drainage system, sanitation, and perennial floods in the settlements. The teams kept in touch through a coordinator from the YouthMappers group, while community leaders were given updates by the Ugandan Red Cross volunteers.

A data collection form used was created using KoBoCollect, an android-based geographic data collection app that facilitates the creation of open-source Mobile Data Collection forms in both online and offline formats (Fernanda et al. 2020; KoBoToolbox 2021). This application was used to build a 12-question survey pertaining to both rubbish dumps and drainages in the settlements (Fig. 18.5). Some of the data collected pertained to attributes such as the geographic location, vegetation close to these features, proximity to roads, and land use. At the end of the fieldwork, the student volunteers uploaded their data to the kobo server for collation and assessment.

Fig. 18.3 YouthMappers listen intently to the mobile briefing provided by partners in the Red Cross

Fig. 18.4 YouthMappers
and Red Cross volunteers
collect data together in the
field

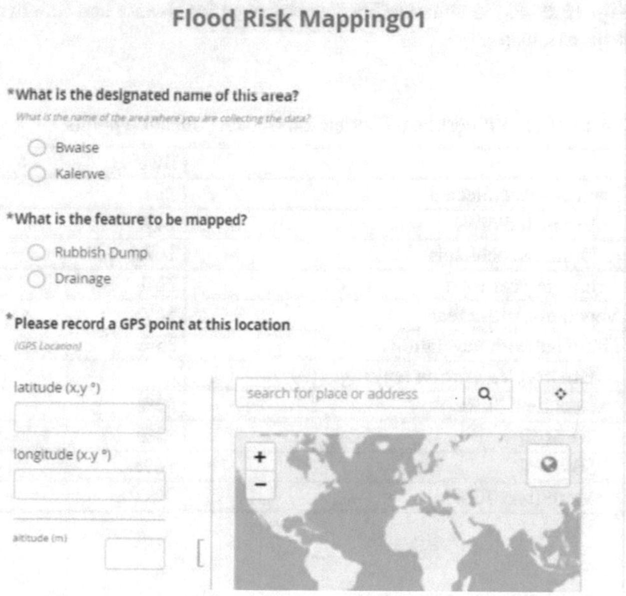

Fig. 18.5 The project
utilizes a KoBoCollect
data collection form

Flood Risk Mapping01

*What is the designated name of this area?

What is the name of the area where you are collecting the data?

○ Bwaise

○ Kalerwe

*What is the feature to be mapped?

○ Rubbish Dump

○ Drainage

*Please record a GPS point at this location

(GPS Location)

latitude (x.y °)

longitude (x.y °)

altitude (m)

search for place or address

2.4   Community Mapping Results

A total of 562 data locations were collected in both settlements, 316 in Bwaise and 246 in Kalerwe. Particularly, 303 data points were located along drainage channels and 259 represent rubbish dumps in the two settlements (Fig. 18.6; Table 18.1).

Of the 167 drainage points collected in Bwaise, 26% were near the road network whereas 74% were in drainage channels; 56% of these had vegetation close to them; 35% of the drainage points had grass cover, with 52% of the points having vegetation covering about 30% of the channel (Fig. 18.7).

The rubbish dumps, on the other hand, were found both along the roads as well as inside the community. Rubbish dumps had been created in various spots all over the settlement, though in most instances, people dump their waste materials at their nearest convenient location.

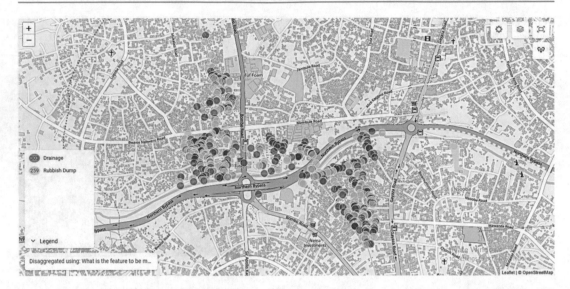

Fig. 18.6 Flood vulnerability data collected for Bwaise and Kalerwe slums are visualized on an OSM building foot-print basemap

Table 18.1 Proportional distribution of data collection points

	Bwaise	Kalerwe	Total
Data points collected	316	246	562
Rubbish dumps	149	110	259
Drainage channels	167	136	303
Drainage near road	26%	41%	
Vegetation close feature	56%	46%	
Drainage with vegetation	35%	31%	
Extent of vegetation coverage (%)			
Less than 30%	52%	57%	
30–50%	33%	31%	
50–70%	16%	5%	
More than 70%	0%	7%	

In Kalerwe, 41% of the drainage points were collected along the main road, whereas 59% were collected within the community (Table 18.1). Forty-six percent were close to vegetation, also predominantly grass and 31% had vegetation covering the drainage with 57% covering about 30% of the drainage channels. The rubbish dumps in Kalerwe were found all over the community both along the roads and in the community similar to the situation in Bwaise (Figs. 18.8, 18.9, and 18.10).

3   Lessons from Fieldwork in Informal Settlements

Bwaise and Kalerwe informal settlements are the worst affected areas due to perennial flooding in Kampala district. Floods are attributed to the state of the drainage system and lowlands, poor waste disposal systems, not only in these settlements but also in Kampala city, as well as increased built-up areas resulting in more impervious areas, less infiltration, and more stormwater runoff (Chereni et al. 2020).

Fig. 18.7 A drainage canal marks a path in Bwaise

Fig. 18.8 Some drainage canals like this one in Kalerwe are fully covered by vegetation

Fig. 18.9 Rubbish is dumped indiscriminately in informal sites in Bwaise and Kalerwe

Kampala is famously known as the city of seven hills, where hilly areas are inhabited by the affluent members of the population. However, the majority of people live and work in the lowland areas of the city. Stormwater runoff from the areas of high elevation or hilly terrain is directed

Fig. 18.10 Rubbish
dumps are situated very
close to the drainage
canals, and some are
evidently blocked,
leading to stagnant water

to the lowland areas such as Bwaise and Kalerwe, leaving the inhabitants vulnerable to extensive flooding and destruction of property.

Additionally, a number of wetlands in and around Kampala have been reclaimed to locate industrial projects such as factories and warehouses. Such activities have also greatly disrupted the hydrology in the area leaving no option but for the stormwater to collect faster into lowlying areas of Bwaise and Kalerwe. This condition is often exacerbated by inadequate waste disposal systems in the city.

3.1 Flood Risk from Waste Disposal in Kampala District

Despite many improvements in the sanitation and waste disposal in the Central Business District and the suburbs, there has been very little improvement in informal settlements such as Bwaise and Kalerwe despite year after year of flooding (WaterAid 2007). Large heaps of rubbish are scattered around the study area, and they are neither properly delineated nor closed off. A lot of the rubbish are either thrown into drainage channels or washed into them, causing blockages and rapid overflow during typical rainstorms. The impact of poor waste disposal, especially plastic

and polythene, is very evident after rain events with the drainage channels filled with water and materials (Fig. 18.7).

Results from the fieldwork confirm the dire need for improvement in the drainage channels in Bwaise and Kalerwe. The drainage system is characterized by blocked, silted sections and presence of vegetation in the channels which makes them shallower. Some drainages have been neglected for a long period of time such that vegetation covers close to 50% of their surface area. This also greatly contributes to the difficulty of stormwater flow and increases the flood risk of the inhabitants. Whereas some efforts have been made to remove the blockages, the debris is often left alongside the channels and drainages, and they are subsequently washed back into the channels during the next rainstorm (Douglas et al. 2008).

3.2 Mitigating Perennial Flooding Using Open-Source Geospatial Data

In order to cope with perennial floods, residents have to scoop water out of their houses using troughs. In more severe cases, residents are forced to vacate their homes temporarily or permanently, especially during the rainy seasons. Other ways in which residents have adapted is through using flood-proofing mechanisms to reduce the impact of water by raising their houses, thickening their walls with more mud, and constructing barriers at the entrances to their houses, to mention but a few built solutions. There have also been some efforts by the members of the communities to desilt choked drainages, but this occurs less frequently.

Flooding not only affects residents' homes but also businesses in the area. In some instances, schools and shops close most of the day and the roads are impassable. Perennial flooding events usually destroy personal property as well as disrupt the livelihoods of residents in informal settlements.

Douglas et al. (2008) proposed a number of solutions to upgrade the slums such as setting up new housing for the residents and relocation, carrying out rehabilitation activities, construction of wider and deeper drainage channels, frequent desilting, setting up a proper waste disposal system and enforcing stricter flood prevention policies and urban planning codes. Solutions require decision makers to have access to up-to-date information in the form of the residents' voices through social surveys and geospatial information which shows the "what" and the "where" aspects of flood risk factors in the communities. A first important step for any mitigation will be to map all of the informal settlements in Kampala, to an acceptable degree of completeness and at a low cost. This can be realized using open geospatial tools, e.g., OpenStreetMap, which are freely available and usable by everyone.

3.3   Mapping with Local Residents of Informal Settlements

OpenStreetMap (OSM) and its many associated tools are very well suited for the collection of geolocated data. OSM is a free editable map of the world composed of data collected and created by mappers from different parts of the world. It emphasizes local knowledge and is largely community-driven providing freely available data for many purposes (OSM 2021).

The data collected during the fieldwork in Bwaise and Kalerwe was based on Kobo Collect, an open-source app which can be used together with OSM. It exemplified how citizen volunteers can contribute vital place-based data to facilitate the accurate assessment of flood vulnerability and mitigation issues in developing countries. Harnessing this geospatial technology opens up the possibilities of efficient decision-making through the use of analytical models – intuitive visualizations which create an awareness of what is happening on ground, especially in informal settlements such as Bwaise and Kalerwe. It also provides an opportunity for proper planning of the drainage system, best locations for relocation, rehabilitation initiatives, and monitoring of waste disposal. This approach allows direct engagement with residents who are suffering the flooding impacts.

4   Toward Informing SDG 11 Through Flood Mapping

The flood risk data collected by the YouthMappers volunteers from Makerere University demonstrates the importance of participatory mapping (volunteer) data to support the effort to meet Sustainable Development Goals aimed at improving the livelihood of urban communities (United Nations 2021). Making progress toward flood risk reductions in vulnerable informal settlements of Bwaise and Kalerwe in Kampala specifically addresses SDG 11 for making cities and human settlements safe, resilient, and sustainable. It also intersects directly with addressing SDG 6, whichpromotes clean water and sanitation.

In particular, SDG 11 focuses on equitable access to adequate, safe, and affordable housing; promoting sustainable and resilient buildings; ensuring sustainable human settlement planning and management; and protecting the urban poor, especially in vulnerable informal settlements. Pursuing SDG 11 will provide good grounds to achieve other sustainable development goals such as minimizing the growth of slums and their associated environmental problems and improving health and sanitation conditions of urban areas, which are directly related to other goals, and a fundamental part of meeting basic development needs.

Overall, the flood vulnerability data collection exercise was a success and highlights the synergy in working with partners already established in the local community and who have garnered people's trust, namely, the Ugandan Red Cross volunteers, to facilitate the process. The exercise provided a vivid picture of what could be achieved if this process is further developed for a longer period of time and covers more areas other than the cases of Bwaise and Kalerwe.

This type of community-based flood vulnerability mapping achieves two main objectives that inform the SDGs. First, it provides detailed geospatial information on the locations and types of obstructions to water flow in informal settlements with limited water and sanitation infrastructure. This is useful information for flood disaster resilience planning by Kampala city planning officials. Additionally, the process of collecting data on rubbish dumps and drainages in itself helps to enlighten the residents of slum communities about the aggravated dangers of indiscriminate rubbish dumps and clogged drainages, given the low-lying nature of their communities. Hopefully, such information may help people rethink their waste disposal habits, to reduce the risk of flooding in their communities.

The results also show the value of taking social-spatial structures into consideration when drawing plans to support proper living conditions, especially in marginalized informal settlements where growth is largely driven by informal social networks and economic migrants. The flood vulnerability mapping activity in Bwaise and Kalerwe could be the genesis of a flood risk and resilience database to support planning and sustainable mitigation efforts that engage residents in marginal or informal slums settlements within Kampala and other major urban areas in developing countries.

The fieldwork also provided a rare learning opportunity for the GeoYouthMappers chapter members, as most of the student volunteers had never been in a slum environment before, let alone interacted with residents of informal settlements. This is a vital social awareness and learning experience for students, many of whom could end up making policy or planning decisions that will affect the livelihoods of residents in marginal informal settlements such as Bwaise and Kalerwe. Personally, it was helpful for us to see how the Ugandan Red Cross is protective of its integrity with the people in informal settlements. That confidence made us, as outsiders, feel welcomed by the people our work aimed to support.

Still, not all challenges can be overcome, despite best intentions. One of the issues we often encountered was residents who approached us asking for help with some basic needs. Sometimes, we YouthMappers felt powerless to help and wished we had been aware of how to address some of the basic needs of residents. This perhaps could have allowed us to achieve multiple goals of community flood mapping and humanitarian assistance.

This activity highlights the necessity as well as the possibility of continual data collection efforts to mitigate perennial flooding in informal settlements such as Bwaise and Kalerwe. We aspire for this exercise to be scaled up on a broader basis for other slums in and around Kampala district, and especially places located in flood-prone areas. It is noteworthy to mention the initial success of one of the student volunteers in the data collection exercise who wrote a grant-winning proposal as part of a YouthMappers Research Fellowship, which will take this initiative from simply collecting data to performing network analysis for access to illegal dumping sites by Kampala Capital City Authority (KCCA) and flood risk analysis. We appreciate the fact that a limited activity such as this may not be sufficient to give a full picture of the state of the informal settlements. Inspired by this start, we have initiated more discussions at the project level on this type of citizen geospatial data collection beyond Bwaise and Kalerwe to include other slums in Kampala district.

In addition to spatial data, we also hope to incorporate the temporal domain and spread over a longer period of time. Coordinating this activity with the visiting professor, student volunteers, the Red Cross Society, and local community leaders gave rise to new insights into managing the interests of different stakeholders and how to ensure that there is effective communication among them to carry out a successful community mapping exercise aimed at improving livelihoods of some of the most vulnerable urban communities.

Acknowledgments Fieldwork was made possible by the Carnegie African Diaspora Program (Institute of International Education / Carnegie Corporation of New York) funded fellowship travel support for Prof. Henry Bulley to visit Uganda during the summer of 2019, as part of a collaborative multi-institutional and multinational project titled, "Addressing perennial flooding in Ghana and Uganda by harnessing capacity and research into integrated landscape-based assessments and geospatial science & technology."

References

Chereni S, Sliuzas RV, Flacke J, Maarseveen MV (2020) The influence of governance rearrangements on flood risk management in Kampala, Uganda. Environ Policy Govern 30(3):151–163. https://doi.org/10.1002/eet.1881

Douglas I et al (2008) Unjust waters: climate change, flooding and the urban poor in Africa. Environ Urban 20(1):187–205. https://doi.org/10.1177/0956247808089156

Fernanda M, Arruda N, Kintu I (2020) How to easily use open source software technologies to face COVID-19 in your region. YouthMappers Blog. Available via YouthMappers. https://www.youthmappers.org/post/2020/07/09/how-to-easily-use-open-source-software-technologies-to-face-covid-19-in-your-region. Cited 28 Dec 2021

Hachmann S, Jokar Arsanjani J, Vaz E (2018) Spatial data for slum upgrading: volunteered Geographic Information and the role of citizen science. Habitat Int 72:18–26. https://doi.org/10.1016/j.habitatint.2017.04.011

KoBoToolbox (2021) Data collection tools for challenging environments. Available via Kobo. https://www.kobotoolbox.org. Cited 26 Feb 2021

OECD/SWAC (2020) Africa's urbanisation dynamics 2020: Africapolis, mapping a new urban geography. West African Studies, OECD Publishing, Paris. https://doi.org/10.1787/b6bccb81-en

OSM (2021) About OpenStreetMap. Available via OSM. https://www.openstreetmap.org/about. Cited 31 Mar 2021

UN Habitat (2003) The challenge of slums global report on human settlements. United Nations Report. https://doi.org/10.1006/abio.1996.0254

United Nations (2021) The 17 goals – sustainable development. United Nations Department of Economic and Social Affairs. Available via UN. https://sdgs.un.org/goals. Cited 12 Mar 2021

WaterAid (2007) Social marketing report: Usage and Attitudes of Sanitation facilities in Kawempe division, Kampala. Available via WaterAid. https://washmatters.wateraid.org/publications/social-marketing-report-usage-and-attitudes-of-sanitation-facilities-in-kawempe. Cited 28 Dec 2021

Sustainable Mobility Through Knowledge Exchange and Collaborative Mapping of Cycling Infrastructure: SIGenBici in Medellín, Colombia

Natália da Silveira Arruda,
Hernán Darío González Zapata,
and Ana Maria Navia Hermida

Abstract

Sustainable mobility is a strategic tool that directly influences the fulfillment of the sustainable development goals (SDGs). We highlight the need to improve cycling as an urban transport option through our research in the city of Medellín (Colombia) and its Metropolitan Area. We present SIGenBici, a collaborative cycling infrastructure mapping project supported by the use of free and open-source software (FOSS) and the generation of open data. This work is aimed at contributing directly to SDG 11: Sustainable cities and communities, and SDG 9: Industry, innovation, and infrastructure. We explain the project development process from the perspective of the mappers, a gender-sensitive approach, and community knowledge transfer. We conclude with reflections on what we have learned throughout this process and future expectations.

Keywords

Sustainable transportation · Cycling · Infrastructure · Participatory planning · Colombia

1 Sustainable Transportation and Open Mapping for SDGs

As in other cities worldwide, in Medellín, the externalities of transportation are manifested in increasingly frequent periods of high levels of air pollution, congestion, travel times, costs, and

N. da Silveira Arruda (✉)
GeoLab, University of Antioquia,
Medellín, Colombia

School of Geographical Sciences and Urban
Planning, Arizona State University, Tempe, AZ, USA
e-mail: narruda@asu.edu

H. D. González Zapata · A. M. Navia Hermida
GeoLab, University of Antioquia,
Medellín, Colombia
e-mail: hdario.gonzalez@udea.edu.co;
ana.navia@udea.edu.co

P. Solís, M. Zeballos (eds.), *Open Mapping towards Sustainable Development Goals*, Sustainable
Development Goals Series, https://doi.org/10.1007/978-3-031-05182-1_19

road incidents. These problems, which have both local and global characteristics and impacts, must be approached from an integrating vision considering their multiple spatial and temporal relationships (Anderies et al. 2013).

Sustainable mobility is an essential element in territorial planning and applied to the urban context it is a strategic tool to break the transportation vicious circle and disassociate population and urban growth from the pressures exerted on the social-environment system. Cycling is a mobility option that, in addition to promoting an active lifestyle, directly influences the fulfillment of 11 of the 17 sustainable development goals (SDGs) stipulated by the United Nations Development Program (WCA 2016). We highlight the need to promote the use of the bicycle as an urban transport mode in the city of Medellín (Colombia) and its Metropolitan Area through the SIGenBici project, a collaborative cycling infrastructure mapping project supported by the use of free and open-source software (FOSS) and the generation of open data. In particular, our work is aimed at contributing directly to two of the Global Goals: Sustainable cities and communities (SDG 11); and Industry, innovation, and infrastructure (SDG 9).

We are keen to respond to the urgency of responding *locally* to the Global Goals. SIGenBici is a project that arises from the collaboration between GeoLab, our research group, and Corporación Colectivo SiCLas. At GeoLab, we are a group of students and professors from the University of Antioquia, whose main campus is in Medellín. GeoLab is characterized by its interdisciplinarity because even though it has been created within the Faculty of Engineering, it receives members from other disciplines interested in learning more about FOSS, the generation of open data, and the application of collaborative methodologies. For this reason, we have different lines of applied research, and SIGenBici was born within the urban sustainability line, coordinated by professors from the area of mobility and transportation. On the other hand, SiCLas is a group of cycling activists from Medellín, who are dedicated to promoting active

mobility centered on cycling through various activities with a social and environmental focus.

We believe that active mobility associated with technology has the potential to offer more equitable access to the territory. The support network to which we have access by being part of YouthMappers allows us to act in a growing international community of "digital humanitarians" that use mapping tools to connect people and places in response to global emergencies (Solís et al. 2020). This multiscalar network encourages the generation of citizen science connected to academic research as a means for creating social innovation (Brovelli et al. 2020).

2 Context of Cycling in Medellín

Medellín is the core city of the Aburrá Valley Metropolitan Area (AMVA, by its acronym in Spanish), as well as the capital of the Department of Antioquia, Colombia. AMVA is composed of ten municipalities located in a narrow valley limited by two plateaus, presenting a canyon configuration. The bottom and flat area of the valley is almost completely urbanized, which has led to a phenomenon of urban expansion toward its slopes (AMVA 2016).

This complex metropolitan system concentrates 58.4% (3,994,645 inhabitants, projection for 2020) of the population in 1.8% of the departmental area, which largely directs socioeconomic flows from rural regions toward this urban conglomerate. The population growth trend and the marked increase in the number of vehicles in the AMVA (an increase in the vehicle fleet of 282% and a motorization rate of 190% between 2005 and 2015) led to a rise in fixed and mobile polluting sources (AMVA 2016). Adding the above to the geographical conditions of the valley, where the mountains act as a barrier to the wind circulation and pollutant dispersion, has increased the frequency of atmospheric contingency episodes over the past decade.

Framed in this context, we proposed SIGenBici to expand the information on the existing cycling infrastructure in the territory and to encourage more people to use bicycles as a mode of urban transportation. The Bicycle Master Plan for the Aburrá Valley Metropolitan Area 2030, developed in 2015, planned that by 2019, 4% of the total trips made in the valley would be on bicycle and that by 2030, these would increase up to 10% (AMVA 2015). To achieve this goal, the region promised to increase the cycling infrastructure of the capital city. However, there is no updated information on the fulfillment of the goal. Additionally, since 2012 there has not been a significant increase in cyclists, and according to the Origin-Destination Survey of 2017 (AMVA 2017), only 1% of people used this mode of transport in the same year, an increase of 0.3% compared to 2012.

The availability of up-to-date and reliable information about the existence, location, and condition of bicycle lanes and other cycling infrastructure is an incentive to the use of technologies for decision-making of trips, as well as the absence of these data may represent a barrier. This is a problem for cyclists who want to travel around the city and who seek to identify the most appropriate route to their destination, as well as information on bike parking, EnCicla stations (our bike-sharing system), among others.

In this sense, cyclists are part of a community that has been growing locally, and they play a key role in the generation and use of open data in the field of mobility. By strengthening this technological connectivity, they become active mappers who have local knowledge and the interest to keep the data constantly updated, since they use it in their day-to-day life.

3 Open Mapping Methods and Activities

We based the development of the project on specific objectives of remote and field data collection, the training of the community of cyclists, the population and spatial analysis of the users of the cycling infrastructure, and the dissemination of results for their use by cyclists and planning

entities. To meet our goals, we applied a mixed methodology divided into three phases combining purely spatial data elements with qualitative surroundings perception, the latter based on concepts of Emotional Cartography (Nold 2009).

3.1 Remote Data Collection

The first phase referred to data collection remotely, with the aim of completing the information about cycling infrastructure publicly available on the OpenStreetMap platform (OSM). In a collaborative tag definition process with SiCLas members, a list of elements was created, complementing the cycling lanes offer with facilities such as bike parking, bicycle repair shops, public bike stations, among others. We realized that the terminology used to refer to cycle paths is not standardized, and changes even within the same region, making it necessary to establish a data mapping standard so that the information was as consistent and uniform as possible. To facilitate this, we organized a step-by-step mapping guide, which was socialized with the university volunteers, SiCLas members, and the general community involved in the project. Then, through a series of workshops on OSM and its editors, the mapping process was explained from the beginning, to empower the community in the use of these tools, inviting them to become active agents, using their local knowledge in the construction of the territory (Fig. 19.1)

To carry out the remote mapping activity, we used OSM Colombia´s Tasking Manager platform, intending to minimize conflicts at the time of editing. Likewise, we relied on Mapillary in some areas that were not so frequented by volunteers. In total, sixteen projects (tasks) were opened, one for each commune of the city of Medellín, so that the volunteer mapper could more easily identify the area that they travel to daily. As a result of these workshops, the information on the bicycle lanes and cycling amenities was published both in OSM and the SIGenBici cycling map (SIGenBici 2021) (Fig. 19.2).

Fig. 19.1 YouthMappers
train and validate in the
GeoLab in Medellín,
Colombia

Fig. 19.2 Bicycle infrastructure is depicted using CyclOSM, making visible features from ramps to repair shops

3.2   Socio-spatial Survey
of Cyclists

In the second phase, we intended to learn more
about cyclists: their profile, the routes they follow,
their motivation, their behaviors, and perceptions.
The main objective of this phase was to capture
information that could be associated with the spatial data reported by the survey respondents,
which would allow generating social analyzes and
identifying route patterns according to characteristics such as gender, level of experience, age,
etc., so that more subjective, emotional, and per-
ceptual aspects of the surroundings could be identified. We also wanted to identify how cycling
routes were perceived in terms of safety, comfort,
and risks, which would help us to identify issues
and opportunities throughout the territory.

To collect such information, a socio-spatial analysis survey was formulated, in which four sections were defined: Information about the subject, transportation mode, identification of routes, and perceptions of the surroundings. The first section allowed creating a general profile of the urban cyclist, by requesting information regarding age, gender, educational level, and occupation, among others. The second section was intended to identify the main mode of transport, as well as the intermodality of the trips, the type of bike used, and the level of experience as a cyclist. The third section covered spatial aspects, including their frequent route tracks and their characterization data (such as the reason for the trip, schedule, and frequency, among others). Finally, the fourth section combined the georeferenced data about the surroundings of the routes with some open comments, used to identify and classify areas, according to the perception of cyclists, as well as cases of theft or vandalism. In addition, it was asked which conditions hinder or discourage the use of the bicycle, as well as which ones optimize and encourage it.

The survey was developed as a specific web form for the project, considering that after reviewing many of the tools available as free software, none could combine the features required for the project with an intuitive way of operating, in a way that was easy to use for the community and volunteers. To build the form, we used languages and libraries typical of web development, such as HTML, CSS, and JavaScript (JQuery and Leaflet libraries), grouping in the same tool the capture of quantitative, qualitative, and spatial data. Additionally, this data development gives the possibility of receiving answers permanently, in order to continue improving the database. After two months of its publication, the number of responses received reached the equivalent of 72% of the sample of cyclists in the official Origin-Destination Medellín Metropolitan Area Survey of 2017.

We also planned a street-level photo capture activity using the Mapillary platform for this phase. Project members and volunteers were trained to capture photos using their action cameras or cell phones, cycling through the cycle paths mapped in the first phase of the project. The sequences of photos published on the platform linked to the SIGenBici project will allow the integration of street images into the SIGenBici cycling map in the future.

3.3 Thematic Mapping and Data Analysis

The third phase included the analysis of the data obtained through the survey, as well as the generation of thematic maps. To achieve this, the responses received were standardized and refined, then associated with spatial information and stored in GeoJSON-type file formats, which represent the base format for viewing on the SIGenBici website. It is important to note that both the project and its tools were designed in such a way that they had continuity over time. Therefore, people from the cycling and noncycling community can continuously add information to the database, contributing to the collection of specific data that are used for thematic maps and spatiotemporal analysis, carried out in QGIS and visualized on the SIGenBici platform.

3.4 Essential Participation of the Community as Researcher

The development of the SIGenBici project has been a learning process for the entire team involved, since many of its activities have been developed in virtual mode due to the restrictions of social isolation experienced in recent months due to the pandemic caused by COVID-19. This has required a constant adaptation of the methodological process, which, being essentially participatory, has presented great barriers in its application. However, the fact that the project was structured with a collaborative approach was what has allowed us to overcome the difficulties presented and move toward its conclusion. We have applied a community-based participatory research model, actively involving the cycling community from the beginning of the project formulation. This has guaranteed us to address the issues of interest to the community and investigate the problems that really need to be addressed (Wilson et al. 2010).

The cycling community is a group that, we could say, can represent the complexity and richness of the city in a reduced version. It is a community of people with different characteristics (considering demographic variables such as sex, age, educational level, and occupation) linked by sharing a common vision and engaging in joint action (MacQueen et al. 2001). Each one of these people who has participated in the different phases of SIGenBici has contributed with unique knowledge, generating a process of knowledge transfer and exchange between the community and the academy (Wilson et al. 2010). The production of relevant research evidence has guaranteed the interest of the community in advancing the project over time. Beyond that, it projects SIGenBici into the future, guaranteeing actions for changes.

4 Main Findings of Cycling Infrastructure

Despite the increase in cycling infrastructure over recent years, taking into account the difficult topography of the valley, there has not been an adequate distribution throughout the city. Cycling infrastructure is established mainly in the southern and flatter area of the city, where the valley is widest, favoring 6 communes of the 16 existing in Medellín. Precisely, these communes are those that present the highest Multidimensional Living Conditions Index (LCI), meaning that their quality of life is higher, as is the ratio of green and recreational spaces per inhabitant. Unlike the six communes with the most built cycling infrastructure, the communes located in the north of the city have few or no kilometers of bicycle lanes. Coincidentally, they also have the lowest LCI in the city and have little public space and urban facilities. In addition to that, in that part of the city, the valley narrows and the slopes increase, which means that the streets do not present good conditions for the use of bicycles.

Regarding the data generation, it was very interesting to note that the participation of

minorities, despite being proportionally small, has made interesting contributions and generated important reflections. Concerning gender, although the fact that female participation was equivalent to only 36%, compared to 62% of male participation, the results presented a female participation three times higher than that identified in the AMVA's Origin-Destination Survey of 2017. The gender option "Other" was mentioned by only 1% of those surveyed, which leads to the recognition of the need to involve the LGBTIQ + community more in these studies to represent their reality. Likewise, it was interesting to review the comments of non-cyclists (14.4%) who participated in the survey, since they provided their perspective about their relationship as external actors and the bicycle users' issues existing in the city, which for them still represent sufficiently large barriers to the point of discouraging them from migrating to the cycling mode.

When analyzing the influence of the cyclist's gender on the preferred routes, we observed that the majority of women choose to follow the bicycle paths and not so much the shared lanes. In the heat maps analyzed, the roads with the highest density (for the female gender) are those with cycle tracks or bike lanes. Women prefer to avoid high-traffic roads, contrary to the behavior of men, who use more alternative routes for their commutes, which are more spread out in the territory (Fig. 19.3).

This preference for bicycle paths or shared routes could also be explained by the self-perception of the survey respondents about their level of experience. The results we obtained indicate that women perceive themselves to be less experienced in cycling than men, despite not having a proportional relationship with the time they have been cyclists. However, neither gender nor cyclist's education nor the type of bicycle used was shown to influence the number of kilometers traveled by bicycle. The difference in average kilometers between men and women is 0.20 km, and the gender "others" has the same average as the female.

Regarding trip purposes, most cyclists use the bicycle for a broad range of motives, such as to

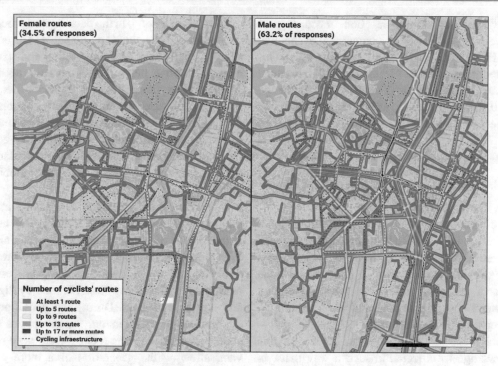

Fig. 19.3 Routes taken by female cyclists are compared to routes taken by male cyclists

go to work, to study, and for recreation, among others. The greatest distances traveled are for leisure rides; in other words, people are willing to take longer trips for recreation or sport, a culture that is very widespread among Colombians. We can deduce that the reasons related to daily obligations, such as work and study, have shorter distances traveled because the cyclists surveyed live relatively close to their travel destinations. The origin-destination map that we generated corroborates the average values traveled and confirms our hypothesis.

It is worth mentioning that both men and women cyclists justified their route preferences by giving greater importance to safety and the existence of cycling infrastructure. In addition, taking into account the level of experience of the cyclists, safety, tranquility or comfort, and less vehicular traffic are important for everyone. On the other hand, cyclists who have a high level of experience prefer direct and fast routes regardless of the existence of cycling infrastructure, a characteristic that is more important for cyclists with a low and medium level of experience.

Finally, we took into account the cyclists' perceptions about the routes they traveled and mapped several points in the city categorized as pleasant, safe, dangerous, with infrastructure problems, or risk of incident. Unfortunately, the last three perceptions coincide in being concentrated along the routes where more road incidents involving cyclists happen in the city. These results have allowed us to identify vulnerable points and dangerous areas where cycling infrastructure could be established in the future. They have also allowed us to discover areas where cyclists feel they are in a safe and pleasant environment, reinforcing the fact that cycling in Medellín has great potential as sustainable urban transport, and it needs to be encouraged through more and better infrastructure, as well as communication and educational strategies.

During SIGenBici development, we had the opportunity to greatly socialize the project through participation in different events, linked both with the use of GIS tools and with the cycling community. We can highlight our engagement in local events like the Medellín Bike Festival and the GIS Day organized by the

University of Antioquia, both occurring in November 2020; and in international events like the HOT Summit 2020, in December 2020. Our presentation at the HOT Summit 2020 was related to our contribution to the Mapillary Challenge 2020, in which we were selected as the #map2020 project of the year.

5 Reflections About Cycling and Participatory Planning

As citizens, we are the territorial actors more immersed in it, and we know first-hand the needs and strengths that exist; however, many times this knowledge is not used, sometimes due to the lack of tools that allow a correct articulation between the government and the community. Other times, it is due to the absence of educational processes that allow the transfer of this knowledge to citizens, in such a way that can generate more articulated and sustainable territories.

Based on this idea, during the SIGenBici project process, we were able to show that barriers that have traditionally existed in the city, associated with social and economic issues, also extend to the offer of cycling infrastructure, where the areas that have been more segregated, are less covered. For instance, given the need to migrate many activities to the virtual sphere, we were able to notice the digital divide that exists within our community, in terms of internet access and mastery of information technologies. This is evidenced by the data reported in the survey, where we receive very little information from some of the areas that have been more vulnerable, in part, due to the way in which the project was disseminated, closely linked to social media networks. There is also an evident relationship with the educational level since a large part of the cyclists surveyed have secondary and higher education levels above the population of cyclists in the Metropolitan Area.

On the other hand, in the data capture process, we noticed that some participants presented difficulties in terms of interpretation and spatial location through maps. This shows the need to generate more literacy spaces on spatial issues, so as to increase community participation in many of the digital initiatives that involve spatial data, which are becoming more and more common, especially on issues of city planning, facilitating citizen participation in urban mobility planning.

Considering this, the idea of thinking about the project as a more robust platform was born, with the aim of giving continuity to SIGenBici over time, preserving the structure already designed and working on improving processes, as well as expanding its dissemination by digital and not digital means. During the definition of this current project version, the comments and contributions of the community were very relevant, allowing the generation of a more user-friendly and intuitive environment, which is still under development. Eventually, we hope that this information can be used in the future, in matters of planning and management of infrastructure at the city level, allowing collective participation in the improvement and creation of facilities, at least in terms of cycling mobility.

References

Anderies JM, Folke C, Walker B, Ostrom E (2013) Aligning key concepts for global change policy: robustness, resilience, and sustainability. Ecol Soc 18(2):8. https://doi.org/10.5751/ES-05178-180208

Área Metropolitana del Valle de Aburrá (AMVA) (2015) Plan Maestro Metropolitano de la Bicicleta (PMB2030). Available from AMVA. https://encicla.metropol.gov.co/Documents/5PMB2030.pdf. Cited 15 Jan 2022

Área Metropolitana del Valle de Aburrá (AMVA) (2016) Plan de Gestión 2016 - 2019: Territorios integrados. Primera edición, 1 October. Medellín, Antioquia. Colombia

Área Metropolitana del Valle de Aburrá (AMVA) (2017) Encuesta de Origen - Destino - Análisis de Viajes.

Available from AMVA. https://www.metropol.gov.co/encuesta_od2017_v2/index.html. Cited 15 Jan 2022

Brovelli M, Ponti M, Schade S, Solís P (2020) Citizen science in support of digital earth. In: Guo H, Goodchild MF, Annoni A (eds) Manual of digital earth. Springer, Singapore. https://doi.org/10.1007/978-981-32-9915-3_18

European Cyclists' Federation, World Cycling Alliance (WCA) (2016) Cycling delivers on the global goals. Available from WCA, 9 February. https://www.world-cyclingalliance.org/wp-content/uploads/2018/12/The-Global-Goals_internet.pdf. Cited 15 Jan 2022

MacQueen KM, McLellan E, Metzger DS, Kegeles S, Strauss RP, Scotti R, Blanchard L, Trotter RT II (2001) What is community? An evidence-based definition for participatory public health. Am J Public Health 91:1929–1937

Nold C (2009) Emotional cartography: technologies of the self. Emotion Cartogr. http://emotionalcartography.net/EmotionalCartography.pdf. Cited 15 Jan 2022

SIGenBici (2021) Mapa cicloinfraestructura. Available from SiCLas. https://www.siclas.org/sigenbici/. Cited 15 Jan 2022

Solís P, Rajagopalan S, Villa L, Mohiuddin MB, Boateng E, Wavamunno Nakacwa S, Peña Valencia MF (2020) Digital humanitarians for the Sustainable Development Goals: YouthMappers as a hybrid movement. J Geogr High Educ, 1–21. https://doi.org/10.1080/03098265.2020.1849067

Wilson MG, Lavis JN, Travers R, Rourke SB (2010) Community-based knowledge transfer and exchange: helping community-based organizations link research to action. Implement Sci 5:33. https://doi.org/10.1186/1748-5908-5-33

Wastesites.io: Mapping Solid Waste to Meet Sustainable Development Goals

Chad Blevins, Elijah Karanja, Sharon Omojah, Chomba Chishala, and Temidayo Isaiah Oniosun

Abstract

The World Bank has conservatively estimated that 33% of global waste is managed in an environmentally unsafe way (Kaza et al. 2018). Waste generation could nearly double by 2050 with generation per capita expected to increase by 40% in low- and middle-income countries, many of which are growing at an unsustainable pace with limited resources dedicated to waste management. YouthMappers creating local geospatial data about sites of illegal trash dumping can play a key role in mitigating impacts and improving waste management, and in turn, impact public health. Several of the UN SDGs are supported by creating and sustaining a clean, healthy environment, particularly in this case, SDG 12 Responsible Consumption and Production and ultimately, SDG 3 Health. A novel tool, Wastesites.io, has been initiated to leverage youth action and connectivity of YouthMappers in order to solve these challenges together.

Keywords

Responsible consumption · Waste · Mapping apps · Kenya · Zambia · Nigeria · Public health

1 The State of the Trash Problem

Virtually every person and human activity generates waste in some form or another. While certain countries have reliable municipal or private

C. Blevins (✉)
Locana, Vienna, VA, USA
e-mail: chad.blevins@critigen.com

E. Karanja
Dedan Kimathi University of Technology, Nyeri, Kenya

S. Omojah
Women in Geospatial+, Map Kibera, Nairobi, Kenya

C. Chishala
Humanitarian OpenStreetMap Team, University of Zambia, Lusaka, Zambia

T. I. Oniosun
Space in Africa, Federal University of Technology, Akure, Nigeria

P. Solís, M. Zeballos (eds.), *Open Mapping towards Sustainable Development Goals*, Sustainable Development Goals Series, https://doi.org/10.1007/978-3-031-05182-1_20

waste services, many do not. Developed countries have national policies and entire agencies dedicated to monitoring and improving the environment, with waste management being an essential service. The impact improper waste management has on human health, water quality, air quality, and overall life on land and water has been well researched and documented. Accurate data are a key component for managing resources and measuring human and environmental health (OECD 2019). Unfortunately, quality data to support proper management is lacking in many places around the world. Timely and accurate data are needed to track the quantity, composition, and type of materials being consumed by humans. Each city and location requires information unique to that location to understand the effect of waste on humans and the environment and offer insight into potential solutions (Ejaz et al. 2010).

1.1 Missing Location Data for Waste Site Management

The absence of detailed data to help manage solid waste is inherently a geospatial problem. All waste is generated in one location then transported through a series of bins, dumpsters, trucks, or transfer stations, ultimately ending up in a landfill, incinerator, or processing facility. The adoption of geospatial technology in this industry is lagging, and without a comprehensive database of supporting infrastructure, it is impossible to plan and develop an efficient system for proper management.

In the United States, it took until the 1950s for the government to begin examining legislation and policy to reduce solid waste emissions (Hickman 2000). Up until then, open burning pits were commonplace across the U.S. and hazardous byproducts from industrial pollution were regularly dumped into local waterways, which is still common practice throughout the world. Public health impacts are well known, particularly the impact on air and water quality. A pin-

nacle turning point in the U.S. occurred along the Cuyahoga River in Ohio which caught on fire several times from oil slicks and pollutants floating on the surface. The economic impact of these fires caught the attention of local and federal government officials and was used to campaign for stricter regulations.

Eventually, in 1965, the Solid Waste Disposal Act (SWDA) was passed as an Act of Congress which is described as "the first federal effort to improve waste disposal technology." The Act set safety standards for landfills and established a framework allowing states to create independent standards unique to their needs. More importantly, the SWDA created a system to gather information about existing infrastructure. Officials needed information such as the location of landfills, open burn pits, and factories. These data allowed insight into the industry, the amount of waste being generated, impacts on the environment, and potential reductions and improvements to the amount of waste being generated. In 1970, the Environmental Protection Agency (EPA) was launched, which helped improve waste monitoring and research. The EPA established environmental baselines to measure the success or failure of different management practices. From there, the Resource Conservation and Recovery Act (RCRA) was created and became law in 1976. It remains the primary law governing the disposal of solid waste. The Act has been amended several times to accommodate for changing demands, amounts, and new types of waste. This example highlights the framing role of policies within waste management which is essential to establishing laws to protect human and environmental health, in turn affecting these SDGs.

Proper waste management takes a series of resources to be accomplished successfully. These resources include government funding, built infrastructure such as paved roads, large trucks, and equipment, along with a population willing to implement sound practices. The fastest-growing cities in the world are all within developing countries and lack resources to accommodate this surging population. As people migrate toward urban areas for work, waste management becomes worse each day. UN-Habitat estimates that 25% of the world's urban population live in informal settlements (Avis 2016). In some cases, city governments attempt to evict residents of informal settlements while other times they are tolerated but ignored with no access to municipal services. The preferred approach is for governments to work with residents to help secure land tenure rights and integrate them into urban society. It is impossible to accurately create municipal budgets and plans to support populations without timely and accurate information.

1.2   The Open Spatial Data Opportunity

Today there is tremendous opportunity to improve waste management policy and regulation throughout the developing world using geospatial tools. Not all waste requires the same regulation or management process. Sources of waste can be classified into four broad categories: industrial, commercial, domestic, and agriculture. This research focuses on creating geospatial data to better manage household-generated domestic waste. It is necessary to understand changes in production and consumption for local authorities to develop strategies for proper collection and disposal of waste. Many countries lack basic information needed to manage waste such as addresses, complete road networks, and the location of waste facilities and informal dump sites. With the exponential increase of solid waste these data are necessary for governments to understand and plan future investments keeping people and the environment clean for generations to come.

Fortunately, maps and geospatial data are more accessible in today's world than any other time in human history. Satellites are constantly orbiting earth snapping images to monitor the changing landscape of our planet, smartphones allow users to easily navigate cities and populated areas discovering places of interest and other essential services. Researchers are constantly finding new methods, tools, and commu-

nities to engage and benefit from technological advancements. Accessible and accurate data are the foundation for improving waste management and planning for future demand.

All the tools and technology needed to support an operational and efficient waste management system exists; however, there are many barriers in employing technology to make smart decisions. Government regulations and financial costs are high on the list. The first step in moving toward technology- and data-driven approach is improving the accessibility and quality of geospatial data for features essential to waste management. It is the first step in addressing this challenge at a global scale.

2   Case Studies of Mapping Solid Waste

YouthMappers around the world have been innovating to help fill in missing data for waste management, using the tools of OSM and local fieldwork to make a difference breaking the consumption cycle, and building better health for their environment and local populations. These cases narrated below represent only a few examples among many and have served as independently designed, serendipitous, and organic movements to use geospatial tools and ideas to address SDG 12 and SDG 3 in different local contexts. They share at their heart, a commitment to responsible production, healthy communities, and youth action.

2.1   YouthMappers in Nairobi, Kenya

Acute growth and rapid urbanization in Nairobi Metropolis, the heart of Kenya's economy, has led to the rise of illegal waste disposal which has become a major threat to the city and environment. Cases of epidemics like cholera have been reported in areas nearby landfills and disposal areas. In 2020 the Nairobi population was 4,735,000, a 3.93% increase from 2019.

Industrialization and urbanization associated with rapid urban growth at this scale are exponentially increasing the generation of solid waste throughout the city. There is a need for the evolution of Kenyan culture to develop spatial data improving economic, political, and social development. There is a growing need for both accurate and open data at both government, private institutions, and the local community level. This project will help Kenya meet Sustainable Development Goals (SDGs) that exist within institutions and government.

Effective solid waste management is essential to maintaining a healthy population and environment. Many low- and middle-income families have little to no municipal services to properly dispose of solid waste. This often leads to unpleasant neighborhood waste piles putting residents at risk of contaminated drinking water, disease-carrying rodents, and poor air quality causing refuse to regularly flow into waterways and forcing residents to organize periodic burns (Fig. 20.1).

The objectives outlined in this project contribute to achieving Kenya's Vision 2030 by promoting a healthier environment, empowering Youth to promote their own social and economic development, building more resilient people and ecosystems in a green growth economy, and creating an enabling environment for private sector investment in solid waste management. Several Sustainable Development Goals (SDGs) are supported throughout this work as we focus on collecting data to create a cleaner and healthier ecosystem.

This overall goal of this project is to increase data availability and accessibility for the government and solid waste industry in Nairobi by mapping formal and informal waste sites. Having data on where resources and threats to the environment exist is the first step toward planning improvements to the existing system. Training and field mapping exercises will help strengthen data literacy while providing youth and community members the opportunity to gain valuable workforce development experience.

2.2 YouthMappers in Lusaka, Zambia

Most cities in sub-Saharan Africa are growing at an unsustainable rate. Lusaka is the capital and largest city of Zambia and happens to be one of the fastest-growing cities in Southern Africa. As of 2019, the city's population was about 3.3 million, up from 2.5 million in 2018. According to Lusaka City Council (LCCS), 70% of solid waste

Fig. 20.1 YouthMappers at the Dedan Kimathi University of Technology help to remove waste at a site near Nyeri, Kenya

is generated from peri-urban areas in Lusaka, and half of it remains uncollected. The uncollected solid waste has a negative effect on the health of the people, at times this solid waste blocks the water deranges hence coursing floods in the city during the rainy season.

Waste collection and management are vital public services for every community and are obligatory for the protection of public health and the environment. Quality waste management services are important to urban management and policies; they fortify thriving local economies and are essential to ensure the protection of public spaces.

Sustainable Development Goals (SDGs) cannot be achieved if waste management is not prioritized. It poses a risk to the well-being of the people, when people have good health the productivity of a country can be attained. Zambia has a vision for 2030 to shift from a developing nation to a middle-income country, for this vision to be achieved a key consideration is to ensure the environment is clean and free from any illnesses that can emulate from a polluted environment.

Solid waste management has been a challenge for the Local authority in Lusaka due to the overpopulation of peri-urban areas. Most of these areas are too big for effective collection of solid

waste and there is a lack of supporting geospatial data. YouthMappers took part in a project to map the zones for solid waste. During the data collection exercise by YouthMappers at the University of Zambia (UNZA) chapter, students were able to map boundaries for the solid waste zones in 6 peri-urban areas, identify the major road networks in these peri-urban areas, and identify the location of dumpsites to study the accessibility of these sites by waste collection vehicles (Fig. 20.2).

Having detailed maps for each zone provides the council with information to enable planning for future waste management efforts. It allows them to study the amount of waste being generated for each area, research different interventions to improve management, and ultimately help improve the health of people and the environment.

2.3   YouthMappers in Akure, Nigeria

YouthMappers in Akure have been shocked by the amount of pollution he found in this big city in southwestern Nigeria. They were surprised to see trash piled high in the streets of neighborhoods across the city. Due to poor infrastructure and a lack of policy and regulation on trash collection, informal dumping sites, and poor waste

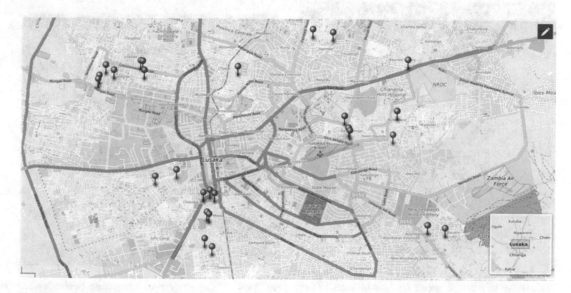

Fig. 20.2 YouthMappers from the University of Zambia pinpoint waste sites in Lusaka, Zambia

management created a large urban waste problem (Blevins and Lefeber 2018). Without anywhere to go with their trash, most households just contribute to the issue, piling waste around their neighborhoods (Oniosun et al. 2020; Oniosun 2017).

Akure's mounting urban waste problem is not just unsightly – it poses significant health, sanitation, and environmental consequences for the community. Though tackling this challenge is no easy task, YouthMappers at the Federal University of Technology were motivated to find a way to clean up their city. As students taking the remote sensing and Geographic Information Systems (GIS) major, the chapter decided to put their mapping skills to work to take on the challenge. Forty students from across the university began mapping urban waste across the city (Figs. 20.3 and 20.4).

Clearly, many of the problems African nations face could be addressed with better information – and mapping adds an important dimension to this. Spatial data can help deal with issues such as desertification, water resources, locust plagues, planning and urban development, and monitoring the latter's impact on biodiversity. But all of these also depend on human activity in the production-consumption cycle not leading to poor environmental sanitation. According to the Active Times,

out of the World's 25 dirtiest cities, 16 of these are from several countries in Sub-Saharan Africa. The initiation of the Urban Waste Mapping project by the YouthMappers chapters in Akure, Nigeria set out to address this. We all know there is trash everywhere in our city, we see it every day, but in order to clean it up, we first have to get a clear picture of the situation. YouthMappers first mobilized students at universities around the world to help remotely create new geospatial data using OpenStreetMap – to accomplish base maps of streets and buildings. Sitting at computers with satellite imagery of Akure as the backdrop, students from around the world digitally mapped roads, buildings, and parks in the city that had never been mapped before.

In the YouthMappers chapter of the Federal University of Technology, Akure decided to focus from the field on identifying illegal dumping sites in the city. Armed with this new base map of the city, YouthMappers hit the streets of Akure on bicycles and motorbikes to identify the location of illegal dumping sites. Students recorded GPS coordinates, descriptions, and photos for each site they found.

About 40 volunteer mappers collecting necessary street-level data. KoboToolBox was used to collect information such as: pictures of the dump-

Fig. 20.3 YouthMappers from the Federal University of Technology take to the streets on their bicycles to map waste in Akure, Nigeria

Fig. 20.4 Illegal waste sites exacerbate public health, infrastructure, and water quality or flooding problems

site, coordinates, site descriptions, type of waste (special waste, liquid waste, hazardous waste, restricted solid waste, general solid waste [Putrescible], general solid waste [Non-Putrescible]), proximity to residentials or water bodies (<10m, 10–30m, 30–80m, 80–150m, >150m), size of site and accessibility (either motorable or not). Navigation through the city for data collection was done using motorcycle/bodaboda/okada while the processed information was made available openly on Umap.

One major issue faced during mapping involved residents of some areas being concerned about why we are taking pictures and collecting data at each site. This provided a good opportunity for environmental outreach, sensitizing people about why they need to keep their environment neat and keep away from illegal dumping of refuse. The students ultimately proposed more suitable locations for disposal facilities away from residential areas and water sources. It wasn't enough for us to simply map illegal dumping sites, we also needed to give solutions to the problem. So the project output was presented to Ondo State Ministry of Environment to inform urban waste policy planning and also serve as a resource for local NGOs that work on urban waste clean-up. It was also later transposed to other cities toward environmental sanitation and became a chapter in a book authored and edited by the YouthMappers leadership.

But perhaps the most important reason to map is to spread the message. If seeing is believing, then looking at a virtual map of overflowing trash sites will help open our eyes to the problem. And, hopefully, it will also inspire people to join us to do something about it so we can look forward to a cleaner city (Oniosun 2017).

3   Wastesites.io: A Solution for the Global Goals

These case studies showcase the passion and power of today's Youth in addressing the SDGs. There is no shortage of volunteers or energy to engage in activities to improve environmental and human health. Over the course of several years,

each of these projects evolved independently from one other, but serendipitously to address the same exact problem: the abundance of waste in the human environment. There are two critical factors that should be considered when looking at this kind of action as potential solutions for the SDGs.

First, youth action around trash occurs in the context of solid waste policy. It is important to recall that when the U.S. first addressed waste at a national level, the first step was enacted by Congress approving the Solid Waste Disposal Act. This in turn set loose a wave of regulation to basically collect and create data surrounding waste management throughout the country at local and regional scales. However, the work of YouthMappers chapters is unfolding in very different policy environments, especially in places where regulation just does not exist or is not enforced. Part of the power of this work is that it can bypass a lack of regulation, by creating data to directly address the problem. Often it takes a grassroots movement to influence national-level policy change. Mapping waste can be seen as a lever of engagement where citizens take responsibility for their own sites and actions by bringing public attention to local concerns, with the added aim of persuading decision makers to enact policy to solve problems in favor of their constituents – and ultimately in support of SDGs.

A second important aspect of this work is related to the potential for scaling up solutions across many different communities facing common problems. Each of these projects may appear small but combined they bring a greater level of attention, a robust model for working through common challenges, and importantly, the connectivity of a network of youth bringing together like-minded individuals. To build upon this youth action and advance the impact and scale of work exemplified by the case studies in this chapter, solutions need to be designed that consider the data-policy context, as well as the opportunity for establishing and nurturing a growing community of practice (Blevins and Lefeber 2018).

Wastesites.io is one such solution potential. Wastesites.io has been conceptualized and initiated to support these efforts by building a virtual community around waste feature mapping. The goal is to provide a platform for capturing data not only to visualize or highlight the problem but also to help solve the problem. Organizing and implementing field-level mapping campaigns takes an immense amount of work and a multitalented team. Wastesites.io aims to lower the barrier of entry for students and fellow community members interested in promoting responsible consumption while fostering a healthy environment. While each case study showcased here was unique, many of the same tools and methods were used to achieve results.

Mapping waste features in developing countries introduces a host of variables which make creating standard collection guidance difficult. The informal waste sector is one example, as seen in YouthMappers efforts. Many times, what someone may view as an informal dumping site is, in a practical place-based sense, the locally-understood formal spot where people have been and will be dumping their waste, often due to lack of alternatives. Basic guidance for designing and applying a mapping schema to cover each scenario is important and small distinctions occur at the hyper local level. While these results differ, what the YouthMappers' experiences underscore is that the process is very, very similar.

Clearly, proper waste management addresses multiple SDGs with goal 12 "Responsible Consumption and Production" likely being the most influential for countries like Kenya, Zambia, and Nigeria. But while waste represents the ending stages of the productive cycle, the question of how YouthMappers everywhere will continue to innovate and evolve such grassroots work could also take a focus on the production side, in higher GDP countries, where YouthMappers are also concerned about what the SDGs mean for them. For instance, while the United States has a high-functioning waste management system, the country overall ranks number one in waste generation (Tiseo 2020). When compared to lower GDP countries, the U.S. creates a disproportionate amount of waste per capita by a factor of more than 10 times. "Reduce, Reuse, Recycle" is a familiar quote used to demonstrate the power of behavior change. The kinds of tracking of data and location that solutions like Wastesites.io could

offer include opportunities for all YouthMappers despite their local policy context, to unite and share around common problems, even if at first, they do not look similar. This could also happen because many of the same solutions for responsible waste management could be put in place wherever there is human production and consumption. One example of this solutions-driven principle is illustrated by the fact that many places around the world are embracing these words by banning the use of plastic bags. It takes the attention of politicians to implement bans like this with small costs for big returns in the solid waste system, and countries with lower GDP are leading the way, having implemented them for years, whereas countries like the U.S. are just now starting to put in partial, local effect. The relevance of a global community of practice cannot be overstated.

In conclusion, the YouthMappers experience with waste sites, represented by these three use cases, may seem small when compared to the global solid waste management crisis. However, each incremental step moving toward a solution is helping improve the health and well-being of local residents while contributing more broadly to a global capacity to address the SDGs. We aim to keep innovating, sharing, and building the network and tools to accelerate this work.

References

Avis WR (2016) Urban governance: informal settlements (Topic guide). GSDRC, University of Birmingham, Birmingham. Available at https://gsdrc.org/wp-content/uploads/2016/11/UrbanGov_GSDRC.pdf. Cited 20 Jan 2022

Blevins C, Lefeber M (2018) After mapping effort, one city in Nigeria cleans up urban waste. Available via USAID Medium, 7 February. https://medium.com/usaid-2030/after-mapping-effort-one-city-in-nigeria-cleans-up-urban-waste-4945348e82e0. Cited 20 Jan 2022

Ejaz N, Akhtar N, Nisar H, Naeem UA (2010) Environmental impacts of improper solid waste management in developing countries. WIT Trans Ecol Environ 142:379–387. https://doi.org/10.2495/SW100351

Hickman Jr HL (2000) A brief history of solid waste management in the US 1950 – 2000. Available from MSW Management, Part 3. https://www.mswmanagement.com/collection/article/13000352/a-brief-history-of-solid-waste-management-in-the-us-1950-to-2000-part-3. Cited 20 Jan 2022

Kaza S, Yao L, Bhada-Tata P, Van Woerden F (2018) What a waste 2.0: a global snapshot of solid waste management to 2050. International Bank for Reconstruction and Development/The World Bank, Washington, DC. Available at https://datatopics.worldbank.org/what-a-waste/ Cited 20 Jan 2022

Oniosun T (2017) Capitalizing on maps to achieve SDG goals in Africa. Available via YouthMappers Blog, 27 September. https://www.youthmappers.org/post/2017/09/30/capitalizing-on-maps-to-achieve-sdg-goals-in-africa. Cited 20 Jan 2022

Oniosun T, Ifeoluwa Adebowale BI, Solís P (2020) Capitalizing on geospatial technologies to solve urban waste in Akure Nigeria. In: Froehlich A (ed) Space fostering African societies. southern space studies. Springer, Cham, pp 31–46. https://doi.org/10.1007/978-3-030-32930-3_3

Organization for Economic Cooperation and Development (OECD) (2019) Municipal waste, generation and treatment: municipal waste generated per capita. Available via OECD Data. https://data.oecd.org/waste/municipal-waste.htm. Cited 20 Jan 2022

Tiseo I (2020) Energy & environment: waste management, global per capita generation of municipal solid waste by select country 2018. Available via Statista 27 October. https://www.statista.com/statistics/689809/per-capital-msw-generation-by-country-worldwide/. Cited 20 Jan 2022

Mapping for Resilience: Extreme Heat Deaths and Mobile Homes in Arizona

21

Elisha Charley, Katsiaryna Varfalameyeva, Abdulrahman Alsanad, and Patricia Solís

Abstract

YouthMappers help discover hidden vulnerabilities to extreme heat in the face of a changing climate by mapping health outcomes compared to energy assistance. What emerged is a pattern of disproportionate deaths by housing type, necessitating innovations in tagging unique mobile home attributes in OpenStreetMap (OSM). The resulting community engagement generated solutions that stakeholders and residents of mobile homes can implement for greater resilience, and a model for connecting SDG 13 (Sustainable Development Goals) for climate action to SDG 3 good health and well-being by looking at the homes where people live.

Keywords

Climate Impacts · Heat · Housing · OpenStreetMap Attributes · Arizona · Health

1 Local Impacts from a Global Problem

In an era where youth are demanding global climate action, there is a profound need to use local microdata in local places to better understand the impacts of climate change in our towns and

E. Charley (✉)
School of Geographical Sciences and Urban Planning, Knowledge Exchange for Resilience, Tempe, AZ, USA

Diné, Navajo Nation, AZ, USA
e-mail: echarle@asu.edu

K. Varfalameyeva
School of Geographical Sciences and Urban Planning, Knowledge Exchange for Resilience, Tempe, AZ, USA

Brest, Belarus
e-mail: kvarfala@asu.edu

A. Alsanad
School of Geographical Sciences and Urban Planning, Knowledge Exchange for Resilience, Tempe, AZ, USA

Kuwait City, Kuwait
e-mail: asanad@asu.edu

P. Solís
Global Futures Laboratory, Knowledge Exchange for Resilience, Arizona State University, Tempe, AZ, USA
e-mail: patricia.solis@asu.edu

neighborhoods. Data gathering is one way the youth can contribute to SDG 13, climate action. Such impacts include how climate change is affecting our health and well-being, which implicates the added opportunity to contribute as well to SDG 3, healthy communities. By creating local microdata in new ways, we can uncover some of the missing insights about what is happening, and show where vulnerable people are falling between the gaps. YouthMappers at Arizona State, who include international students who come from around the world to study geography and urban planning, have contributed to a discovery of hidden health vulnerabilities to the climate in places we live and work in the Southwest United States. Our connections link the global and the local in unexpected ways.

The most surprising fact that we learned through mapping locally about global problems is that local people's health is impacted by climate heat extremes in different ways, depending on their housing shelter characteristics. Namely, our team's research uncovered that although only 5% of housing in Maricopa County, Arizona, is in a particular kind of low-cost housing called mobile homes, approximately 30% of indoor heat-related deaths occur in these types of homes. Mobile homes are typically constructed with lightweight building materials and can be manufactured elsewhere and moved to lots (where they often become immobile over time). These parking areas are sometimes referred to as trailer parks and can include homes on wheels known as Recreational Vehicles (RVs). Often people from other parts of the country migrate from cold weather during the milder local winter months.

Especially the older mobile homes, which were built with lower energy standards as well as RVs that are not fully heat-ready, represent options that are affordable to struggling families or the elderly, but they do not withstand well during the Arizona summers, that are growing increasingly hotter each year. Thus, the residents of mobile homes in Maricopa County, Arizona are disproportionately affected by heat (Fig. 21.1).

Mobile home residents are vulnerable specifically because they are extremely exposed to heat due to the high density of mobile home parks,

Fig. 21.1 People who live in Recreational Vehicles (RVs) or mobile homes like these in Mesa, Arizona, are disproportionately exposed to extreme heat and suffer heat-related illness and death at rates up to six times higher than similar residents in other types of housing

poor construction of dwellings, lack of vegetation (and thus lack of shade trees), sociodemographic features, and not being eligible to get utility and financial assistance. According to the Centers for Disease Control and Prevention (CDC), more people die in the United States from heat than from all other natural disasters combined. According to the Environmental Protection Agency (EPA), more than 1,300 deaths per year in the United States are due to extreme heat. Arizona, California, and Texas are the three states with the highest burden, accounting for 43% of all heat-related deaths.

2 YouthMappers Making Vulnerability Visible on the Map

During the summer of 2019, the YouthMappers at ASU hosted a series of collaborative mapping exercises in-person and online. Collectively as Arizona State University (ASU) students working as interns and research assistants in the research unit called, Knowledge Exchange Resilience (KER), we contributed our efforts in developing mapping data specifically for Recreational Vehicle (RV) and mobile home units across a number of trailer parks in Mesa, Arizona, east of our campus. This effort was launched to advocate for the gap in mapping data displaying RV and mobile homes as primary residences for long-term tenants within urban mobile home parks. Normally these areas on the map are not even marked for internal roads and typically not buildings, as they represent a neighborhood of privately owned property where lots are rented to mobile homes and RVs as tenants who own their unit but not the land beneath it. Our area of focus specifically marks spaces that were identified with a higher frequency of mortality rate resulting from heat-related death and illnesses.

2.1   Extreme Heat Deaths

In the summer of 2019, we requested and received data from the Maricopa County Public Health Department broken out by trailer parks versus all morbidity cases that happened indoors. They reported that year, 25% of all indoor heat deaths occurred in mobile homes, but in some years that rate reached as high as 40% (MCPHD 2006–2020). The data provided by the public health department was overlaid with other data we collected from organizations that provided utility assistance in form of various payment options. These utility assistance programs serve primarily low-income household residents occupying single-family homes and apartments, in the form of funding to offset the cost of electricity to run air conditioning. These services are provided through the local utility companies including Salt River Project power and water (SRP) and Arizona Public Service Electricity (APS). However, RV and mobile home dwellers are rarely ever eligible for the forms of utility assistance programs at that time, due to various policies implemented by the utility companies and certain federal (national US) guidelines.

2.2   Mapping the Pattern Beyond Indicators

The KER team began to analyze the provided data through GIS mapping and identified a pattern of hot spots of areas of indoor heat-related deaths. When compared to where utility assistance was provided, there were new patterns to analyze with stakeholders to triage where there might be a higher frequency of heat-related deaths with limited or no utility assistance recipients.

Phillips et al. (2021) recount our process of data acquisition and meaning-making that unfolded in our community geography research setting. The major discovery from this process was the extent to which people who reside in mobile homes were disproportionately at risk (MCPHD 2020). Many of the stakeholder organizations that provide support to people suffering health issues due to extreme heat were well aware of individual vulnerabilities like age or household income vulnerabilities. On the whole, this stakeholder community also utilized indicators, tracking county-level statistics through a Social Vulnerability Index. However, the spatial analysis done prior to KER and YouthMappers had not yet revealed the danger that the housing type implied (Wang et al. 2021). We pointed out that the SVI was missing any variables related to housing at all (Varfalameyeva 2020).

2.3   Gathering Data on Unseen Climate Vulnerable Locations

There are 92,031 mobile homes in Maricopa County, which is 5.2% of the total housing stock. These homes are located in approximately 686 mobile home parks spread throughout different parts of the urban core of Maricopa County, usually on the outskirts or in less valuable areas.

Our targeted area for mapping coincided with the densest location of heat-associated deaths where little to no assistance was being provided, namely in the stretch of surrounding mobile home parks located along Main Street in Mesa, Arizona. The satellite image below displays a wave of white and silver metal roofs reflecting

Fig. 21.2 The concentration of trailer parks in Mesa, Arizona is visible on satellite imagery due to the reflection from metal roofs, paved lots, and a lack of trees (Google 2021)

from mobile homes and RVs. We highlighted the area of interest in red. It was very alarming to see the overwhelming density of mobile homes and RVs expanding along Main Street, which are typically hidden from street view, and are not represented on most maps because they are private property (Fig. 21.2).

Typically, mobile home parks in the US have an owner and landlord that manages these private development properties. As a result, mobile home building footprints or lots are not commonly identified on maps. The unique challenge of this matter required a mapping platform allowing local advocacy and access – but also global visibility. We utilized OpenStreetMap (OSM), which is an online open-source mapping engine that enables mapping by looking at satellite imagery as well as mapping using local tools and information. This allows local to global access in obtaining or generating maps to communicate local knowledge.

2.4 Labeling Attributes – The Challenge of a Novel Housing Type

The mobile home parks in this area, were not identified and labeled within OpenStreetMap, let alone each mobile home lot, so we as students and YouthMappers co-created mapping data

identifying each park, and then each location for mobile homes and RVs. Co-generating and sharing our research efforts on the mobile home parks were key in our mapping data so they could be validated by other students, verified with city and county property data, and interpreted with community residents in order to generate solutions that make their homes more resilient to extreme heat.

We soon identified the lack of available categories, legends or labeling, that were adequate for identifying RVs and mobile homes as primary homes within OpenStreetMap. We wanted to create mapping data to specifically tag to identify long-term mobile home and RV dwellers in mobile home parks. However, OpenStreetMap was limited to labeling mobile homes and RVs as recreational units only. This definition does not fit the local circumstances well.

The challenges we experienced are utilizing the existing tags (labels) in OpenStreetMap identifying RVs as "tourism=caravan_site" which is fitting for tourist RVs but not for permanent living like for mobile homes. The tourism label also can be mistaken when people who migrate annually for the winter months consider themselves as residents, not tourists or even visitors (Varfalameyeva 2021).

After careful consideration, and consultation with knowledgeable people from the broader

Tag:building=static_caravan

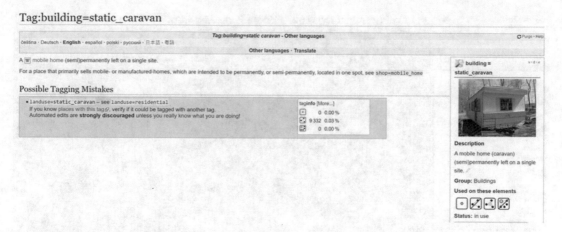

Fig. 21.3 Because of the unusual housing type, YouthMappers at ASU developed a special tagging convention for the project to identify these building features on OpenStreetMap

OSM community, we decided to utilize the tag "building=static_caravan" to label permanent living RVs, (which again are not the same as people who use RVs for travel or vacation). The mobile home tags were labeled as "A mobile home semi-permanently left on a single site" (Varfalameyeva 2021). Developing two specific tags would allow us to develop a data set that distinguishes the built environment within mobile home parks. This will tie later into the specific distinct solutions sets for making these dwellings more heat resilient. We can better manage the data and specifically identify the qualities of the two dwellings during our mapping process from the start. The static_caravan tag begins to characterizes semi-permanent mobile homes positioned on lots (Fig. 21.3). (OpenStreetMap 2021) This tag is our mapping method and tool to apply to a long-term mobile home unit (Fig. 21.3).

2.5   **Results of Data Collection Campaigns**

The image below is a real-time OpenStreetMap displaying our collaborative mapping data efforts in the red box which lists the contributors #YouthMappers, #HeatMappers, #ArizonaState, and #teachosm-project-834. We feel a sense of accomplishment about the amount of mobile home

and RV data our team has been able to create and share with the community. The research focus area display blue polygons, polylines, and points on the map it describes the mobile homes positioned on the lots within the mobile home parks. This data has become a fundamental feature of ongoing research in ways that improve our ability to analyze and solve for heat. As a visual layer, we are able to utilize this OSM layer in ArcGIS Online as a basemap, which is useful for everything from presentations with community members to share quick analysis of overlap with other layers that may relate to social vulnerability (such as factors in the SVI). We can also tally units, calculate roof versus impervious surface space, and import real data into landscape design projects to seek how to optimize shade. These have been the foundational data for theses and student honors projects, as well as for visualizations informing local and national press articles. This particular collaborative mapping generation is serving as a beneficial pilot experience for our team members' future academic research focus on rural Indigenous communities (United States) sharing similar gaps of limited mapping data. Other research questions are to look at the dynamics of geospatial climates of indoor versus outdoor temperatures in mobile homes and RVs within existing landscapes, from vegetation to limited or no vegetation, to test the effects of possible solutions (Fig. 21.4).

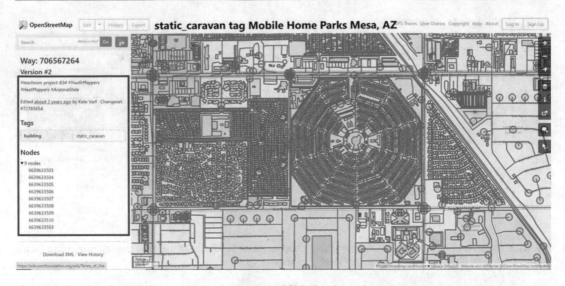

Fig. 21.4 Density of mobile homes are clearly visible on OSM after the project

Fig. 21.5 The basemap
display vulnerability
scales,
specifically mobile
home dwellers.
OpenStreetMap is an
open mapping
platform that provide
access for our team to
inform ongoing research
solutions, climate
change mitigation

3 Community Implementation with YouthMappers

There is an urgency of representation and communicating local knowledge, especially in the context of global conversations that may seem distant or remote. Mapping data allows our team to communicate the scales and impact of heat-related deaths within the focus area and have dialogue about how climate change is impacting Maricopa County residents. ASU students and YouthMappers emerge as the global citizens who are able to identify and learn about these communities through collective mapping generation efforts intersecting climate action. We think our efforts of advocating for mobile home park residents can impact future adaptation or remediation within local development, and become a model for resiliency planning. In addition, advocating affordable housing and immediate shelter resources to mitigate the impacts of extreme heat (Holmes 2021; Kear et al. 2020; Kutz 2020; Leahy 2020; Peterson 2021) (Fig. 21.5).

3.1 Global Partners, Local Stakeholders, and Mobile Home Residents

YouthMappers collaborative mapping exercises have been a wonderful experience and we (ASU students) have developed a lot of applied methods for our own academic studies. Most of our research is advocacy through the agency of community and local knowledge, including local residents of parks, and with the county-wide organizations that serve and support them. We enjoyed the real-time web interactions with global collaborators and student chapter members during the live YouthMappers collaborative exercises and varying remote campaigns. Many of the ASU students had never participated in a real-time global hands-on mapping workshop and this was an exciting learning experience. The events were successful and the generated mapping data continues to serve the mobile home communities, the KER research team, and the Maricopa County Public Health Department as well as others.

The Mesa mobile home parks mapping data continues to be developed further in various quantitative data and attributes specifically, communicating ground truth for new questions or providing deeper understandings of on-going issues. Our co-generation data efforts helped us present the vast density of urban mobile home developments in Maricopa County. The mapping exercises developed tangible data for mobile home residents and local leaders to participate in spaces and conversations about problem solving.

3.2 Toward Heat Resilient Solutions

The most motivating outcome of this work has been to co-develop heat solutions with residents and local stakeholders. Through months-long series of meetings convened by KER in 2020 and 2021, we researched numerous solutions across different domains that could help build the heat resilience of mobile home residents. As a result, we generated 50 heat mitigation solutions for mobile home residents, at various budget levels and depending on locally available resources (Varfalameyeva et al. 2021). The 50 solution goals are to raise awareness of the health impacts of extreme heat and recommend additive built material, legal, and policy education for mobile home residents (Fig. 21.6).

From our studies we learned most Maricopa County mobile home parks/property conditions vary from no vegetation to some vegetation, abandoned rental lots, added carports, added

Fig. 21.6 The OSM data on mobile home sites created by YouthMappers at ASU serves as a basemap for new heat-ready trailer park designs

storage units, and an abundance of impervious surfaces. Identifying, collecting, and mapping these hyperlocal forms of microdata sets would provide the varying scales and interrelated impacts of the built environment in mobile home parks. These forms of data are necessary to create site-specific designs, generate cost-benefit analysis, and enlist the investment of local cities and actors to propose solution agendas to protect the health and wellbeing of mobile home residents.

3.3 SDGs as Link from Global to Local Climate Action for Health

Branching from our initial mapping data set, we share this understanding with the entire YouthMappers network to emphasize the different patterns of structural variability. The exposure to extreme heat varies, so local health impacts due to climate change depends on the type and quality of housing / shelter, and demographics of residents in the mobile home parks. In the future, we hope to inspire cross-chapter comparisons of how the built environment impacts the structural performance of housing under conditions of climate change, connecting the concepts of the SDGs, especially 13 climate action and 3 health and well-being through the idea of housing as a right to shelter that saves lives.

References

Google. (2021). Google Map data, Imagery layer. Maxar Technologies, U.S. Geological Survey, USDA/FPAC/GEO. Available from maps.google.com. Cited 18 Oct 2021.

Holmes C (2021) Arizona Researchers look for ways to decrease heat deaths in trailers. ABC 15 News, 23 April. https://bit.ly/32Y9y5y. Cited 4 Jan 2022

Kear M, Wilder M, Hondula D, Bernstein M, Solís P (2020) Self-isolating from COVID-19 in a mobile home? That could be deadly in Arizona. Arizona Republic, 3 May. https://bit.ly/2WsytKN. Cited 4 Jan 2022

Kutz J (2020) Extreme heat is here, and it's deadly. High Country News, 1 September. https://bit.ly/3gVo172. Cited 4 Jan 2022

Leahy S (2020) As summer arrives, how will the most vulnerable escape deadly heat and COVID-19? National Geographic, 16 June. https://on.natgeo.com/2YISFdX. Cited 4 Jan 2022

Maricopa County Public Health Department (2006–2020). Annual final reports. Available from https://www.maricopa.gov/1858/Heat-Surveillance. Cited 4 Jan 2022

OpenStreetMap. (2021). Tag:building=static_caravan, OSM Wiki. Available from https://wiki.openstreetmap.org/wiki/Tag:building%3Dstatic_caravan. Cited 18 Oct 2021.

Peterson K (2021) Extreme heat is killing people in Arizona's mobile homes. Washington Post, 2 July. https://wapo.st/3yooHe0 (English) and https://bit.ly/3hlhHZW (Spanish). Cited 4 Jan 2022

Phillips LA, Solís P, Wang C, Varfalameyeva K, Burnett JL (2021) Hot for convergence research: a community-engaged approach to heat resilience in mobile homes. Profession Geogr. https://doi.org/10.1080/00330124.2021.1924805

Varfalameyeva K (2020) An illusion of affordability: the economic costs of heat exposure for mobile housing in the phoenix metropolitan area. Unpublished master's thesis, Arizona State University

Varfalameyeva K (2021) How to define recreational vehicles where they serve a different purpose? Available from YouthMappers Blog, 29 October. https://www.youthmappers.org/post/2019/10/22/how-to-define-recreational-vehicles-where-they-serve-a-different-purpose. Cited 4 Jan 2022

Varfalameyeva K, Solís P, Phillips LA, Charley E, Hondula D, Kear M (2021) Heat mitigation solutions guide for mobile homes. knowledge exchange for resilience solutions series. Arizona State University, Tempe. Available from https://keep.lib.asu.edu/collections/160080. Cited 4 Jan 2022

Wang C, Solís P, Villa L, Khare N, Wentz E, Gettel A (2021) Spatial modeling and urban analysis of heat-related morbidity in Maricopa County, Arizona. Journal of Urban Health. https://doi.org/10.1007/s11524-021-00520-7

Mapping for Women's Evacuation Plans During Climate-Induced Disasters

Airin Akter and Mobashsira Tasnim

Abstract

Increasingly disasters like floods, droughts, cyclones, and heat waves are recurring in the wake of climate change impacts. Women and children are among the most vulnerable, particularly in developing countries. At the same time, young people are among the most eager to respond with solutions to protect populations affected by disasters. In the low-lying riverine country of Bangladesh, YouthMappers are working to create fundamental maps to support the information needs of women who seek safety from climate-inducted disasters, but currently must rely on their husbands for evacuation. Initiatives like our open data mapping that entails participation of young women to alleviate such vulnerabilities toward disasters contribute to global goals for climate (SDG 13) and gender (SDG 5).

Keywords

Climate change · Women · Disaster vulnerability · Emergency response · Bangladesh · Gender inequity

A. Akter (✉) · M. Tasnim
Alumna, Department of Geography and Environment,
University of Dhaka, Dhaka, Bangladesh

1 Youth and Women Affected by Climate-Induced Disasters

Disasters are engendered from extreme climatic events when anomalies in normal temperature occur, initiating heat waves and a more intense precipitation than normal. This causes a higher rate and intensity of floods, landslides, avalanches, and soil erosion with associated damages. Climate-induced disasters like floods, droughts, cyclones, and heat waves are the most recurrent disasters and bring about deaths, whereas women and children have died mostly. They also dismantle the economy which disproportionately affects the young generation of a country.

What happens when these forces come together? The answer is found in the example of Bangladesh where the percentage of workers seeking jobs is the highest, and the percentage of dependents is the lowest. Climate impacts are among the most severe and gender inequality is high. Considering this scenario, young people,

P. Solís, M. Zeballos (eds.), *Open Mapping towards Sustainable Development Goals*, Sustainable Development Goals Series, https://doi.org/10.1007/978-3-031-05182-1_22

mostly women and children, are highly affected by the process of these forces converging.

Climate change, the focus of action for SDG 13, apparently aggravates the frequency and intensity of disasters, raising economic costs associated with mitigating it. The threat is greatest in the so-called developing countries as there is an unequal impact of climate change within high-income countries and low- and middle-income countries like Bangladesh. The disasters and their costs inevitably affect gender equality and result in being evaluated as the single biggest threat to development in the future. If true, the young generation would be the hardest-hit victims of this compounded threat. It is worth noting that young people largely did not create the problem of climate change, and especially those of us living in Bangladesh are disproportionately not the ultimate cause of increased sea level rise or more frequent cyclones as a result of oil consumption.

So why are youth, especially young women, so willing to respond to a problem we did not create? Maybe because we have the most to lose. We are the ones who need to design our escape from this problem.

2   Present Scenario of Climate Change in Bangladesh

Bangladesh is our home. Being a low-lying riverine country of South Asia with unique geographical vulnerability to climate-induced disasters, this place where we were born, where we live, is regarded as one of the climate change "Hot Spots" suffering frequent climate change-induced disasters, impeding sustainable development goals. However, the present scenario of climate-induced extreme weather events exposes the coastal region of Bangladesh to cyclones, the most prominent ones causing immense loss of lives, properties, and economic infrastructure. These are our mothers and fathers, cousins, our family homes, and our livelihoods. These extreme weather events are creating vulnerabilities categorized as biophysical and socioeconomic vulnerability among coastal communities, and

young women are at utmost risk of not only losing their lives but also because they are exposed to socioeconomic insecurities at critical moments.

2.1   Impact of Climate Change-Induced Disasters on the Coastal Area of Bangladesh

In the past 20 years, Bangladesh has managed to face this problem and reduce the death rates significantly, though the rate is still alarming for Bangladesh. The Bay of Bengal is a perfect breeding ground for tropical cyclones where on average 12–13 depressions form annually, leading to at least one or two powerful cyclones (Paul and Routray 2010; Alam and Rahman 2014), for example, cyclones Sidr (2007), Rashmi (2008), Bijli (2009), Aila (2009), Mahasen (2013), Komen (2015), Roanu (2016), Mora (2017), Titli (2018), Fani and Bulbul (2019), and Amphan (2020). Now, tropical cyclones have become an annual occurrence.

In 2020, The cyclone Amphan washed away embankments, mud homes, and fisheries, which thousands of families relied upon for their livelihoods, and sanitation infrastructures collapsed. On 15 November 2007, a huge cyclone brutally struck the coast of Bangladesh which caused thousands of deaths where unfortunately most of the casualties were women and children.

One major reason for the large number of casualties is the lack of implementation of disaster awareness programs along with proper planning, in our viewpoint. We mention proper planning since, at that time, disaster awareness programs were offered in generalized ways. According to Mallick B (2014), Community-based disaster management must take into account people's capacities, perceptions, and resources, as well as their participation. So, we set our eyes on this perspective that women and children need different interventions compared to men because of different vulnerabilities of women in remote areas, such as: physical weakness, health conditions, and dependency on their husbands or a male representative of their family for preparedness and mobility. Some women have less information about disasters and have less decision-making power given that they are unaware about the disaster and uneducated about preparing for them (Kibria 2016). This grave scenario is still the same after a decade.

In this YouthMappers project, we visited one of the unions from the Bagerhat district where women find it difficult to understand the warnings of disasters and are not really aware about the programs. They still depend on their husbands for evacuation. We are motivated by the thought that young women need to be more cautious as we need to face the problem together and make our world more resilient toward climate-induced disasters.

2.2   Gendered Impacts of Climate Change

Climate change-induced disasters strike the coastal areas almost every year in Bangladesh and various research data show that women die more often than men during disasters. Vulnerabilities toward the natural hazards for men and women are determined by the differences in biological and social responsibilities. Men are considered to be the income source of a family, basically covering all the expenses for the loss after natural hazards. As men are the provider of economic safety and most of the families in Bangladesh are headed by men, they take all of the decisions to respond along with the preparedness in advance for any uncertain climate change-induced hazards.

On the other hand, women are usually expected to be the nourisher of children and carry out all the household chores. Hence, women depend on their husbands or a male representative of their family when it comes to taking any major decision. Moreover, women's access to information and control over taking any precautions for any natural hazards are less compared to men. Women have less disaster information and have less decision-making power. They may even

be unaware about a coming disaster which makes them more vulnerable compared to men.

2.3   Role of Gender in Emergency Response and Evacuation Planning During Cyclones

Evacuation Planning is one of the basic components in the phase of preparedness and Emergency Response. Studies into the role of gender in evacuation efforts have found that women may be more likely than men to evacuate their homes as hurricanes approach. In developing countries, socioeconomic conditions are different and so can be the determinants of evacuation behavior. One important factor causing low evacuation here is identified to be the failure to perceive the potential risk from warning messages by the general public due to both lack of awareness and knowledge gap between them and the experts.

According to Mallick and Vogt (2009), a study conducted after Cyclone Sidr among 124 sample households found that two-thirds did not evacuate. They decided to stay at their home but sent some of the elderly and children to the safe shelters or a relative's house. Research indicates that the overall evacuation rate during Cyclone Sidr in the study areas was only 33% (Paul and Routray 2010: 96). Even a recent study after Cyclone Roanu shows that the evacuation rate was less than 16% (Rezwana 2017). Among the overall number of fatalities, Ikeda (1995) sees a gender effect in terms of loss of lives.

According to Rezwana (2018), women and children do not know from where and which way to follow as they are mostly dependent on the male head of household about any evacuation issues during cyclones. Again, it has been seen that women tend to put the lives of their children and family members before their own. They may grow too busy saving their children and even livestock and household items before making arrangements for themselves. Therefore, they can end up being the biggest sufferers during a crisis (Collins and Edris, 2011).

3 Voices of Women and Men Facing Climate-Induced Disasters

For our work, we have found most of the women from Rayenda union, a highly disaster-prone area, are indifferent about evacuation plans. Some of them are still unaware about what preparation should be taken during climate-induced emergencies. As women and children suffer the most due to disasters, and still there is lack of implementation of evacuation planning, the participation of YouthMappers using open data like OpenStreetMap (OSM) might be helpful to increase the evacuation rate, especially among women as well as their resilience against natural disasters like cyclones.

We interviewed a woman from Rayenda Union named Rumi. She told us, before Sidr, there were not enough cyclone shelters in the union. However, after the deadly consequences of Sidr, authorities tried to build more cyclone shelters in the area. We asked her about plans to evacuate and how she would know, when and if she has to go to a cyclone shelter. She told us, she would just know with a smiling face and mentioned that she will wait for her husband and ask neighbors about the situation. We then asked her, "What will you do if your husband isn't at home?" She was surprised at the question and said, "I may wait for him."

We also interviewed one of the survivors of Cyclone Sidr, Afzal kha, a fisherman from Rayenda union, Shrankhola, Bagerhat, Bangladesh who lived near the riverbank of Boleshwar river with his wife and two baby girls. Usually, Afzal kha catches fish from the Bay of Bengal, to sell them in the local market. Afzal kha was stuck out at sea while his family perished at the terrible hand of Cyclone Sidr. His wife was caught dead on a date tree wrapped by her Sharee, a traditional Bengali cloth. His two baby girls' dead bodies were found in the paddy field. After a couple of days, Afzal khan returned from the sea and found this dreadful consequence of his family. The whole family lost 26 members and most of them were women and children.

Afzal kha still lives along the same riverbank with all the difficulties. We were there to interview one of the female members of his neighbor's house. As he told us during the interview, "I lost my family during Sidr, and the key scenario is still the same despite having many cyclone shelters in our village. Women and children are still not trained enough to take steps forward during disasters. People are not as afraid of cyclones nowadays because most of them happen in daylight. If it somehow attacks in the night, that would be more deadly." Every year, during cyclone season, women and children face the same problem and are still not getting enough training or resources to prepare against climate-induced disasters in Rayenda Union (Figs. 22.1 and 22.2).

4 Integration of Open Data and Youth Participation in Emergency Response Management Through YouthMappers

The integration of open, publicly created, and publicly available data and YouthMappers can enhance the capacity-building program for women and children which promises to reduce the vulnerabilities of women. Technology-based capacity development could be a door for women to build resilience. The following case study is showing that women in the local areas of Bangladesh need capacity building and the traditional process has not been working. The combination of open data and youth could be a revolutionary change inside the traditional Emergency Management System (EMS) of the country.

Fig. 22.1 A roundtable discussion with women from Rayenda union, a highly disaster-prone area, informs YouthMappers work in Bangladesh for evacuation mapping

Fig. 22.2 Evacuation from cyclones, flooding and hurricanes requires planning, mapping, and training with families in communities to build knowledge and resilience

The growing availability of mobile applications and technology-based emergency management in cities has had a remarkable impact on the overall emergency management system. However, remote populations of Bangladesh are still facing problems in navigating mobile apps or technology-based approaches. These coincide in the remaining places where women are deprived of higher education, health facilities, and basic needs. In this regard, the technology-based emergency management approaches are quite impractical in the context of a remote area in Bangladesh. Bridging between technology and manual approaches needs to be emphasized to obtain better outcomes in primary level preparedness and the open data and participation of YouthMappers aim to make this happen.

4.1   YouthMappers and OSM

YouthMappers introduces voluntary geospatial initiatives building on OpenStreetMap (OSM) as a platform, where women are leading information and analysis programs, such as under the YouthMappers campaigns of EverywhereSheMaps and LetGirlsMap. These campaigns are being regarded as an active way to engage young women in not only collecting, gathering, and analyzing data but also encouraging more women in this profession so that we can make systems change, which further may be used in mapping structures, identifying evacuation paths and engaging more women in this process to abridge women's vulnerabilities to potential disaster risk.

Cyclones are considered one of the major natural hazards in Bangladesh. Every year, many people die or lose their livelihoods in the coastal and offshore islands of Bangladesh due to cyclones and storm surges. The government has taken many initiatives such as the National Plan for Disaster Management (NPDM) to build resilience for sustainable human development and reduce the vulnerabilities of communities from disaster-prone areas.

But most of the people from remote areas in Bangladesh are not as socioeconomically well

off to facilitate the technology-based emergency management system. It is not possible to fully prevent natural hazards but preparedness can mean less damage. Upgrading awareness-raising initiatives according to the demand and situation can curtail the damages even further. Considering the socioeconomic conditions of remote area populations, manual approaches with proper guidance are needed. An awareness program and training designed in the native language with the help of OSM platform might change the situation through the leadership of YouthMappers.

Under the YouthMappers Research Fellowship program, we have experience working in a cyclone-prone coastal area, Rayenda, Sharankhola, Bangladesh. Rayenda inhabitants, vulnerable to frequent cyclones, are more concerned about their livelihoods than preparing themselves for uncertain natural hazards. Studies found that, in a household, women consider their cattle's safety before taking decisions on evacuating themselves to the shelters. Apart from that, lack of interest in attending awareness-raising programs has been perceived among women and children. During our fellowship research, we tried to figure out why women and children are more vulnerable and how to help them regarding the preparedness for cyclone disasters.

One woman from the union described her situation during Cyclone Aila. Though she managed to reach the cyclone shelter, she arrived asking "Where is the nearest cyclone shelter?" She had the idea to seek out safe shelter but did not know the exact name or location, and that made her more confused. She told us, "I managed to save myself because Aila (another major cyclone in Bangladesh) stuck during daylight, otherwise me and my family would have died during cyclone Aila."

Under the program, we trained some women from the union with a Bengali (Native Language of Bangladesh) Version of Evacuation Planning created by using OpenStreetMap (OSM) data. Figures 22.1 and 22.2 show the training initiated by the YouthMappers with collaboration of local authorities. Figures 22.3 and 22.4 exhibit part of our Bengali Version Evacuation Planning for Rayenda, Sharankhola Bangladesh.

Now, YouthMappers chapters from different universities in Bangladesh can play a vital role to continue this process and expand to other areas. Our training program covered a small portion of Upazila. Considering the whole coastal area, YouthMappers could put their eye on mapping all cyclone shelters and finding out the shortest routes. We can exhibit it in a Bengali version which is easier to explain to local people.

Arranging such a program is not easy, yet we must try. Contacting local government could be a plus point in this regard as the government has many programs to provide proper guidelines and building resilience. It will not only leverage the conduction of the training but also catches the attention of government authorities to take into account how women are more vulnerable and differently affected during disasters. Making preparedness decisions, and managing everything to evacuate on time is rather an issue of women's empowerment. The more women we train with proper guidance, the better feedback on preparedness we can get. It could arm a new generation with environmental knowledge as mothers are the first teachers for the children. Once we succeed in training women, we are on our way to ensure family safety (Figs. 22.3 and 22.4).

4.2   Climate Change Education and Emergency Response Awareness

Environmental education and awareness programs on natural disasters can easily reduce the possible loss as well as make people more resilient for an uncertain future of climate-induced disasters. In developing countries like Bangladesh, a native language-based preparedness program can play a vital role in inaugurating climate change and its impact. Why native language? In most remote areas, poverty devours the possibilities of getting proper education, and women and girls are the main victims. To introduce climate change and its impact we need to understand the present educational conditions of the region and take measures according to these needs.

4.3   Youth Advocating Climate Action That Inspires Us

With the UN-led celebration of the International Year of Youth from August 2010 to August 2011, there was a renewed interest in young people and the vital role they can play in important issues, such as disaster risk reduction (DRR). This cele-

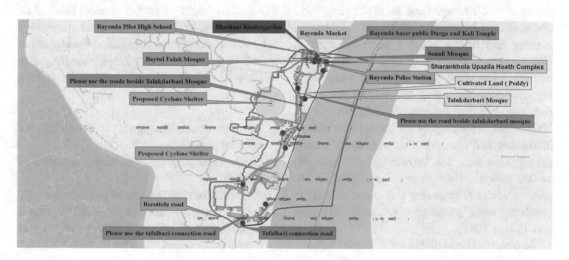

Fig. 22.3 A final evacuation plan map has been produced for Rayenda, Sharankhola Bangladesh

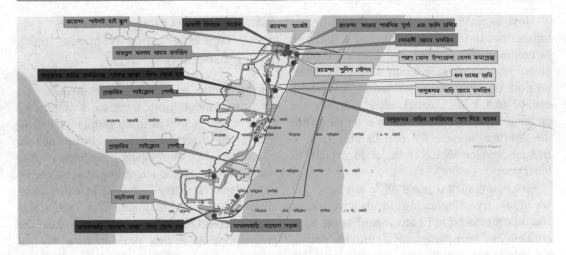

Fig. 22.4 The Bengali version is shown of the resulting Evacuation Plan for Rayenda, Sharankhola Bangladesh

bration not only created a major framework but also inspired us to work in the field of disaster risk management with the understanding that any comprehensive disaster risk management can be improved when it is incorporated with dynamic and lively young women in its programs. The first campaign of the United Nations International Strategy for Disaster Reduction (UNISDR) designated as "Disaster Prevention, Education, and Youth" (UNISDR 2000), focused on the eminence of recognizing youth participation.

Young women are enthusiastic when it comes to sharing and contextualizing knowledge about disaster risk management incorporating local and scientific knowledge which aids in decision making and brings about a potential change. However, the younger generation, especially women, have often been neglected while sharing and communicating knowledge through a risk communication network (Mitchell et al. 2008). But we can play an impressive role in increasing and disseminating awareness among people who lack this knowledge, for instance, hazard education programs (Finnis et al. 2010) science clubs (Fernandez and Shaw 2014), or youth-centered participatory videos (Haynes and Tanner 2015).

At a regional and global level, YouthMappers, a network of students, takes initiatives like EverywhereSheMaps, LetGirlsMap, Leadership Program, and Research Fellowships to create leaders from different regions of the world. It also focuses on the women leading programs to impact on the development of women around the world by giving a platform to the young women which has been experiencing a tremendous growth in the sector of young women development.

As the repercussions of climate change are not distributed homogeneously, some would have to confront a direct impact while some would have an indirect impact slowly. But we need to walk this arduous path together to create changes and make our world live up to the goals of the UN SDGs. How might we further connect SDG 13 on climate action, and SDG 5 on gender equality? Why are young women considered the key actors regarding the discussion of Emergency Response in Bangladesh? There are several opportunities, such as to:

- Raise awareness among women from different cultural views
- Grow understanding regarding the issues of women
- Act with considerably more reach in the context of village women, as most of the women in villages feel uncomfortable to talk freely with men in Bangladesh because of their religious views
- Promote how young women can be examples of decision-making for other women

Youth climate activists are trying to transform the world by putting their step outside their horizon, and that inspires us to work under the umbrella of YouthMappers to make this initiative to bring changes along our horizon as well.

future of our country, albeit a climate insecure future. Our initiatives like open data mapping with the participation of young women to alleviate our vulnerabilities toward disasters will make us more resilient.

5 Conclusion

Gender equality and climate change are inevitably considered to be among the major concerns to ensure sustainable development. There are close relationships among gender impact, climate change, youth, and development which need to be understood to make sustainable change in the world.

Bangladesh has been working on disaster management for a long time but gender aspects of vulnerability still need more attention. Integration of gender perspectives with climate change and disaster risk management can help ameliorate vulnerabilities as well as it can be considered as the stairway to sustainable development. Moreover, educating women can be another approach toward change because intensive preparedness plans for disaster risk will not be effective if women are not educated to understand them, and trained young women can be an asset in this regard. Different types of approaches include native language-based evacuation plans along with local community engagement. As our experience shows, open data platforms like OSM are a convenient technology for people, and if young women could get training, they easily can identify their evacuation path. We are working to engage more women in this platform so that they can be a part of this intervention, including aims to create a women's club in remote areas where the awareness training would be given on a monthly basis. Gendered impact can be addressed by the open mapping explained through native language, where young women themselves would lead this process to profoundly reduce the disproportionate vulnerabilities of women compared to men.

We, young women, have been experiencing tremendous growth and contributing toward the

References

Collins EA, Edris A (2011) Cyclone disaster vulnerability and response experiences in coastal Bangladesh. Disasters 34(4):931–954. https://doi.org/10.1111/j.0361-3666.2010.01176.x

Fernandez G, Shaw R (2014) Participation of youth councils in local-level HFA implementation in Infanta and Makati, Philippines and its policy implications. Risk Hazards Crisis Public Policy 5(3):259–278

Finnis KK, Johnston DM, Ronan KR, White JD (2010) Hazard perceptions and preparedness of Taranaki youth. Dis Prev Manag

Haynes K, Tanner TM (2015) Empowering young people and strengthening resilience: youth-centered participatory video as a tool for climate change adaptation and disaster risk reduction. Children's Geogr 13(3):357–371

Ikeda K, (1995) Gender differences in human loss and vulnerability in natural disasters: a casestudy from Bangladesh. Indian Journal of Gender Studies 2, Page: 171–193

Alam K, Md., Rahman H (2014) Women in natural disasters: A case study from southern coastal region of Bangladesh. International Journal of Disaster Risk Reduction 868–882. https://doi.org/10.1016/j.ijdrr.2014.01.003

Kibria G (2016) Why are women more vulnerable to climate change? Climate change implications on women with reference to food, water, energy, health, and disaster security. Research Gate Online Publication. 10p. https://doi.org/10.13140/RG.2.1.2577.9683

Mallick B (2014) Cyclone shelters and their locational suitability: an empirical analysis from coastal Bangladesh. Disasters 38(3):654–671. https://doi.org/10.1111/disa.12062

Mallick B, Vogt J (2009) Analysis of disaster vulnerability for sustainable coastal zone management: a case of Cyclone Sidr 2007 in Bangladesh. IOP Conf Ser Earth Environ Sci 6. https://doi.org/10.1088/1755-1307/6/5/352029

Mitchell T, Haynes K, Hall N, Choong W, Oven K (2008) The roles of children and youth in communicating disaster risk. Child Youth Environ 18(1):254–279

Paul SK, Routray JK (2010) Household response to cyclone and induced surge in coastal Bangladesh: coping strategies and explanatory variables. Natural Hazards 57:477–499

Rezwana N (2017) Refusing safe shelter during cyclone warning: dilemma or limitations? A case study of recent Cyclone Roanu in Bangladesh. Unpublished paper, DRTMC, Dhaka University

Rezwana N (2018) Disasters, gender and access to healthcare: women in coastal Bangladesh. Routledge, London

UNISDR (United Nations International Strategy for Disaster Risk Reduction) (2000) Mobilizing LocalCommunities in Reducing Disasters. ISDR, Geneva.

Sustainable Development in Asia Pacific and the Role of Mapping for Women

23

Celina Agaton

Abstract

In the archipelago of the Philippines, surrounded by the Pacific Ocean, sustainable development takes many forms, but livelihoods are always shaped by our ocean surroundings. This chapter explores a collection of research that addresses concerns that emerge when advancing SDG 14, to conserve and sustainably use the oceans, seas, and marine resources for sustainable development, with an eye toward the particular role of women in the creation of development, SDG 5. This includes their household contributions, as well as what they give to promote knowledge, policy, and programming and how the household and enterprise needs of women are critical to life in this region. The use of free and open-source tools through the Open Knowledge Kit Regeneration Program presents universal benefits to address the combined devastation of the pandemic, climate change, and of marginalized communities, especially women. Easy to use data collection, analyses, and modeling tools remove barriers to participation and the creation of knowledge.

Keywords

Pacific · Sustainable development · Gender equity · Transportation · Safety · Mobility · Violence · Philippines

1 Women's Sustainable Development in the Asia Pacific Region

Across the Asia Pacific, micro, small, and medium enterprises (MSMEs) make up the majority of economies, up to 99.5%. Although 60% of MSMEs are owned or led by women, studies show that women have the least access to household essentials like phones, water, paved roads, and Internet, but also essential business infrastructure like logistics, equipment, processing facilities, training, and finance—everything necessary to succeed in life in general, but in the space of our archipelago, these infrastructures take on a new meaning. Meanwhile, although we

C. Agaton (✉)
MapPH, Manila, Philippines

P. Solís, M. Zeballos (eds.), *Open Mapping towards Sustainable Development Goals*, Sustainable Development Goals Series, https://doi.org/10.1007/978-3-031-05182-1_23

know that women are responsible for the lion's share of household tasks, their significant economic contributions should make their needs a priority, and yet women remain mostly invisible to public and private sector planning and investments. There are still many data gaps in the multitude of ways women are crucial to the well-being of their communities here in the Pacific Region (Australian Aid, 2017; FAO, 2011; Philippine Department of Trade and Industry, 2019).

Between land and ocean, here the pandemic and climate crisis present unique opportunities to make the critical shift to local data collection and lifesaving local employment models. With travel restrictions, Indigenous and local women can safely participate in data collection and analyses using new, low-cost, low-equipment methods, trained remotely online using Facebook Messenger's free messaging and video platform. The interest and capacity to participate in data collection and knowledge sharing are the only prerequisites for participation. The higher value placed on local knowledge versus technical ability changes the outcomes and increases the quality of the recruitment and retention of new mappers. The priority is to help communities make the most informed decisions about their own communities for the long-term, versus contract revenues.

Many of the communities have experienced stress and trauma from violence and conflict, with illegal activities escalating during the pandemic due to the loss of patrols. The devastating impacts of the pandemic on local organizations further allow international organizations to increase their Western influence over communities and institutionalize their bias. Lastly, along with the increasing impact of the pandemic and climate change, many communities here in the Pacific deal with multiple hazards on an annual basis (Fig. 23.1).

2 The Gender Gap in Mobility

In 2019, as part of a World Bank and Australian Department of Foreign Affairs and Trade Strengthening Road Connectivity to Support Agriculture and Regional Development in Mindanao project, we completed a gender gap mapping study in the conflict region of Mindanao, Philippines. The household survey was designed to obtain the experiences and perceptions of women and men with regard to the impact of current road and public transportation conditions on their health, well-being, and livelihood in the six provinces in 18 municipalities and 54 barangays or villages classified as rural agricultural commu-

Fig. 23.1 Poor road infrastructure makes it extremely difficult for women to complete daily household and livelihood tasks, especially because they travel mostly on foot

nities surrounding the three major ports in Mindanao, namely Davao City, Cagayan de Oro, and General Santos. The informants were family members not less than 18 years of age at the time of the interview. The use of mobility provides a tangible, quantifiable metric with which to analyze differences between the lives of women versus men.

This study utilized a descriptive research method in which the identified variables were measured as they existed. Tests of significance were utilized to answer three questions: (a) How big is the gender gap if there is? (b) Is the gender gap significant? and (c) Is gender the reason for the gap? The results of the study will serve as a baseline study and shall be compared with the findings of the series of studies that may be periodically conducted in the future. The minimum required sample size was determined using Slovin's formula. Multistage sampling and systematic sampling techniques were utilized to ensure the representativeness of the final sample. The OpenStreetMap free and open-source platform's six satellite imagery sources and Google Maps were utilized to determine the existence and non-existence of roads and the location of the barangays or villages from the población or town center. The computed sample totaled 540, comprising 277 males and 263 females of rural households. However, the survey realized only a total of 512 respondent households composed of 257 males and 255 females. The response rate was 94.4.

A large, extended family characterized the households covered in the study. In addition to the members of the nuclear family (husband, wife, sons, and daughters) are other household members who are either the couple's parents, in-laws, brothers, sisters, nieces, nephews, cousins, and grandparents. The data also disclosed that there are more female household members (51.1%) than their male counterparts (48.9%) although the difference is not substantial. The average household size is five (5) which is not consistent with the national average of 4.3, based on the Philippine Statistics Authority 2015 Census of Population. Consistent with the patri-

archal setup of a Filipino family, the majority of the households are mostly headed by the males who are husbands/fathers. Nonetheless, about one in every 10 households is headed by females who are mostly widows. There is a high proportion of female spouses belonging to the fertility age.

Household heads and spouses have low educational attainment. The low level of education may be dictated by poverty, among which are geographical location, the condition of the roads, and access to public transportation. Although gender gap in education exhibited by the household heads and spouses may not be significant, it is worth noting that 2 in every 10 women pursued college education compared to their male counterparts. Concerning Technical-Vocational schooling, more male household heads/spouses than their female counterparts took the skills-focused courses. Another feature of the households is that they are culturally and linguistically diverse. Coming from the same ethnic background tends to be the norm when it comes to marriage. Nonetheless, many seemed to be more open to intermarry with someone coming from another ethnic background; notably a spouse who is indigenous to the place. The language spoken at home depends on the ethnic origin of the household head and spouse. However, due to the deep influence of the migrant settlers, the respondents of the study also speak, or at least, understand well the language of the migrant settlers.

3   Key Findings and Analysis

3.1   Transportation Access and Mobility for Household Essentials

3.1.1 *Chores considered traditionally as "men's work" that entails heavy lifting, long walking and time spent, and food production are generally performed by the male family members while women's household tasks are associated with housework, childcare, school care, and medical care for the family.*

This is exemplified by the huge proportion of the males doing the following household tasks: fetch water, gather/collect firewood, go to the rice/corn mill, purchase cooking gas, gather food in the forest, and take care of the crops. The tasks of the women, on the other hand, involved bringing the children to school and back to the house almost daily. Women going to the sari-sari or convenience store daily more so than the men (16.6% v 6.25%) and to the wet market once a week (20.3% v 13.9%). It is obvious from Table 1 that women's productive work is also associated with raising a few heads of livestock (pigs, goats) and poultry (chicken) in the backyard. Women are also predisposed to visiting relatives, neighbors, and friends more than men.

3.1.2 *Women travel longer distances, longer travel times at higher travel costs to perform household tasks.*

Women may be homebound because of responsibilities like food preparation, laundry and house cleaning, and child/school/medical care of their family; however, they travel more distance (kilometers), spend more travel costs, and spend more travel time than the men. Also, the women are taking on more roles in petty vending/ microenterprise and microservices activities thereby contributing to the households' income. Almost the same proportion of men and women do household chores on a daily and weekly basis.

Significance tests were made to determine whether or not the number of times a specific household essential/task is performed (daily, per week, or monthly) varied by gender. These tests were particularly applied to six household essentials namely: (a) fetch drinking water; (b) fetch water for domestic use; (c) gather firewood; (d) buy LPG/gasul; (e) fish in the river/creek/sea; and (f) gather food in the forests.

3.1.3 *The prevailing road condition makes it difficult for people to get to their destination, making it extremely difficult to accomplish their household tasks.*

They described the condition of the road as steep, narrow, too grassy, filled with stone rocks, muddy and slippery when it rains.

3.1.4 *The location of household essentials determines the gender of the doer.*

The farther the location of the household essentials the men are the doers; in contrast, the closer the location the doers are the women. This is illustrated by the Bukidnon households, where 22.9% of the males reported to be the doers, while 52.8% are women for household essentials that are located less than a kilometer away from their respective homes. This holds also with the households in Sarangani and South Cotabato. This could be explained by the fact that domestic chores (house cleaning, washing of clothes, food preparation, childcare) are located within the house premises and are traditionally assigned to the women.

Two significant differences were found in terms of distance from the house to the location of the household essentials/tasks. The location of the area where drinking water is fetched varied significantly by gender. Women, in general, live farther from the location where drinking water is fetched (M = 649.17 meters, SE = 149.77) compared to the men (M = 281.09 meters, SE = 50.26). This difference, −368.08 meters, is significant $[t(180.83) = −2.28, p<.05]$.

3.2   Transportation Access and Mobility for Healthcare Services

3.2.1 *Women take on more tasks related to healthcare services compared to men.*

The traditional role of women associated with healthcare services is evident in the study. All tasks related to access and mobility for healthcare services are completed by the women. This is evidenced in the data presented in the table below (Table 4). Considering the devastating impact of the current pandemic, these further aggravate the burdens on women.

3.2.2 *More men than women travel more distance but the latter spend more travel time and travel cost for healthcare services.*

Although the men (11 kilometers) travel more in terms of distance than women (10.04 kilometers) the difference (1.33 kilometers), however, is not substantial. The women spend more travel time (54.52 min v 49.34 min) and travel costs (98.08 pesos v 91.27 pesos) than men (Table 6).

3.2.3 *More sick male family members are brought to the hospital than the women patients while more men than women decide on emergency medical transport.*

For emergency medical services, male patients are more likely to be brought to the hospital than female patients, and female patients are more likely brought to the health center than their male counterparts. The males are also more likely the decision-maker as to the emergency medical transport than their female counterparts (Table 6).

Health Clinic Female respondents visit the health center more frequently (M = 1.24, SE = .08) than the male respondents (M = 1.13, SE = .12); the difference, however, is not significant. There is a significant association found between the gender of the person who goes to the health center (Rural Health Unit or Barangay Health Center) and the mode of transportation used to reach the nearest facility. Women are more likely to walk than men to reach the nearest facility (59.7% vs 42.1%) while men are more likely to use a motorized vehicle than women (57.9% vs 40.3%). These responses are significantly different.

Across gender, women are significantly more likely to use tertiary and national roads than men.

Mental Health There is a significant association between the two variables, where the distance of

the residence of women is significantly farther from the nearest mental health facility (mean rank = 18.07 kilometers) compared with men (mean rank = 10.94 kilometers).

Pharmacy Based on this survey, women live farther from the nearest pharmacy (M = 11.71 kilometers, SD = 1.90) than men (M = 9.66 kilometers, SD = .87). However, this difference (2.04 kilometers) is *not significant*. Results also showed that women have more frequent visits to the nearest pharmacy (M = 1.12, SD = .06) than men (M = 1.10, SD = .08) but this difference (−.02) is not significant. Similarly, women have longer travel time to the nearest pharmacy (M = 58.29 min, SD = 3.16) than men (M = 51.35 min, SD = 4.28) but the difference (−6.94 min) is *not significant*. In terms of transportation cost, men reported spending more (M = 100.77 pesos, SD = 12.66) than women (M = 99.20 pesos, SD = 8.02) but this difference (1.56 pesos) is *not significant*.

Interestingly, a significant association is found between the gender of the person who usually avails of pharmaceutical services and the mode of transportation used to reach the nearest facility. Across gender, women are more likely to walk than men to reach the nearest facility (14.0% versus 7.2%) while men are more likely to use motorized vehicle than women (92.8% versus 86.0%). These responses are *significantly different*. Further, a *significant relationship* is also found between the gender of the person who usually avails of services in the nearest pharmacy and the type of road used to reach the nearest facility. Across gender, women are more likely to use tertiary, secondary, and national roads than men.

Dental Services In this study, the residence of men is farther from the nearest dental facility (M = 13.78 kms, SE = 1.64) than the women's (M = 8.29kms, SE = 1.08). This difference, 5.49 kilometers, is significant. For the other scale variables, the men have higher mean scores than women, but the difference is *not significant*. Interestingly, however, a significant association is found between the gender of the person who usually avails of dental services and mode of transportation used to reach the nearest facility. Across gender, women are more likely to walk than men to reach the nearest facility (17.8% versus 2.3%) while men are more likely to use motorized vehicle than women (97.7% versus 82.2%); these responses are *significantly* different. However, the relationship between gender and the type of road used to reach the nearest dental facility is not significant.

Emergency Medical Transportation/Emergency Medical Services In this study, men registered higher mean scores than women for the following variables: distance from residence to the nearest facility, number of times the person visited the facility, and cost of transportation. The difference is not significant, however. Women had higher mean scores for transportation cost than men, but the difference is also *not significant*. No significant association is also found between gender and mode of transportation to reach the nearest facility, on one hand, and the type of road used to reach the nearest facility, on the other hand.

3.3 Transportation Access and Mobility for Livelihood and Microenterprise Purposes

3.3.1 *Four in every 10 household respondents are engaged in agriculture.*

Note that while the research areas are considered agricultural communities, only about 41% of the household respondents are engaged in agriculture. Major crops raised include corn, rice (upland and lowland), banana, coconut, and abaca. Their secondary crops are pepper, cacao, root crops and vegetables, and fruit trees. Livestock and poultry are raised in the backyard by a few households. Most farms are located within the barangay, to which they just go on foot (males = 70.2%; females = 65.6%) while others ride on a single

motorcycle to reach the farms. For those who have farm animals (horse, carabao, cow) and turtle tractor (*bao-bao*), the men, in particular, use these as their modes of transportation. The roads going to the farms are that of the unclassified type (male = 75.4%; female = 77.5%) and are mostly not in good condition, especially during the rainy season.

3.3.2 *Generally, they only have one source of livelihood; others, though, have diversified their sources of income.*

Eight in every 10 households have only one source of income. They generally subsist on agriculture-related activities. They are the crop producers (corn, rice, vegetables, coconut, abaca, banana, coffee, cacao, peanut, rootcrops, vegetables, and other unspecified crops) and livestock and poultry raisers. Nonetheless, 3 in every 10 households have diversified their sources of income by taking in jobs related to microservices (construction worker, farm laborer, welder, *habal-habal* or modified motorcycle driver, carpenter, public servant, employee, teacher, salesgirl, manicure, pedicure, and wedding organizer, among others). Others have additional income derived from micro-enterprise/petty vending activities (sari-sari store, buy and sell, *kakanin/* ice cream/ice drop vending). Only a few are engaged in fishery-related livelihood. Homebased livelihood includes a sari-sari store, manicure and pedicure, dressmaking, and hog/poultry raising (Fig. 23.2).

3.3.3 *More women are employed in microenterprise and micro-services than in agriculture.*

While agriculture is dominated by men, in contrast, petty vending/microenterprise are noticeably dominated by women. They manage a sari-sari store, process food, vend food delicacies (*kakanin*), and handicraft, and engage in the buy-and-sell business of used clothes (*ukay-ukay*), and raise livestock and poultry in the backyard. It is also noticeable that more women than men are employed in micro-services activities.

Province	Agri-farming		Fishery		Petty vending/ Micro-enterprise		Micro-services[*]	
	Female	Male	Female	Male	Female	Male	Female	Male
Bukidnon	22.2	8.3	0.0	0.0	13.9	13.9	30.6	11.1
Davao Oriental	20.0	22.9	0.0	2.9	14.3	8.6	8.6	22.9
Davao del Sur	13.5	35.1	0.0	0.0	20.3	21.6	4.1	5.4
Misamis Oriental	13.4	22.0	0.0	1.2	6.1	3.7	22.0	31.7
Sarangani	18.8	27.5	1.3	8.8	5.0	5.0	21.3	12.5
South Cotabato	14.8	22.7	1.1	0.0	1.1	1.1	34.1	25.0
Total average, %	16.2	24.6	0.5	2.3	8.9	8.1	20.8	18.7
Total	40.8		2.8		17.0		47.3	

Includes formal employment like teacher, government employee, government official, salesgirl, etc.

Fig. 23.2 Compared to men in agriculture, more women were employed in micro-enterprise and microservices

Fig. 23.3 Compared to men, women took on majority of tasks related to banking, finance, government and business services

Activities related to agriculture livelihood services	In percent	
	Female	Male
Procurement of fertilizer and pesticide	31.6	68.4
Bring farm produce to the processing center	23.1	76.9
Bring farm produce to baf sakan / drop-off center	18.9	81.1
Deliver farm produce to the market	22.4	77.6
Deliver farm produce directly to the buyer	17.2	82.8
Go to the bank/financing or lending institutions	78.6	21.4
Go to municipal, provincial or city hall	58.1	41.9
Logistics	50.0	50.0
Go to the Dept of Trade & Industry (DTI)	100.0	0.0

3.3.4 *Women's role in activities related to agricultural livelihood services can be said as minimal.*

This is evidenced by the little role the women have in the procurement of farm inputs like fertilizer and pesticides as there are more men than women (68.4% v 31.6%) assigned to do the task, as shown in Table 8. The nominal role of women versus the dominant role of the men is further highlighted in the chores of bringing the farm produce to the processing center (76.9% v 23.1%); to the *bagsakan* or drop-off center (81.1% v 18.9%); to the market (77.6% v 22.4%); and bringing their farm produce to the direct buyer (82.8% v 17.2%).

Interestingly, the activities in which the women are dominant are in going to the bank/financing or micro-lending institutions (78.6% v 21.4%), and in going to the municipal/city/provincial hall (58.1% v 41.9%). Presumably, the women's going to the bank/financial/micro-

lending institutions is for financial assistance for the family's agricultural production activities and/or for the family's domestic needs (e.g., children's education, medical care) (Fig. 23.3).

3.3.5 *Women have less access to production support services than men.*

Only 25% of peasant women have access to production support services. The data further shows that the farm technology is the least the women have access to, like hand tractor, thresher, weeder, and harvester thus implying a backward technology in crop production. This is compounded by having less access to the other production support services like production capital, irrigation, seeds/seedlings, fertilizer and pesticides, and pest management.

3.3.6 *Males spent more travel time than the females going to the farm, but the females spent more on travel cost than the males.*

On average, the males spent 1 h and 7 min compared to the females' 48 min in going to the farm. However, the females could reach the farm faster than the males; the former's walking time is 44 min compared to the males' 57 min presumably due to the distance of the farms to their homes. In terms of travel cost, the women spent an average of P140 while the men P58 (Table 9).

3.3.7 *Location of agriculture livelihood services is paved with unclassified and/or tertiary roads.*

Data shows that the farmers have to traverse the unclassified road and/or tertiary road when bringing the farm produce to the processing center, *bagsakan*/drop-off center, to the market, and bringing the farm produce directly to the buyer. The condition of the road would also depend on the season, and whether the activity is being done during the dry or wet season. Walking/head loading/backloading the farm produce from the farm to the location of the agriculture livelihood services is done if these are located within the barangay. Hiring a motorcycle (*habal-habal*), multicab, jeep, or elf-truck is done depending on the weight and number of sacks to be transported. Outside the barangay, the road typifies that of the secondary and national roads. Travel cost depends on the distance but ranges from as low as P20 to as high as P300. Travel time outside the barangay ranges from 30 min to 2 h.

3.3.8 *Six in every 10 household respondents are engaged in non-agricultural activities.*

Table 10 shows that almost 60% of the household respondents have sources of income that are not agriculture related. Furthermore, there are more women whose sources of income are derived from petty vending/micro-entrepreneurial activities; however, there are more men than women whose income is sourced from selling their labor or services (Fig. 23.4).

3.3.9 *Overall, women have a longer average travel time and higher travel costs than men.*

As shown in the table, the women, on average, spend 48 min of travel time compared to the men's 46 min. These are particularly demonstrated by the women from Davao Oriental, Bukidnon, and Davao del Sur. They incur 96.03 pesos per trip higher than the men's 79.09 pesos, as revealed by the data from Bukidnon, Davao Oriental, Davao del Sur, and Misamis Oriental. The women are mostly involved in the tasks related to petty vending/microenterprise and micro-services which require them to frequently take trips either by walking or motorized riding (single motorcycle).

3.3.10 *More women than men take the unclassified and tertiary types of roads.*

The data show that more women than men take the unclassified and tertiary types of roads in going to the location of their livelihood. The table further shows that while more men than women walk on the unpaved road going to the farms and fishing grounds, there are more women involved in microenterprise and micro-services livelihood and walk traversing the unclassified and tertiary road to the location of their livelihood in conducting their economic activities (Table 12).

3.4 Transportation Access and Mobility for Education

3.4.1 *Children's education is affected by the current road and transport situation.*

More than half of the respondents affirmed that the road and transport situations in their community have created barriers or challenges to their children's education, in which, according to them, the female students (54.1%) are more affected than the male students (22%). But female or male they are all affected as school children.

Fig. 23.4 For livelihood tasks, women spent longer travel times and costs, and traveled mostly on local roads compared to men

Table 10. Non-agriculture based main livelihood or source of income by gender (multiple responses), in present

Livelihood	Male	Female	Total
Fishing	1.5	1.3	1.4
Petty vending/micro-entrepreneur	7.1	15.0	10.7
Micro-services	48.2	40.9	44.9
Employment/formal sector*	0.5	4.4	2.2

*Includes formal employment like teacher, government employee, government official, salesgirl, etc.

Table 11.1 Travel time and travel costs by gender

Province	Total travel time (minutes)		Total travel costs (pesos)	
	Female	Male	Female	Male
Bukidnon	53.2	39.3	130.00	123.00
Davao Oriental	72.7	40.0	116.67	16.67
Davao del Sur	53.2	40.3	130.00	123.00
Misamis Oriental	54.5	60.0	114.50	44.86
Sarangani	25.0	50.0	35.00	92.00
South Cotabato	31.2	44.5	50.00	75.00
Total	48.3	45.6	96.03	79.09

Table 11.2 Travel cost, in pesos

Province	Low	High	Average
Bukidnon	10	300	70
Davao Oriental	70	70	70
Davao del Sur	20	300	125
Misamis Oriental	10	300	70
Sarangani	10	30	67.50
South Cotabato	10	150	69.16

Table 12. Percent of respondents by gender and type of livelihood and type of road

Livelihood	Tertiary		Secondary		National		Unclassified	
	F	M	F	M	F	M	F	M
Agri-farming	15.9	15.2	6.6	5.3	0.0	4.6	27.8	24.5
Fishery	0.0	0.0	0.0	0.0	0.0	0.0	90.0	9.1
Microenterprise	30.2	18.9	11.3	3.8	5.7	0.0	20.8	9.4
Micro-services	10.8	10.8	10.8	10.8	2.4	6.6	30.1	17.5
Total	15.2	13.4	8.9	7.3	1.8	4.7	27.3	21.3

F=female; M=male

Table 13. Percent of respondents by gender who foresee an increase in their income with improved transporatation options

Province	Yes		No		Not sure	
	F	M	F	M	F	M
Bukidnon	50.0	15.5	5.2	1.7	22.4	5.2
Davao Oriental	49.1	43.5	0.0	1.9	0.0	3.8
Davao del Sur	40.0	41.1	1.1	2.2	6.7	8.9
Misamis Oriental	36.2	54.3	0.0	0.0	5.3	3.2
Sarangani	36.9	32.1	2.4	1.2	11.9	15.5
South Cotabato	44.0	52.0	2.7	1.3	0.0	0.0
Average %	42.7	39.7	1.9	1.4	7.7	6.1

F=female; M=male

3.4.2 *Girls have limited mobility or have mobility restrictions and are mostly accompanied by a male family member if they go out.*

Four in every 10 respondents said that girls in the family have limited mobility "all the time" and 3 in every 10 respondents also said "sometimes" while another 3 in every 10 respondents said that girls in the family have no mobility restrictions. If girls go out, they are accompanied by a male family member (54.9%), a female family member (27.2%), or any male or female family member (17.8%).

3.5   Safety and Security

3.5.1 *Adult men and teenage boys are mostly involved in road-related accidents.*

Adult men are almost always involved in road-related accidents in the study areas as compared to adult women (Table 16). The same is true with teenage boys being primarily involved as compared with teenage girls. Young boys and young girls also figured in road-related accidents. It can be assumed that these road-related accidents are caused by the kind of road and the kind of users and uses that ply these roads. Asked about any concerns regarding safety when using transportation, both women and men identified over-speeding of vehicles, overloading of passengers, and overloading of cargo (farm produce, etc.). Unlit roads are also mentioned which implies many accidents do happen in the evening. Poorly maintained vehicles and poor traffic signaling in addition to roads with obstructed views/foliage/overgrowth contributed much to the road-related accidents. Since the most common transport vehicle in the rural barangays is the *habal-habal or modified motorcycle*, the commuters are also concerned with the drivers without proper registration papers and are not wearing helmets. Six in every 10 female and male respondents described their roads as generally accident-prone (Table 17) (Fig. 23.5).

3.5.2 *Both men and women experiencedverbal and physical abuse, andsexual harassment-during commutes but more women than men report the incidence to authorities.*

Table 16. Road-related accidents in the past year by gender

Mostly involved in the accidents	%
Adult men	46.5
Adult women	15.8
Teenage boys	25.0
Teenage girls	5.5
Young boys	3.5
Young girls	2.7
Total	99.9

Table 17. Concerns regarding sagety and security by gender, in percent

Concerns	%	
	Female	Male
Over-speeding of vehicles	69.7	78.0
Overloading of passengers	66.2	68.2
Overloading of cargo (farm produce, etc)	65.7	60.9
Unlit roads	69.9	63.1
Roads with obstructed views, foliage	59.6	59.4
Sheltered and lit waiting areas	46.4	40.4
Accident-proneroads	66.5	62.3
Safe pedestrain paths	49.1	45.6
Fear of sitting next to unknown passengers	62.0	59.9
Poorly maintained vehicles	64.4	60.2
Poor traffic signaling	60.2	61.2
Lack of reliable information (i.e., bus stop schedule of arrival/departure of the bus, jeepney, etc)	44.3	36.1
Vehicles without registration papers	45.9	44.6
Motocycle drivers without ahelmet	52.5	47.5

Fig. 23.5 More men and boys are likely to be involved in road accidents, and more women reported security and safety concerns

Women and men report a fairly similar experience of verbal abuse/confrontations with the driver/conductor or co-passengers during commutes. Men have a higher frequency of experience of physical abuse, about 5 times more in a year, compared to women. Findings show that women are more likely to report the incidence of physical abuse than men (Fig. 23.6).

As shown in Table 18, it appears that more male respondents have experienced verbal and physical confrontations with the driver/conductor or co-passenger. In most incidents, these are not reported to the proper authorities by the adult men; in contrast, one or two female respondents involved in the incident found their way to the authorities to report the incidents of verbal and physical abuse.

In general, women reported a higher frequency of experiencing sexual harassment ($M = 1.68$ or about 5 times in a year, $SE = .22$) compared with men ($M = 1.67$ or about 5 times in a year, $SE = .36$). However, the difference ($-.24$) is not significant [t (38.72) $= 1.86$, $p>.05$]. Noticeable though is that none of these respondents have reported to the proper authority the sexual harassment they experienced during their commutes.

Asked if they knew someone who experienced sexual harassment during their commutes, results show that women expressed greater affirmation to this question compared to men (20.2% v 18.8%) although the study shows there is no significant difference between gender and knowledge of someone who experienced sexual harassment during their commutes [Pearson chi-square analysis, $x^2(1) = .18, p>.05$]. Among respondents who reported that they knew of someone who experienced sexual harassment during their commutes (n = 99), response patterns are quite similar for both men and women. It is worth noting that men are more often observant of this type of abuse experienced by another person during their commutes ($M = 2.43$, $SE = .22$) compared to women ($M = 2.19$, $SE = .19$). The difference, however, is not significant. Of the 99 individuals having observed someone being sexually harassed during their commutes, none of them reported the incident to the proper authorities.

Table 18. Experiences during commutes by gender, in percent

Experiences	%	
	Female	Male
Verbal abuse/ confrontations with driver/conductor/ co-passengers	17.9	19.0
Physical abuse/confrontation with driver/conductor/ co-passengers	2.8	4.8
Sexual harassment	6.2	9.2
	Mean score	
Sexual harassment	1.68	1.67
Respondents' knowledge of someone who experienced sexual violence during his/ her commutes, in percent		
Adult men	15.0	12.6
Adult women	6.4	9.0
Teenage boys	4.0	8.0
Teenage girls	1.5	1.5
Young boys	6.5	4.5
Young girls	0.0	0.0

Table 19. Concerns about personal safety, in percent (multiple responses)

Concerns	Female	Male
Personal safety and travel	45.9	46.2
Felt not safe waiting along the street for their commute	37.1	36.5
Road accidents	56.2	59.8
Harassment	17.7	13.0
Accidents, harassment, and theft	7.3	6.5

Table 20. Respondents' restrictions on time of day of travel by gender, in percent

If there is a restriction	Female	Male
Yes, with restrictions	24.4	19.8
None, no restrictions	27.9	27.9
Total	52.3	47.7
Overall total	100.0	
These restrictions are the following:		
Curfew time: be home before 6 pm	2.5	4.0
Curfew time: be home between 7.10 pm	12.1	7.6
Curfew time: before 9 pm	4.5	2.0
Curfew time: be home by midnight	0.5	0.5
Curfew time: be home by 1 am	0.0	0.5
Daytime travel only, absolutely no evening	23.2	21.2
If it is evening and raining	2.5	2.5
Not in the late afternoon because there'd be no more trips	1.5	1.5
Whenever the weather is bad, no travel at all	6.1	6.1
Avoid travelling if there are checkpoints for unregistered vehicles	0.5	0.5
Total %	53.4	46.4

Fig. 23.6 More women reported concerns with personal safety and reported restrictions on time of day travel compared to men

3.5.3 *Both women and men are concerned about their safety during commutes.*

Men are slightly much concerned about personal safety and travel than women (46.2% versus 45.9%) but this difference is *not significant*. Asked if they feel safe waiting along the street for their commute, there is a higher percentage of women who felt they are not safe (37.1% versus 36.5%). However, the difference is *not significant*. Women and men who felt unsafe waiting along the street for their commutes (n = 188) register the same response patterns in the following order as follows: accidents (56.2% for women and 59.8% for men); harassment 17.7% for women and (13.0% for men); and accidents, harassment and theft (7.3% for women and 6.5% for men).

Mostly, women feel unsafe as they fear road accidents, the verbal/physical abuse perpetuated by the driver/conductor/co-passenger, possible sexual harassment, and theft that may ensue while waiting along the street. Likewise, they feel unsafe because the streets have no lights and are quite far from houses and that when it is dark, wild animals, snakes, and harmful insects usually make an appearance frightening the residents.

3.6   Well-Being

3.6.1 *Are the household respondents satisfied on the impact of their current transportation options on the six dimensions of social life or wellbeing namely, household tasks, family relations and support networks, health care, education, and livelihood?*

Comparing the mean scores of women and men, it appears that they share the same response patterns. In other words, they are "satisfied a little" on five out of the six dimensions. It is on the dimension of livelihood that women and men are "not satisfied at all" on the impact of their current transportation options on their livelihood (1.89 versus 1.73). Interestingly, more women are not

satisfied at all on the impact of the current transportation options on health emergency and livelihood. Inferential analysis, however, reveals that the degree of satisfaction on the impact of current transportation options did *not significantly* vary by gender across each of the six dimensions of social life.

On the impact of travel time, more women than men are most likely "not satisfied at all" on their impact on the six social dimensions (1.87 versus 1.84). The degree of satisfaction on the travel time significantly differed only on healthcare. About travel cost, the data shows that more women than men are not satisfied at all (1.86 versus 1.84) on its impact on the six dimensions of social life. The inferential analysis, however, shows that women and men *differed* only on two dimensions namely, family relations and support networks and healthcare.

4   Lessons Learned

... the full and complete development of a country, the welfare of the world and the cause of peace require the maximum participation of women on equal terms with men in all fields. – Declaration on the Elimination of Discrimination Against Women

The Philippines, a signatory to the Declaration on the Elimination of Discrimination Against Women, has the Magna Carta of Women and several issuances to fast-track gender mainstreaming in infrastructure projects, among others, aimed at showing results in the quality of lives of women and men, girls and boys. Irrespective of the type of infrastructure project, the men and women are unfortunately wrongly viewed as a homogenous population. The Philippine Commission on Women concludes that "having the same needs, vulnerabilities, access, and opportunities in deciding what facilities are needed where, how they will be maintained, how much they should cost each user, and the like. By recognizing the differences among those affected by infrastructure, projects can help achieve better gender equality results."

4.1 Transportation Access and Mobility for Household Essentials

Undertaking household essentials/ chores is doubly difficult for the women than the men because the former have to travel longer distances, travel more frequent and longer trips, and spend more for traveling. The prevailing road conditions and mode of transportation to get to their destination makes the journey extremely difficult to accomplish essential and daily tasks. The roads that these women travel on a daily basis are generally the tertiary and unclassified types and the condition is wanting—steep, narrow, too grassy. Landslides are a common occurrence in some parts of the surveyed villages and the roads are muddy and slippery during rainy days and the eight-month rainy season from May to December. The locations of some of the household essentials are far for the women who have to walk a distance. The women who perform the tasks of going to the rice mill or corn mill, going to the farm, and collecting or harvesting vegetables are more likely to walk than their male counterparts. The frequency of fetching drinking water is higher for women—who live farther from the water source—than the men. Thus, the condition and the type of the road and the mode of travel of the residents are significant when the distance from the house to the location of the household essentials/chores is considered.

4.2 Transportation Access and Mobility for Health Care Services

The traditional role of women associated with healthcare services is evident in the study, that is, all tasks related to access and mobility for healthcare services mainly rest on the shoulders of women. Although more men travel longer distances, the women, however, spend more travel time and travel cost for healthcare services because they have more frequent visits to the pediatric facility, health clinic, and the pharmacy or drugstore. It is also important to note that women than men are more likely to pass through tertiary and unclassified types of roads. The women are also more likely to walk while the men are more likely to use motorized vehicle in going to the health facility. Improvement in the transportation access and mobility for health services shall bring in significant impact on the health of the people and particularly women who are less likely to receive medical services, emergency healthcare, and be brought to the hospital.

4.3 Transportation Access and Mobility for Education

The road and transport situation are barriers to the children's education. Children's tardiness to class is pervasive because of the road condition and scarcity of vehicles and travel costs. On rainy days, the road is full of mud holes that vehicles encounter frequent breakdowns. It is also during the rainy days that travel cost becomes expensive. Other than scarcity of vehicles and road blockage, some vehicles are also no longer fit for commute endangering the riding school children.

4.4 Transportation Access and Mobility for Livelihoods

Depending on the distance of the farms from the house which are generally located within the barangay, the women were more likely to walk. The men are also more likely to walk to their farms or take a motorcycle ride. The men's travel time is longer than the women's because their farms are either located in the fringes or outside the barangay where they live. Comparatively the women are also more likely to travel on tertiary and unclassified types of roads going to the farm. The same can be said with regard to the location of the agriculture livelihood services. Although the following are more performed by the men, they have to pass through the unclassified and/or tertiary types of roads when bringing the farm produce to the processing center or to the *bagsakan*/drop-off center, to the market, or directly to the buyer.

The majority of the rural households deriving their sources of income from non-agriculture activities reveals there are constraints that muddle the development of agriculture in the sampled areas. Beyond roads and transportation is the landownership problem, low farm productivity, limited connectivity to the market, particularly the lack of quality rural transport such as farm-to-market roads, and the lack of adequate and timely market information support services, among others. All these make farming a "non-bankable" undertaking, discouraging farmers from increasing productivity or even making farming a source of their livelihood. The problem is more profound on the part of the rural women who appear to be marginalized in agriculture. The study further revealed that only a few women have access to production support services particularly farm technology, production capital, and farm inputs (seeds, fertilizers, and pesticides), to name a few. Because there seems to be no "space" for the women in agriculture, many of them instead sell their personal services (laundry, manicure and pedicure, massage, etc.) or derive their source of income from petty vending (food vending, buy-and-sell) and, or microentrepreneurial livelihood (sari-sari or convenience stores, livestock, and poultry raising) which are mostly home-based. In contrast, the men would do additional farm work and find other means to augment the family's coffers.

Should transportation options be improved, more women than men have anticipated a reduced time to accomplish livelihood-related activities. More women than men also foresee an increase in their income, better access and mobility to the location of their source of livelihood, and faster marketing of the family's farm produce. Rural folks would then be encouraged especially the young to go into farming. If the road and transportation situation is improved, it would set off possible business opportunities for residents and migrants alike, consequently creating jobs for the locals and more people paying taxes which the barangay government can judiciously use to finance its public services. If road and transportation options are improved, more

women than men can now find time to rest, social, and leisure activities.

However, the Food and Agriculture Organization study pointed out that "the labor burden of rural women exceeds that of men… contribution of women to agricultural and food production is significant…invariably women are overrepresented in unpaid, seasonal and part-time work, and the available evidence suggests that women are often paid less than men, for the same work." Future empirical studies that delve into the essential contributions of women to agriculture and rural enterprises would help in contextualizing gender-sensitive and gender-responsive policies and programs relative to rural infrastructure and livelihood.

4.5 Safety and Security Issues

The roads in the sampled areas are generally described by the residents as accident-prone caused by the kind of road and the kind of users as previously described. The current state of road conditions affects the residents, especially the women's transportation access and mobility for household essentials, healthcare services, education services, and livelihood and microenterprise purposes. More women than men are worried of their personal safety and travel as they feel unsafe waiting along the street for their commute, the road accidents, possible verbal and physical abuse, theft, and sexual harassment. The residents complained about the roads that are almost impassable during the rainy days. They are alarmed by the road accidents that occurred quite frequently—the habal-habal modified motorcycle drivers and passengers who, more often than not, do not wear helmets, the over-speeding, overloading of passengers, overloading of cargo, the poorly maintained vehicles, poor traffic signaling, the unlit roads, and the overgrowth foliage that obstruct the driver's view of the road ahead. The residents, too, are concerned about public transportation vehicles without proper registration papers and motorcycle (habal-habal) drivers who do not have a government-issued license to drive.

4.6 Gender-Based Violence Issues

Outside of the protective walls of their homes, gender-based violence during commutes lies in wait. Verbal and physical abuse and sexual harassment do happen to both men and women. Most perpetrators of verbal and physical abuse are the drivers, conductors, and even their co-passengers. Experiences of sexual harassment during commutes have a higher frequency among women than men. It is unfortunate though that very few sexual harassment incidents are reported to the proper authorities. If ever a report to the proper authorities is made, more women than men have the courage to do so. This is why girls have limited mobility or mobility restrictions and are chaperoned by a male family member should they go out. To protect their girls from harm, parents have set some rules relative to commute. In fact, more women than men have imposed restrictions on their time-of-day travel. More women than men, too, have restrictions on the mode of transportation.

Thus, gender equality outcomes are in order for any infrastructure program:

- Improving women's access to safe, reliable, and affordable public transport services and essential infrastructure
- Improved capacity of local women and local infrastructure agencies to participate in local data collection, planning, design, implementation, and monitoring programs that address gender issues and the concerns of different groups of women and men users
- Increased employment of women at all levels (planning, construction, technical, and management) in infrastructure projects or services

- The development of a gender and equity index with which to gauge the approval of a government or donor infrastructure program

In the Philippines, micro, small, and medium enterprises (MSMEs) make up 99.5% of the economy, with women representing 51% of registered MSMEs. Public and private infrastructure and investment planning decisions need to reflect the reality that women are central to the economic and social prosperity of their communities. With the devastating impact of the current pandemic, and the overwhelming burden of caregiving, household and livelihood tasks on women, and their reduced access to healthcare, the participation and stewardship of local women are critical to achieving sustainable development goals.

Acknowledgments This 2019–2020 study was funded by the World Bank and the Australian Government Department of Foreign Affairs and Trade. A parallel study on the role of free and open-source software, cross-sector community-based and women-led local and remote mapping programs was used to analyze rural agriculture infrastructure, commodity flows, logistics, road planning, and road safety.

References

Australian Aid (2017) Women and Entrepreneurship in the Philippines.https://investinginwomen.asia/knowledge/university-sydney-july-2017-women-entrepreneurship-philippines/

Food and Agriculture Organization of the United Nations (FAO) (2011) State of Food and Agriculture team. The role of women in agriculture. Available via FAO. http://www.fao.org/docrep/013/am307e/am307e00.pdf. Cited 31 Jan 2022

Philippine Department of Trade and Industry (2019) MSME statistics. https://www.dti.gov.ph/resources/msme-statistics/

Sustainable Coastal Communities in the Anthropocene: Lessons from Crowd-Mapping Projects in Colombia

24

Yéssica De los ríos-Olarte,
Maria Fernanda Peña-Valencia,
Natalia da Silveira Arruda,
and Juan Felipe Blanco-Libreros

Abstract

We emphasize the need to contribute to the promotion of sustainable coastal communities from the recognition of a new era like the Anthropocene and the role of maps and open data. We want to tell you about how engaged YouthMappers contribute to efforts to achieve the SDGs under the demands of this new era and how maps can be a language that promotes the understanding of key factors of the Anthropocene such as via socioecological systems (SES) and connectivity. Examples of our work with coastal communities showcase addressing SDG 14, life below water, as well as SDG 13, in an era requiring climate action. With our voices, we want to call attention to the importance of observing the world without separating nature and human societies like two different entities, but to see them instead as a tightly interconnected tissue.

Y. De los ríos-Olarte (✉) · M. F. Peña-Valencia
University of Antioquia, Medellín, Colombia
e-mail: yessica.delosrios@udea.edu.co;
fernanda.pena@udea.edu.co

N. da Silveira Arruda
University of Antioquia, Medellín, Colombia

School of Geographical Sciences and Urban
Planning, Arizona State University, Tempe, AZ, USA
e-mail: narruda@asu.edu

Keywords

Coastal communities · Anthropocene ·
Climate change · Community engagement ·
Colombia

1 Coastal Communities and SDGs

Everything is dark, you can hear the laughter, moans and comments of the people who are in the same boat. We are all underneath a black plastic that protects us from the seawater that splashes inside. After several minutes of a very hectic trip, calm comes. We are poking out our heads little by little and then it is possible to see that we are approaching one of the mouths of the Atrato River. The blue water of the sea mixes with a brown water

J. F. Blanco-Libreros
Instituto de Biología, Facultad de Ciencias Exactas y
Naturales, University of Antioquia,
Medellín, Colombia
e-mail: juan.blanco@udea.edu.co

P. Solís, M. Zeballos (eds.), *Open Mapping towards Sustainable Development Goals*, Sustainable
Development Goals Series, https://doi.org/10.1007/978-3-031-05182-1_24

laden with river sediments. Pelicans fly close to us as we enter the river channel and leave the sea behind. The partially cloudy sky allows the sunlight to pour in with intensity and reflect in greens on the tall aquatic plants that mark our path. We can only go forward in what seems like a maze that does not end. Suddenly, little pink, yellow, orange-colored houses appear, which seem to compete for which will be the most striking, and in the midst of the wonder for all the color a man says: ¡Welcome to Bocas del Atrato!

What should we say about the implications of sustainable development for coastal communities in this new age? What is the relationship of this understanding with the achievement of SDGs about life under water? What are the limitations of being located in the coastal belt that does not belong completely to either the marine or the terrestrial? What implications does this have to the key ecosystems such as mangroves? What is the significance of SDG 14 to an "amphibian community" affected by aquatic and terrestrial pressures/issues? We are going to address all these questions through the perspective of our experiences in "Coastmap," a project that allowed us to rethink ourselves from the maps, like a part of a system connected to social and environmental aspects, in an era of climate change that requires action (SDG 13).

2 Anthropocene, Socioecological Systems, and Connectivity

First of all, it is important to understand key aspects of these times. It is evident that unprecedented population growth and human activities (construction of artificial structures, land cover change, etc.) are acting as a global force, reshaping the Earth (biosphere, atmosphere, hydrosphere, and lithosphere). Such dramatic transformations have promoted an increase in the understanding of the interdependence between human societies and the surrounding environment, leading to the proposal of the *Anthropocene Epoch* as a potential addition to the geological timeline (Folke et al. 2021; Ellis et al. 2020).

Regardless of whether this new geological epoch naming will be formally adopted, the anthropogenic biosphere is a new reality with important implications for human well-being in the face of global phenomena, demanding a transformative shift to reach sustainable futures (Folke et al. 2016. To achieve this objective, the conceptual framework of socioecological systems (SES) seeks to address the challenge of matching the dynamics of institutions with the dynamics of ecosystems for mutual socio-ecological resilience (Folke et al. 2016.

The SES approach emphasizes that the people, communities, economies, societies, and cultures are embedded parts of the biosphere and shape it, but at the same time are shaped by, dependent on, and evolving with the biosphere. Then, more than just an interaction, people are inhabitants of the biosphere together with all other life on Earth from the local to the global, consciously, or unconsciously (Ellis et al. 2020; Folke et al. 2016). For instance, urban areas and other forms of densely populated human settlements, alongside villages, have become one of the most salient features of the anthropogenic biosphere, particularly since the second half of the twentieth century, known as the Great Acceleration (Ellis et al. 2020; Steffen et al. 2015). In addition, human transformations have resulted in a highly interconnected world with new cross-scale interactions linking people and places in new ways. Linkages can be described as uni- or bi-directional, and positive or negative from the perspectives of the donor and beneficiary areas. The connectivity (or tele-connections) (sensu 7) of urban and rural communities with their surroundings is a major feature that needs to be scientifically understood, but it must also be used in humanitarian and socio-economic development projects. In this sense, we have been addressing two forms of connectivity: spatial and technological. The spatial connectivity refers to the physical linkages where the spatial scale is a key factor defining the strength of the relationships maintained, serving as a conduit for matter and energy. Technological connectivity allows people to communicate quickly and to exchange knowledge between distant parts of the world, where the distance is no longer a barrier to overcome in order to establish a network.

3   YouthMappers in the Anthropocene

The interaction between spatial and technological connectivity is approached by the YouthMappers (YM) workflow, through a model where actions transcend geographical restrictions to operate in other physical locations by learning about communications and geospatial technologies, encouraging people to become global citizens. Such engagement mobilizes "digital humanitarians" (Solís et al.2020) that include students, volunteers, and professionals, who want to make a difference by working online to respond to the needs of international humanitarian campaigns or field actions (Steffen et al.2015).This working model adjusts to the current times allowing individuals, groups, private organizations, and government agencies to face the challenges of the Anthropocene by contributing to the achievement of the Sustainable Development Goals. As a group of students and professors, we recognize that our academic community is part of a complex SES, that interacts, influences, and is affected by other communities, institutions, and the natural environment itself.

3.1   The Coastmap Project

Coastal communities located on the shoreline (the imaginary line between inlands, sea, and the intertidal zone) are directly connected with the sea and the land in a space which belongs to both and at the same time to neither.They are another unit of coastal SES; however, they have high levels of vulnerability because that is where the relations between land and sea converge and where decision making and the management of marine and terrestrial resources and spacesare mani-

fested. Some coastal communities in the Gulf of Urabá belong to the coastal SES unit we are going to talk about in the context of the Coastmap project, which we will describe as follows.

3.2   A Participatory Approach to Mapping Coastal Communities

The Coastmap is a humanitarian research program initiated in 2017, comprising three phases so far, and it has the objective of mapping fishing coastal communities using a participatory methodology,to ultimately contribute to their resilience against flood hazards induced by climate change and environmental variability.In this same sense, this project highlights the importance of the surrounding coastal wetlands, particularly mangrove forests, as bioshields and sources of additional ecosystem services. Is developed as a collaboration between the YouthMappers chapter GeoLab and the research group PEEP (Ecosystem Processes at the Landscape Scale, by its Spanish acronym), that are part of the Faculty of Engineering and the Institute of Biology of the University of Antioquia, respectively. For more than 10 years the PEEP group has worked in the Urabá gulf studying the different coastal dynamics that influence the wetlands located in that area, while GeoLab promoted the interdisciplinary use of GIS and open data for research through different lines of work (e.g. environmental) and collaborations.Therefore, identifying the close links between local communities and ecosystems in the Urabá gulf, and recognizing the importance of including local actors in the different projects,allowed PEEP to join forces with the GeoLab chapter to initiate the Coastmap project.

In the southernmost part of the Greater Caribbean is located the Urabá. It is characterized by the inflow of many rivers of different sizes, among which the Atrato River stands out as one of the largest deltas and highest freshwater discharge in Colombia. It is also the life support for unique and extensive mangrove forests dispersed along the coastline of the entire Gulf. The municipality of Turbo is the main administrative unit in the southern part of the Gulf, where a compact coastal city and different fishing communities are found (approximately 880 fishermen). The program was initiated in 2017 in Barrio Pescao (Called Bahía El Uno due to its location on a homonymous embayment) in the peri-urban with funding from HOTOSM, expanded to Bocas del Atrato (2018) and Puerto Cesar (2020) in rural areas with funding from the YM Research Fellowship (Fig. 24.1). The work in Puerto Cesar was still in progress at the time of writing, due to delays caused by the COVID-19 pandemic.

3.3   Phased Methodological Processes with Community Participation

The growing (both broader and easier) access to technological tools (e.g. laptops and tablets, smartphones, wifi, open software, cloud storage, and cloud computing), allows anyone to be a potential user and a source of data, empowering citizens to contribute to scientific research and humanitarian aid with different levels of participation or leadership. In Coastmap, we applied a community-based participatory approach to incorporate the empirical and local knowledge of the communities into the different phases of our humanitarian mapping project. The participatory methodology of Coastmap involves joint work between the academic community and fishing communities, using different free and open-source tools (Fig. 24.2).

As a first step, we conducted remote mapping activities using OpenStreetMap and the Tasking Manager platform to understand the territories. This technological tool allows us to generate open data overcoming the spatial distance, and at the same time, it creates the conditions to connect us with local actors in the verification process of the data in the field. High school students, teachers, and social leaders support the field activities by verifying local information while conducting home-based interviews.These socio-economic surveys aim to provide data to understand the impact of flood hazards, perceptions about the linkages with climate change and variability,and knowledge about the ecosystem services provided by mangroves and other coastal wetlands.

In the first phase of Coastmap Urabá, in El Uno bay, we conducted interviews engaging students from a local high school, who were trained in the use of smartphone GPS and Field Papers.In the following phases, we have also been using

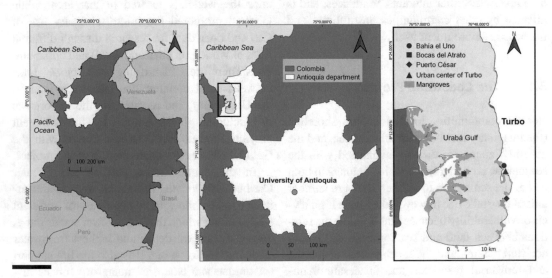

Fig. 24.1 The study area is shown for the country of Colombia, department of Antioquia, and gulf of Urabá

Fig. 24.2 The methodology for Coastmap engages YouthMappers, the academic community and fishing communities, using different free and open-source tools

Kobo Toolbox for the creation of the survey, and Kobo Collect for its application in the field. We engage volunteers in diverse mapping activities realized through the conduction of surveys, description, and verification of elements of housing and other infrastructure in their territory, demonstrations with drone operation, and training on the use of smartphone GPS, KoBo Toolbox and Field Papers.

We have been organizing the field activities annually, in a way that allows us to celebrate the International Day for the Conservation of the Mangrove Ecosystem with local communities. For us, it is very important to be present in the territory, interact with the local community, and undertake knowledge exchange dialogues and recreational activities. The celebration of this particular date is a practice that allows us to incorporate local milestones into the project schedule while generating awareness on the importance of coastal wetlands andimproving the ecological understanding of their territories.

The proof-of-concept methodology carried out during the second phase of the program in Bocas del Atrato allowed the updated and field-verified OSM outputs using visits, drone-based orthophoto mosaics, and digital elevation models (Fig.24.3). These products also aided the flood-

ing risk analysis and supported the knowledge exchange between our working group, leaders, and focal groups within the fishing communities. We focus on guiding communities in the spatial recognition of their territory and context, transcending the geographic discipline. For instance, a school is not just a polygon or a physical structure, but a bridge to education and a gateway to the outer world. The surrounding mangrove ecosystem is not just a group of trees but a fish habitat and nursery as well as a protective barrier against winds and storms. An established route to a fishing ground is not just an invisible path, but a safe and profitable choice for subsistence (efficiency in catching fish and higher remuneration in the sale). Consequently, Coastmap stands out in the capacity of OSM helping to identify spatial features but also for understanding relationships relevant to humanitarian aid.

Cartographic resources resulting from the first and second phases of Coastmap in the community of Bocas del Atrato include: (A) Updating streets and buildings in the OpenStreetMap platform, where a total of 24,264 buildings and 397 kilometers of roads have been drawn throughout the project, of which only 174 buildings correspond to Bocas del Atrato; (B) digital elevation model generated from orthophotogra-

Fig.24.3 Analysis for the project combines layers of OSM outputs, digital elevation models, and orthophoto mosaics

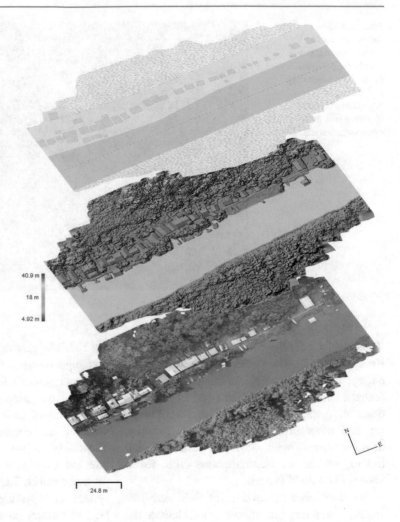

phy; (C) orthophotography created from images taken on drone flights during fieldwork.

In the third version of Coastmap, developed recently, we were working through the promotion of geographic literacy. We think that mapping should initiate from the recognition of the territory using maps as a tool for this learning. The concept of geoliteracy helps also to study the territory across SES as a theoretical framework that contributes to expanding it beyond their residential zone, where the relationships with other communities and key ecosystems start to be appreciated as a part of the system. Geoliteracy promotes a vision of territory through establishing a zone of influence and a network of interconnected sites, which means taking the connectivity into account.

4 Anthromes

Taking into account the importance of the SES in the anthropocene, we can also speak of the anthromes as a SES on a broader scale. According to Ellis et al. (2020), the anthromes are heterogeneous landscape mosaics composed of used lands and ecosystems. We can analyze the different communities immersed in the big coastal landscape of the Urabá gulf based on their structure and relations as coastal anthromes from our experiences and information compiled in this project.

The gulf is located approximately 340 km away from the city of Medellín by road in the southernmost part of the Greater Caribbean, where stands out the discharge of the Atrato River as a large freshwater source and support for development of mangrove forests and other wetlands.If we orient ourselves in the southern part of the Urabá gulf we can see the first anthrome; the urban center of Turbo highlighting as a unit with a compact coastal city.It can be classified as an urban anthrome with high population density, access to land and sea transportation, and with the presence of small patches of highly degraded mangrove forests. Despite being the main administrative unit with the characteristic economic dynamics of urban centers, a part of its population is dedicated to artisanal fishing.Fishermen must travel to areas where mangroves still have the capacity to sustain the fishery resource, while for the rest of the population, the port condition, services sector, and monetary exchanges seem to distance them from the recognition of the importance of preserving the coastal area and the traditional practices of fishermen.

Bahía El Uno is located in the north of the head municipality of Turbo approximately 20 minutes by road. In this anthrome, mangrove forests border the coastal lagoon, but the establishment of crops and pastures for livestock have fragmented both the coastal forests and the rest of the natural cover inland. From participatory mapping, we know that the inhabitants (approximately 450) are highly connected with the center of Turbo as a source of work through different activities, and that fishing accounts for only 11 percent of existing occupations. On the other hand, south of the municipal seat of Turbo is Puerto Cesar, where we can see through that the adjacent mangrove coverage is greater than in Bahía El Uno, and that the residential zone is embedded in an area with small patches of crops and livestock. Also, the inhabitants are connected to Turbo, in this case by sea, and connected by road to a half-hour drive from another small economic center east of the coast, called Currulao.

In contrast, Bocas del Atrato is a township anthrome located in the western side of the gulf, in front of the urban center of Turbo and connected to it only by sea.The community of Bocas del Atrato is embedded in the middle of a huge extent of mangrove in Atrato river delta, therefore this anthrome is almost completely wild with little direct human influence.From our proj-

ect, we know that in this socioecological system of the 356 inhabitants, 48 percent are dedicated to fishing, the rest of the occupations vary mainly between housewife, business owner,and various trades that are intimately related to products or services of the center of Turbo.

We can then observe that the communities studied interact in different ways with their natural and economic context and are reflected in the anthromes at different scales. The structures of these groups of populations change in terms of the cover of mangroves and connectivity (Fig. 24.4) and at the same time, conditions the way in which the local community could respond to coastal hazards. With the maps and open data, we can understand more how it could intervene and what information is important to consider for future planning in the communities that we studied.

Classification of the communities with respect to the anthromes proposed by Ellis et al. (2020) and their spatial connectivity with the municipal seat of Turbo.In this case, connectivity is defined by the access and main route of exchange of people, goods, and services.Each map figure has a diameter corresponding to 700m, the base map is OSM and mangrove forests are in green.

5 Rethinking SDGs 14 and 13 for the Coastal Anthropocene

Coastal SES are complex systems with interconnected reliance across multiple scales; they do not have a delimited area and are affected by social and ecological elements, both marine and terrestrial. How far does the coastline reach? The coastal uplands and river basins despite not having coastal physical characteristics and being farther away from the coast, influence the coastal zone; therefore, it could be considered also as part of the coastal SES unit (Hossain et al. 2020).

Coastal communities located on the shoreline (the imaginary line between inlands, sea, and the intertidal zone) are directly connected with the sea and the land in a space which belongs to both and at the same time to none.They are another unit of coastal SES; however, they have high lev-

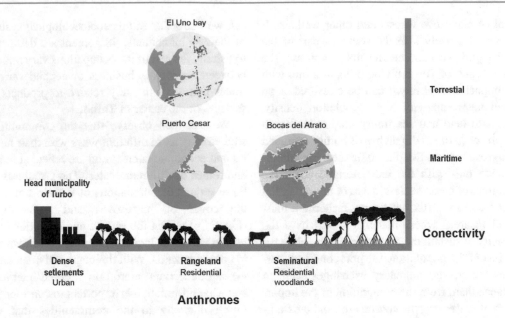

Fig. 24.4 A scheme resulting from the research classifies communities with respect to anthromes vis-a-vis spatial connectivity

els of vulnerability because that is where the relations between land and sea converge and where decision making and the management of marine and terrestrial resources and spacesare manifested. For example, in Bahía el Uno, the structure of this anthrome with the cover of mangrove in the lagoon can avoid marine flooding because these forests act as barriers. But, due to the fragmentation of adjacent lands of the watershed of the Turbo river the population lacks protection as it is in the middle of the floodplain, being affected by torrential rains or inter-annual climatic events such as El Niño.Then, how far is the coastline of Bahía el Uno?

The implications of being a coastal community are commonly ignored, as can also be observed in the objectives of SDG 14. According to Friess et al. (2019), these alone do not emphasize coastal ecosystems, but only marine life. Therefore, the separation between the marine and terrestrial zones leaves out, as usual, the coastal socio-ecosystems (formed for both areas),which not only deprives them of the attention and importance they have for marine resources but can also indirectly affect them negatively. The sustainable development of these amphibian communities must start by avoiding this separation.

For SDG 13, where climate action is called for,we want to carry this logic further to start to think about how the climate is connected to the coastlines, and also to incorporate the call for action of this goal in terms of the new era of the Anthropocene. That is why the voice of local and academic communities is important to promote SES as a cross-cutting element for any decision-making in coastal areas. However, claiming these spaces and having the ability to face the challenges that arise as a consequence of marine and land and even climate connectivity is only possible from a recognition of the territory.Geographic literacy and participatory mapping emerge in this sense as a language to connect communities with the geographic, social, and climatic context,as well as with the challenges that being a coastal SES gives them.

In the end, promotion of geographic literacy must be based on a diagnosis of the community. One of the most important challenges of the third version of the project has been to find the best way to address geographic literacy in the community.In this sense, interdisciplinarity (approaching the community with different elements such as arts and support of personnel with emphasis on the social part) is necessary to approach the community from its culture and context.Mapping projects should not ignore the history and way of life of the communities, but work hand in hand with them so that the mapping exercises remain in the collective memory.

For forward-thinking conversations about any or all of the SDGS, it is thought provoking to reflect on the challenge that it may not be possible to contribute to the resilience of a community (of inhabitants or academics) without a whole understanding of where they are living. The role of international entities, institutional government, and universities is to connect this knowledge to help to improve the capacities to respond to the harms that they face as a coastal community. The use of open data and maps is completely useless if it remains only in academic or governmental institutions.

References

Ellis EC, Beusen AHW, Goldewijk KK (2020) Anthropogenic biomes: 10,000 BCE to 2015 CE. Land 9(5):1–19. https://doi.org/10.3390/land9050129

Folke C, Biggs R, Norström AV, Reyers B, Rockström J (2016) Social-ecological resilience and biosphere-based sustainability science. Ecol Soc 21:41. https://doi.org/10.5751/ES-08748-210341

Folke C, Polasky S, Rockström J, Galaz V, Westley F, Lamont M (2021) Our future in the Anthropocene biosphere. Ambio 50:834–869. https://doi.org/10.1007/s13280-021-01544-8

Friess DA, Aung TT, Huxham M, Lovelock C, Mukherjee N, Sasmito S (2019) SDG 14: life below water – impacts on mangroves. In: Katila P et al (eds) Sustainable development goals: their impacts on forests and people. Cambridge University Press, Cambridge, pp 445–481. https://doi.org/10.1017/9781108765015

Hossain MS, Gain AK, Rogers KG (2020) Sustainable coastal social-ecological systems: how do we define "coastal"? Int J Sustain Dev World Ecol 27(7):577–582. https://doi.org/10.1080/13504509.2020.1789775

Solís P, Rajagopalan S, Villa L, Mohiuddin MB, Boateng E, Wavamunno Nakacwa S, Peña Valencia MF (2020) Digital humanitarians for the sustainable development goals: YouthMappers as a hybrid movement. J Geogr High Educ 1–21. https://doi.org/10.1080/03098265.2020.1849067

Steffen W, Broadgate W, Deutsch L, Gaffney O, Ludwig C (2015) The trajectory of the anthropocene: the great acceleration. Anthr Rev 2(1):81–98. https://doi.org/10.1177/2053019614564785

Collaborative Cartography Making Riparian Communities Visible in Tefé, Amazonas, Brazil

Ana Luisa Teixeira, Silvia Elena Ventorini,
Évelyn Márcia Pôssa, Francisco Davy Braz Rabelo,
Leonardo Cristian Rocha, Mucio do
Amaral Figueiredo, and Paula dos Santos Silva

Abstract

The authors explore the invisibility of riparian communities along the river channels of the Amazon basin and the utility of collaborative mapping as a methodology for increasing their visibility on publicly available maps with the objective of contributing to the recording of their history and presenting a collaborative cartographic product that can be useful for guaranteeing territorial rights and support the creation of public policies more suited to the riverside realities. These efforts carried out by the YouthMappers chapter at the Universidade Federal de São João del Rei and Centro de Estudos Superiores de Tefé - Universidade do Estado do Amazonas support SDG 15 – Life on Land and SDG 16 – Peace and Justice and Strong Institutions.

Keywords

Land · Justice · Invisibility · Riparian communities · Collaborative cartography · Brazil · Amazonia

A. L. Teixeira · S. E. Ventorini (✉) · É. M. Pôssa ·
L. C. Rocha · M. do Amaral Figueiredo
Departamento de Geociências, Universidade Federal
de São João del - Rei, São João del-Rei, Brazil
e-mail: sventorini@ufsj.edu.br; rochageo@ufsj.
edu.br; muciofigueiredo@ufsj.edu.br

F. D. B. Rabelo · P. dos Santos Silva
Centro de Estudos Superiores de Tefé /CEST,
Universidade do Estado do Amazonas, Tefé, Brazil
e-mail: frabelo@uea.edu.br; paulasantos@ufsj.edu.br

1   Introduction

The native peoples and riparian communities, who live on the banks of the river channels of the immense Amazon basin, govern their way of life according to the hydrological and geomorphological dynamics of the place, oftentimes intensified by the effects of El Niño and La Niña (Piedade et al. 2005). Along the Middle Solimões River, many of the riparian communities lack records about their ancestral knowledge, cultures, territories, etc. This is because the riverside dwellers transmit their knowledge and stories orally and, although this is a legitimate source of information, as it is immaterial, it can be lost in time (De Magalhães Lima and Ferreira Alencar 2001).

The riparian population is scattered throughout the vast Amazonian landscape, that is, it is a regional phenomenon. However, it is the local scale that gives exposure to riparian geographic spaces due to their reduced dimensions. All of this constitutes challenges for a regional cartography of this phenomenon and contributes to the imprecise knowledge about their real spatial distribution. The invisibility of these communities in official cartographic or reference documents limits them from the right to maps that document their territories and their histories. Furthermore, it hinders the creation of adequate public policies by the State to address their needs. And it is in this context that many of the communities in the municipality of Tefé are in the central region of the Amazon (Fig. 25.1).

According to the last Demographic Census (2010), the municipality of Tefé had 89 communities with approximately 15,663 inhabitants (IBGE 2012). The socio-economy of these communities is characterized by having a peasant orientation with production aimed at meeting the needs of family consumption (Adams 2006; De Souza Costa and Coelho 2020). The communities closest to the city of Tefé and those connected by land have a larger population due to the availability of commerce, education, and health. The communities farthest from the city of Tefé use river transport and lack basic infrastructure, such as education, health, basic sanitation, etc. The further away from the urban center, the scarcer the resources and the longer it takes to travel to access them.

2   Invisible Communities

Such communities do not have reference mapping whose function is to present the location of natural, artificial features, and intangible limits in

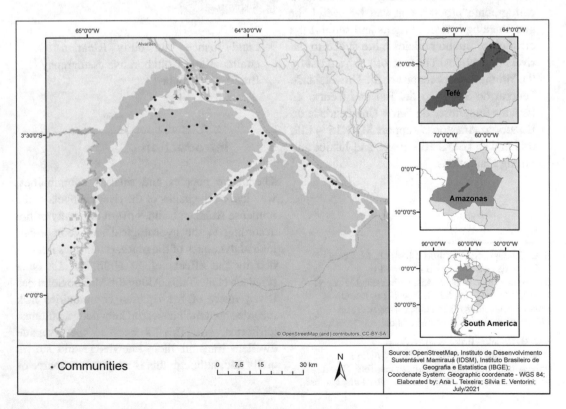

Fig. 25.1 Riparian communities are located in the municipality of Tefé, Brazil

fractions of the Earth's surface, elaborated following a Cartographic Accuracy Standard – PEC, edited by the Technical Standards of National Cartography (Machado and Camboim 2019; Oliveira 1988). This type of material constitutes one of the important elements for mapping at the local level. Oliveira et. al. (2013) considers that official maps still express the interest of the powers in the dominance of knowledge of the territory and the nation. If the State and companies use mapping to legitimize actions and control the territory, making visible only what interests them, why not appropriate such instruments to document the use of the territory by traditional populations, in order to contribute to the perpetuation of their cultural practices and for their own survival (Mascarello et al. 2018)?

Sant'Anna and Young (2010) highlight the vast scientific literature that denounces territorial conflicts between the various actors that form the Amazonian framework: indigenous peoples, settlers, riverside dwellers, quilombolas,[1] land grabbers, miners, and farmers, among others. The violence generated by direct conflict results in injuries and deaths and the loss not only of land, but also of cultural, social, and identity references.

Many traditional communities struggle to guarantee their land rights, and, in this scenario, the map appears not only as a record of their locations but as an instrument that can be used to claim territorial rights (Araujo Junior 2020). Moore and Garzón (2010) describe that the maps were already part of reports and allowed the identification and proof of the traditionality of the occupation, in order to guarantee rights that were being denied in the market economy and administrative practices (Oliveira et al. 2013).

---

[1]Translator's note: *quilombolas* are the people of maroon communities, in the Portuguese nomenclature.

3 Increased Visibility Through Collaborative Mapping

In this context, collaborative, participatory, and social mapping emerge as alternatives to the invisibility scenario of traditional Amazonian populations. Specifically, for the mapping of riparian communities, the mappings carried out through the collaborative project OpenStreetMap (OSM) stand out as promising.

As stated earlier, the riparian dweller is a regional phenomenon, however, the scale that gives them exposure is local. In this case, the OSM platform appears as a solution, as it offers regional coverage of images that allow local analysis (images with high spatial resolution). They allow the exposure of a local phenomenon, which, later, can be transposed to a regional scale. As it is collaborative and has volunteers from all over the world, the territorial extension of the mapping is no longer a limitation.

Given this scenario, Unificar Ações e Informações Geoespaciais – UAIGeo, the YouthMappers chapter at the Universidade Federal de São João del Rei, presents an experience of collaborative mapping of riparian communities in Tefé, with the objective of mapping their spaces in the territory, contributing to the recording of their history and presenting a collaborative cartographic product that can be useful for guaranteeing territorial rights and also to support the creation of public policies more suited to the riverside realities.

3.1 Methodology

The mapping of Amazonian riparian locations in the municipality of Tefé had as initial reference a vector base of the communities' locations, provided by the Mamirauá Sustainable Development Institute, whose source is the Municipality of Tefé. From that, we verified which communities were mapped on OpenStreetMap (OSM) and Google Maps platforms. A search was made for communities in the Census Sector Network

(IBGE 2012), in Higher Education Institutions and, finally, in academic publications.

With the locations available, the project *Mapping Amazonian riparian locations in the northern municipality of Tefé – AM* was registered in the Humanitarian OpenStreetMap Team (HOT) Tasking Manager and made it public so that volunteer collaborators from all over the world can map and collect organized and consistent data on the spatial distribution of communities, taking them out of the invisibility of digital maps.

At first, the vector base of reference of the riparian localities provided only the location. In the Tasking Manager project, employees were instructed to map the buildings of the communities. On OSM, buildings were marked as "buildings" which, ontologically, are related to any covered construction, usually delimited by walls, with the purpose of housing human activities. It is noteworthy that the most common buildings found in riparian communities are made of wood, due to the abundance of such material (Fig. 25.2).

We verified the accuracy of the data (name and location) based on the analysis of alphanumeric information, collected at the Mamirauá Sustainable Development Institute, data from the IBGE Demographic Census (2010), and the Department of Education (using the name and address of the schools). On OSM, the buildings in each community were quantified, considering two distinct temporal scenarios: one in 2010, based on Bing's image collection; and another in 2017, based on Maxar images.

3.2   Discussion

The vector base of reference of the riparian sites represents 72 communities. When transposing the data from this base to the OSM, a displacement was found, which was corrected during the transposition process. The location data come from a physical cartographic document of the City of Tefé. The transcription to the digital environment was made by the Mamirauá Sustainable Development Institute, by scanning, georeferencing, and punctual vectorization in a Geographic Information System (GIS).

Despite the inconsistency of the data, the map was essential to carry out the visual analysis of the communities on the OSM Platform and to define and register the project area in the HOT Tasking Manager. It is noteworthy that this was the only map found with the representation of most communities. There is no Census Sectorial Network in the communities and the Tefé City Hall and Higher Education Institutions do not have digital maps.

Analyzing OSM contributions, it was found that 16 (18%) of the 89 communities were mapped, that is, 73 communities were invisible in all sites with different themes and that use a *base map* with data from the OSM. Through the HOT Tasking Manager, 90 contributors mapped 10,780 buildings in the municipality of Tefé, in a period of approximately three months. The city of Tefé is part of the area specified in the project, hence the expressive number of buildings.

As you move away from the urban area, communities tend to decrease, for example: 49 com-

Fig. 25.2 Examples of buildings in riparian communities are characterized by proximity to water

munities are formed by less than 25 buildings, three with less than five houses, 20 with 26 to 50 houses, and 17 communities with more than 50 houses. The most densely populated communities are found in the outskirts of the city of Tefé, due to the easy access to services and resources. Another point is that the closer to the city, the higher the number of communities.

The time analysis of the number of buildings in 2010 and 2017 indicated that in the period of eight years, the communities closest to the urban center of Tefé showed an increase in the number of buildings. On the other hand, in some more distant communities, there was a reduction in the number of buildings (see Table 25.1).

Quantitative changes define a peculiar characteristic of riparian dwellers in this region: population moving between communities, detectable by the number of buildings. In the region, wooden buildings prevail, as mentioned above. When they migrate from one community to another, the riparian dwellers take their homes with them. This phenomenon is described in the work of Alencar (2010).

This movement of the population happens due to the fact that many communities, because of the natural dynamics of river erosion, commonly called "fallen lands," are forced to move communities around, as occurred with the São Luiz do Macari community, in which the erosion quickly reached (approximately 40 m a year) the area where the community was located, destroying it. In this case, the population migrated and formed a new community elsewhere.

The research on scientific bases indicated investigations of participatory and social mapping in some communities in the researched area,

Table 25.1 Change in number of buildings

Community	Number of buildings	
	2010	2017
Flora Agrícola	105	145
Bacuri	35	53
Boa Vista	15	20
Santa Cruz	24	20
Bela Vista do Sapiá	17	9
Vila Trindade	19	15

but not covering all those mapped in this research. Among these are Oliveira et al. (2013), who carried out land-use mapping in the Tefé National Forest. Mapping, therefore, is crucial to make communities visible and spatialize them. Based on it, it is already possible to identify the location and number of homes in each community, which until then were invisible, in addition to the data being free for local studies in the future.

4   Riparian Communities Seen on the Map for SDGs

The mapping activity and research analysis presented in this chapter arise from the concern when verifying the invisibility of riparian populations on digital maps. The question that guided the research was: why not take ownership of maps to document the use of territory by traditional populations, in order to contribute to the perpetuation of their cultural practices and their own survival (Mascarello et al. 2018)? Collaborative mapping on OSM through the project registration in the HOT Tasking Manager allowed volunteers from all over the world to support students and faculty at the Universidade Federal de São João del Rei map the area. Such collaboration is extremely important when local populations and the government itself do not have access to technological, financial, and human resources.

The internet system in Tefé is precarious and has a high cost for the population and the further away from the city, the more limited the signal is. The volunteer and humanitarian work from the project registered in the HOT made it possible for the communities to be visible in several Brazilian governments and international institutions databases such as the *Iniciativa para Mapeamento de Uso e Cobertura de Solo (MapBiomas)*, *Infraestrutura Nacional de Dados Espaciais (INDE)*, *Agência Nacional de Águas (ANA)*, – *Centro de Sensoriamento Remoto (CSR Maps) da Universidade Federal de Minas Gerais*, Doctors Without Borders, World Bank Group, United States Agency for International Development,

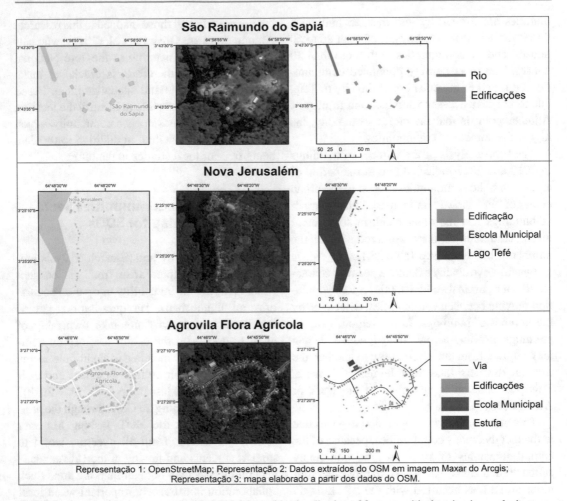

Fig. 25.3 Communities represented with OSM data enable visualization of features critical to riparian analysis

United Nations, Clinton Health Access, among others.

In addition, Geographic Information Systems have plugins that enable the visualization and download of OSM data, such as QGIS and ArcGIS, allowing researchers and professionals from different areas and sectors (public and private) to use the data to generate new mappings. Figure 25.3 illustrates examples of representations of communities that can be generated from OSM data.

Thus, it is concluded that the challenge of mapping riparian communities on a local scale gave them exposure within a regional space, and ultimately, seen for the SDGs, as well as guaranteed the right to the map.

References

Adams C (2006) Sociedades caboclas amazônicas: modernidade e invisibilidade. Annablume

Alencar EF (2010) Dinâmica territorial e mobilidade geográfica no processo de ocupação humana da Reserva de Desenvolvimento Sustentável Amanã-AM. Sci Magaz UAKARI 6(1):39–58

Araujo Junior EM (2020) Cartografia Social nas Narrativas dos Territórios: O Caso das Populações Ribeirinhas na Amazônia Legal. Int J Prof Bus Rev 5(2):153–162

De Magalhães Lima D, Ferreira Alencar E (2001) A lembrança da História: memória social, ambiente e identidade na várzea do Médio Solimões. lusotopie 8(1):27–48

De Souza Costa D, Coelho AA (2020) Os benefícios sociais e a socioeconomia de comunidades rurais do município de Tefé, Amazonas. Oikos: Família e Sociedade em. Debate 31(2):283–312

Instituto Brasileiro de Geografia e Estatística (IBGE) (2012) Censo Brasileiro de 2010. Rio de Janeiro, IBGE

Machado AA, Camboim SP (2019) Mapeamento colaborativo como fonte de dados para o planejamento urbano: desafios e potencialidades. URBE. Revista Brasileira de Gestão Urbana 11

Mamirauá Sustainable Development Institute. https://www.arcgis.com/apps/webappviewer/index.html?id=0167a84111944b8bbe15dc2f3405a4c6

Mascarello M, dos Santos CF, de Oliveira Barbosa AL (2018) Mapas...Por quê? Por quem? Para quem? Revista Movimentos Sociais e Dinâmicas Espaciais. Recife 7(1):126–141. Disponible in https://periodicos.ufpe.br/revistas/revistamseu. Access 2 Aug 2021

Moore, E., Garzón, C. (2010). Social cartography: The art of using maps to build community power. Race, Poverty & the Environment 17(2):66–67

Oliveira CD (1988) Curso de cartografia moderna. Rio de janeiro. IBGE, 1:991

Oliveira MG, Suertegaray DMA, Pires CLZ (2013) Mapeamento participativo e uso do SIG: FLONA de Tefé-AM Anais XVI Simpósio Brasileiro de Sensoriamento Remoto - SBSR, Foz do Iguaçu, PR, Brasil, 13 a 18 de abril de 2013, INPE. Disponible in http://marte2.sid.inpe.br/col/dpi.inpe.br/marte2/2013/05.28.23.12.31/doc/p0269.pdf. Access in 27 jul.2021

Piedade MTF, Schoengart J, Junk WJ (2005) O manejo sustentável das áreas alagáveis da Amazônia Central e as comunidades de herbáceas aquáticas. Uakari 1(1):29–38

Sant'Anna AA, Young CEF (2010) Direitos de propriedade, desmatamento e conflitos rurais na Amazônia. Economia Aplicada 14(3):381–393. https://doi.org/10.1590/S1413-80502010000300006

Open Mapping with Official Cartographies in the Americas

Vivian Arriaga, Adele Birkenes, Daniel Council, Mason Jones, Enith K. Lay Soler, John Sawyer McCarley, Emily Wulf, Calvin Zhang, Jean Parcher Wintemute, Nancy Aguirre, and Patricia Solís

Abstract

Governments need data to be effective and accountable to the SDGs. Spatial data are key to meeting these goals, and youth open mapping represents one way to support official cartographies, especially to advance SDG 16. We explored how youth could add value by integrating OSM data into public agency maps while offering education opportunities (SDG 4).

Keywords

Public institutions · Volunteered geographic information · Official cartography · Belize · Jamaica · Colombia · Costa Rica · Dominican Republic · Mexico · Panama

1 The Case for Open Spatial Data in Public Institutions

Public government institutions are responsible for providing many critical services to all citizens. They also are the entities that report to the United Nations on official statistics about the Sustainable Development Goals (SDGs) (Stuart

V. Arriaga (✉) · P. Solís
Arizona State University, Tempe, AZ, USA
e-mail: varriag1@asu.edu;
patricia.solis@asu.edu

A. Birkenes
Vassar College, Poughkeepsie, NY, USA
e-mail: abirkenes@alum.vassar.edu

D. Council
Ball State University, Muncie, IN, USA
e-mail: dcouncil@bsu.edu

M. Jones
Texas Tech University, Lubbock, TX, USA
e-mail: mason.r.jones@ttu.edu

E. K. Lay Soler
State University of New York, Buffalo, NY, USA

J. S. McCarley
Clemson University, Clemson, SC, USA
e-mail: jmccarl@g.clemson.edu

E. Wulf
University of Southern California,
Los Angeles, CA, USA

C. Zhang
University of Chicago, Chicago, IL, USA
e-mail: zhangcal@uchicago.edu

J. P. Wintemute
Emeritus, Department of the Interior, United States
Geological Survey, Reston, VA, USA

N. Aguirre
Pan American Institute of Geography and History,
Emeritus, National Geographic Institute Agustín
Codazzi, Cali, Colombia

© The Author(s) 2023
P. Solís, M. Zeballos (eds.), *Open Mapping towards Sustainable Development Goals*, Sustainable
Development Goals Series, https://doi.org/10.1007/978-3-031-05182-1_26

et al. 2015; UN 2014). To accomplish this mission, and remain both effective in delivering services and accountable, including indicators like the global goals, our public institutions need to leverage spatial data (Brovelli et al. 2019). They do this typically through a national spatial data infrastructure (NSDI), which is created and managed by official government cartographic agencies like national mapping institutes.

But it is often difficult and expensive to maintain comprehensive NSDIs for many countries, especially lower-income nations. It is a related challenge to do so in ways that offer open public access to this public institution's information.

On the other hand, crowdsourced maps produced by volunteers contain important information for building NSDIs and supporting the efficient services and accountable metrics related to them. But unfortunately, this open volunteered spatial information like that found on OpenStreetMap (OSM) is not frequently well integrated into official cartographies. Some of the barriers to this include that local and national public institutions may not trust data quality, official standards may not coincide, and sometimes even volunteered data is still incomplete.

We developed a project to explore these kinds of challenges and barriers using YouthMappers as a testbed for finding opportunities to link OSM data with official cartographies within a group of seven countries in Central America and the Caribbean. These case studies fit into a larger study focused on expanding our knowledge concerning the challenges and conditions of acceptance of volunteered geographic information being incorporated into the official framework of geographic data, users, tools, and applications of the Western Hemisphere (Wintemute et al. 2021). Our main goal was to connect with government geographers and cartographers to explore how OSM and YouthMappers could support building effective, accountable, and inclusive institutions, especially related to their NSDIs, meeting SDG 16. In the process, by engaging a team of YouthMappers and involving and establishing local chapters, our work offered significant networking and learning opportunities. The placement internships within hosting public agencies

serve as an exciting potential educational model for advancing SDG 4 as well.

1.1 How Do Governmental Institutions Benefit?

Despite the increased availability of internet mapping sites (Google Maps, Google Earth Engine, Microsoft Bing, Nokia Here) that also provide current and historical high-resolution satellite imagery for public access, and data creation tools (such as OpenStreetMap), many governments find difficulty in using or integrating it with official data. If some of the challenges and barriers to using them could be discovered and mitigated, and if the open mapping communities could leverage moments of opportunities to participate with their national mapping agencies, the public as a whole would benefit from more efficient and effective services, and students could gain valuable educational experience. This is particularly true in countries where public access to geospatial data is restricted, or the costs are currently out of reach of government budgets.

It is clear that officials rely on official cartographic data to make informed decisions about a wide range of applications – from infrastructure development to governmental services to locating schools and health care facilities or responding to disasters like droughts and floods. Benefits also include support for economic development and planning, land tenure and taxation, renewable energy resources, and public actions addressing virtually all of the SDGs.

1.2 Where and Why the Americas?

This region is of particular interest because this part of the world has significant experiences with cross-border harmonization of official fundamental spatial datasets to build upon. We also bring our YouthMappers experiences with universities and students to build volunteered open mapping datasets. With funding from the National Science Foundation's broader research project on these questions, the students were able to immediately connect to important officials who are members of the Pan American Institute of Geography and History (PAIGH). PAIGH is a specialized organization of the Organization of American States and provided the context and relationships needed to engage geographic institutes of the specific countries in the region. PAIGH was the platform for accomplishing cross-border spatial harmonization efforts that provided both information and a collaboration context for our work (Norori et al. 2013). As interns, we were competitively selected from among YouthMappers chapters located throughout the United States (eligible to receive NSF stipends), and we choose individual countries to carry out the case studies that supported the larger research project.

1.3 Why and How to Engage YouthMappers?

Youth voices tend to attract the attention of decision makers in ways that other types of research may not be able to draw. But they also might reinforce the distrust that officials have for the quality and completeness of volunteered spatial data. In addressing the issue of trust, student volunteers need to put in place processes to produce data with controls for accuracy and complexities required to successfully support public uses of the data or to meet government standards. Programs such as YouthMappers offer a peer-review data training, collection, and validation system that has proven to work, which ideally can increase the level of trust to work with youth on data campaigns. Although the resulting data might not be fully integrated or taken up as is into the official cartographies, they could serve as useful additional sources for validation, triangulation, or public consultation in order to support decision-making or provision of services. They could also serve as an independent source for measures of accountability, such as to ensure inclusive public services or just outcomes. Doing this does require partnerships across government institutions, especially between cartographic agencies and universities, which may already be

public. Engaging local university students means building educational opportunities that simultaneously address global education goals (Fig. 26.1).

In this project, a competitive call for university research interns resulted in the selection of eight students with Latin American cultural knowledge, Información Geográfica Voluntaria (VGI) experience, and a minimum level of conversational Spanish fluency. While the research interns were being selected, the principal investigators (PI) presented the project to the directors of the National Geography Institutes within Central America, Mexico, Colombia, and PAIGH observer nations in the Caribbean to request their participation in the case studies.

2 A Summary of Six Case Studies

The broader research project was carried out in three main phases: a roundtable forum, user surveys, and then individual country case studies to pilot ideas (Wintemute et al. 2021). Our YouthMappers portion of this effort focused on the case studies. There were two core requirements for each case. The first element was to conduct a series of interviews with both government officials and volunteer mapping participants in the country. The second was to implement a jointly agreed upon a pilot project that yielded some tangible deliverable that would benefit the hosting government agency and ideally involve local chapters of YouthMappers – whether they were already existing or could be gathered and encouraged to establish new chapters at national universities. At first, we expected to spend six weeks in our assigned country but because of COVID-19, the effort had to shift to virtual activities online.

Directors of the National Geography Institutes of Belize, Colombia, Costa Rica, the Dominican Republic, Jamaica, Mexico, and Panama all agreed to participate in the case studies and (virtually) host YouthMappers interns from the United States. The seven countries involved showed differences in the level of involvement and awareness, availability of data and coordina-

Fig. 26.1 YouthMappers researcher Vivian Arriaga presents the project to delegates at a scientific diplomatic meeting of official government cartographic and geographic representatives in the Dominican Republic

tion of VGI from OSM, varied presence of existing YouthMappers chapters and NGOs, and unique opportunities for how the connection between volunteered and official data might be possible in the national context. Below are summaries of that context, the activities, and main takeaways from each case.

2.1   Belize and Jamaica

This case study united the efforts of interns for two countries, in an exchange of training and experiences: in Belize, the host was the Land Information Centre, Ministry of Natural Resources, and in Jamaica, the host was the National Spatial Data Management Branch in the Ministry of Economic Growth and Job Creation.

As for official cartographies, data in Belize is collected by governmental employees and is typically only shared within agencies, private enterprises, and institutions of higher education through the Belize NSDI, via a website that allows registered government and NGOs viewer access in a geoportal and simple download access of non-georeferenced pdfs of base cartographic layers. Official Jamaica's Land Information Council National Spatial Data Management Division coordinates the NDSI policies, training, and access. Limited land information data is available through eLandJamaica web map. Users can view data and purchase proprietary products for download. Jamaica's NSDI is well-developed

but most of the data is sold to recover production costs.

As for OpenStreetMap contexts, in Belize, the OSM community is very small with limited data compared to other countries. Here one finds an average of just under 15 active OSM members at any given time. In Jamaica, there are over 100 senior OSM mappers who frequently work on mapping their own communities and abroad. Additionally, YouthMappers chapters, such as at the University of the West Indies in Mona, Jamaica, regularly contribute to local and foreign projects.

In connecting the dots, discussions with employees of the Belize Land Information Council illuminated for us that VGI has been a topic of discussion, but no project has yet emerged because of time and resources and fear of data quality. A unique example of a government-supported VGI network in Jamaica is their National Emergency Response GIS Team (NERGIST). This is a volunteer group of government employees trained and coordinated by the Land Information Council to collect data during natural disasters, a highly effective response effort.

Our interns in Jamaica and Belize designed a plan to virtually host a joint multi-sector workshop with OSM volunteers, university students, and government officials across both countries. The objectives were (1) to introduce YouthMappers and OpenStreetMap; (2) to conduct a mapathon for data in Belize that supports the management of COVID-19 near Orange Walk; and (3) to seed exchange between geographers in the two countries and between prospective geography students and government geographers within each country. More than seventy people participated in the two-hour session which included local guest speakers, and training on data collection, standards, validation, and satellite imagery interpretation. Jamaican participants remotely mapped buildings and roads, while Belizean locals tagged grocery stores, town halls, and other entities. In all, nearly 40 percent of the task was completed, with 80 percent of changesets later validated without errors; 15 percent required very minor corrections (mostly squaring building edges), and 5 percent were significantly in error and were corrected or deleted by YouthMappers interns (Fig. 26.2).

The experience demonstrated the possibility for cross-nation collaboration and sharing of expertise in terms of official use of volunteer spatial data, at least for uses that deal with emergencies and public health. It also generated enthusiasm for student exchanges.

2.2   Colombia

Multiple host institutions in Colombia offered a platform for the internship, including the Instituto Geográfico Agustín Codazzi (IGAC), Universidad Pedagógica y Tecnológica de Colombia (UPTC), Unidad de Planificación Rural Agropecuaria (UPRA). Of the seven countries involved, only Mexico and Colombia have legislation to allow free and open access to official geographic spatial information. They allow the public to directly download data in a GIS format from their open web portals. In Colombia, VGI success is most visible in citizen or participatory science initiatives. There are over 100 initiatives of citizen science projects focused on collecting data for environmental factors such as forestry, flora, fauna, and air and water quality. YouthMappers are also very strong in Colombia, where there have been chapters formed at seven universities across the country. As a result, the OSM network is inclusive of students in a significant way.

Through conversations sparked by our internship pilot case study, we grew interest on the part of the national geographic institute to apply open mapping applications for geographic names and cadastral data updates in collaboration with student mappers. A longstanding formal agreement between IGAC and UPTC provided a unique opportunity for us to offer new trainings and advance a focused dialogue around deepening connections between student volunteer mapping to official data of Colombian government agencies. We thus facilitated two workshops to encourage greater participation in the very active OSM and YouthMappers networks. The workshops increased the university students' familiarity with iD Editor and KoBo Toolbox. Examples of YouthMappers projects were shared by the YouthMappers regional ambassador and posted to our YouTube channel and the Resource Library as resources for Spanish-speaking students and faculty at other institutions. Another result of this work was that UPRA and our team collaborated to design a new project, titled: "How to Incorporate VGI in the Official Cartographies of Agricultural Landscapes of Colombia: A Methodology Developed for the Rural Agricultural Planning Unit."

2.3   Costa Rica

In Costa Rica, the Instituto Geográfico Nacional of the Registro Nacional (IGN) hosted the internship. IGN manages the Sistema Nacional de Información Territorial (SNIT). Here users can integrate the base cartographic, orthophotos, and thematic data layers in their Open Geospatial Consortium web-based software applications. They can use MapServer, ArcGIS Server, and GEOServer for visualizing, but not downloading data. At the same time, the OSM community here has mapped the majority of the road networks as well as tourist and travel facilities, including with assistance from outside of Costa Rica, making these layers largely complete and of high quality. This is partly due to the fact that OpenStreetMap Costa Rica has developed and documented editing guidelines created by experienced OSM con-

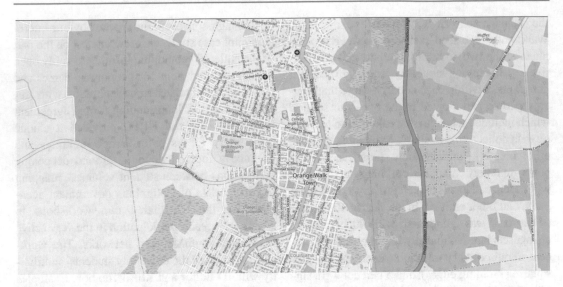

Fig. 26.2 Building footprint data is visible for Orange Walk in Belize, site of a mapathon conducted by the YouthMappers during the project

tributors around the world. There is an active student community at the University of Costa Rica through their YouthMappers chapter, as well. The application of data around international tourism represents a unique and worthwhile opportunity window for connecting volunteered and official spatial data.

Before our case study, IGN had not considered integrating crowdsourced data into their official geospatial database. From our interviews, it was obvious that many OSM volunteers had perceptions of the government as being uncommunicative and uninterested. No framework or platform to encourage government and community discussion and collaboration yet existed. YouthMappers interns and facilitators thus held an initial scoping and exploration meeting where members of the IGN and experienced OSM mappers developed a shared vision for integrating OSM into official government datasets. Despite that there are still challenges to overcome to realize this goal, we were able to ignite these connections through a focused dialogue between two previously detached sets of actors of Costa Rica's mapping landscape.

2.4 Dominican Republic

The one country without an existing NSDI, the Dominican Republic, was developing a spatial data information network when the pandemic hit, and had to put these efforts on hold. Furthermore, there is no YouthMappers chapter in the country, although there are students at universities who are eligible to establish them. Nevertheless, the Instituto Geográfico Nacional José Joaquín Hungría Morell (IGN-JJHM) hosted an internship with our project to continue to progress their thinking about the role of volunteered information in this process.

NGOs in the Dominican Republic have created environmental datasets and vulnerable population data, yet there is no general repository for government or public access to these datasets. So we had to work from the ground up on ideas for simultaneously developing VGI and NSDI in parallel tracks. The relatively recent launching of the IGN-JJHM in 2014 was a challenge but also the fact that the institution is in the early stages offers an incredible chance to introduce awareness of open mapping for both government officials and university students just as this information could be incorporated into institutional strategies and activities. We thus offered four workshops (virtually, and in

Spanish) to various government and academic geographers around: Información Geográfica Voluntaria (VGI), YouthMappers: Usos y Beneficios, YouthMappers: Como Usar OSM, and lastly, YouthMappers: Cómo Realizar un Mapatón. These presentations aimed to inform individuals and groups and provide a set of reference resources specifically tailored to the Dominican Republic, including videos, scripts, instructions, and other necessary material to properly convey information and provide the institution with materials they can easily understand and use in future workshops. Eventually, developing a YouthMappers chapter at a prominent university will complement these efforts to broaden awareness on the topics, including the uses, benefits, and limitations of VGI, crowdsourcing, and OSM, for the upcoming careers of geographers, both in academia and government agencies. This experience showcases how in locations where these ideas are only emerging, the network of youth can support this emergence, and ultimately demonstrate the power of involving local youth in national efforts at open data that could improve official cartographies even from their starting points.

2.5 Mexico

The Instituto Nacional de Estadística y Geografía (INEGI) in Mexico allows free and open access to official geographic spatial information. Collaboration among government agencies, geospatial research centers, and VGI communities have produced an abundance of crowdsourced applications. As a host to our project, we focused together to prepare a detailed report of recommendations to improve and expand the volunteer mapping programs to include greater participation of women, stronger ties to the YouthMappers chapter at the Universidad Autónoma del Estado de México (UAEM) and leveraging a special connection to GeoChicas, an international women's mapping community within OSM that is exclusively active in Latin America.

We began with an information gathering component, including conducting research and interviews with key INEGI personnel and then various actors in the volunteer mapping landscape. This led us to conclude that there exist three different pathways where VGI could expand relative to official cartographies in the country.

First, a top-down approach is possible in a place like Mexico, because several high-level leaders at INEGI expressed an interest in VGI data, and they already operate in a relatively open environment in terms of their NSDI. They have a vision to implement new projects and can serve as champions within the public government agency. They know what specific integration avenues could ensure that the efforts directly speak to official cartographic needs and thus shortcut to success. Direct links to youth and students through mechanisms such as internships could be facilitated.

Another approach to expansion is through a topical focus on inclusion, namely centered on women and on feminist issues. INEGI incorporated a consultant position with GeoChicas. This opportunity to expand participation, representation, and activity is a rich source of ideas on the public utility of INEGI's data. Female students and youth will be among the important communities to engage in through this approach.

A third path that represents a significant opportunity for the growth of YouthMappers chapters across Mexico is through place-based science that merges official data and participatory cartography. The current Director of Geography at INEGI originally founded the CentroGeo scientific think tank which focuses on the need for integrating geographic knowledge and communication in the country at the community scale by applying contextualized science and public participation at the core of nearly every project. One example, in its sixth year, invites the public to suggest updates to place data. This may include updating official names or changing their use designations. While these efforts have had challenges, especially meeting coverage goals, they have expanded their reach by utilizing students in universities. Students are rewarded with community service hours, and have become the largest citizen contributor group to the project. Participatory Cartography's INEGI-university relationship was replicated in INEGI state offices to build collaboration with locally qualified universities. The process includes state offices reviewing and transferring approved data to headquarters. The goal is to identify areas of activity where INEGI data does not exist, which state personnel can field-verify. In the future, the network and structure of more YouthMappers chapters could accelerate these kinds of programs.

2.6 Panama

Panama hosted an intern with the Instituto Nacional Geográfico Tommy Guardia, in the Autoridad Nacional de Administración de Tierras (IGN-TG). Panama does have an NSDI framework and geoportal. IGN-TG allows public downloads of topographic maps in pdf format, but does not make the geospatial data public in GIS format. In Panama, crowdsourced data is present within a small community of independent OSM mappers and mostly by the very active University of Panama's student chapter of YouthMappers. YouthMappers UP generates volunteer data through collaborative projects that are geared toward academic and research projects that have multiple users and stakeholders, and the support of both public and private sectors. As one of the first international chapters of the YouthMappers network, UP has ample experience driving new data campaigns that have important end uses. However, initiatives from the other direction: from government agencies outward to incorporate VGI have not been happening.

Our internship confirmed that Panama has a wide range of stakeholders in academia and government agencies around spatial data. Many organizations produce geographic data, yet there are many silos, and few collaborate by sharing data or incorporating crowdsourced data. For example, Panama's official geographic institute, Tommy Guardia, has the cartography expertise, but lacks the resources to collect data extensively and intensively. YouthMappers has experience in mobilizing volunteers to collect data but expressed the need for more training to maintain the quality of data collected and to utilize more advanced tools and techniques.

Through the conversations sparked by our project, we honed in on a fundamental issue to resolve before greater collaboration could unfold. Different government organizations use different data formats, platforms, and even contradicting attributes for the same objects, hampering basic operations like merging across official datasets. However, these problems that are barriers to more integration of VGI are the same as those that collaboration through VGI might be able to solve.

We chose to explore how to begin this process by focusing on street names in Panama. Here, one street might have an official name differing from its actually used name. Official data did not contain names for 272 of the 613 streets in a pilot project zone. Other streets have multiple names. Updating this complex data platform requires money, effort, and personnel to develop concerted efforts to standardize nomenclature (Fig. 26.3).

Due to the pandemic, the government-mandated quarantine prohibited the ability of government officials to accomplish fieldwork to verify geographic names. Our YouthMappers intern demonstrated a Python program that uses volunteer data from Open Data Kit (ODK) to pinpoint inconsistencies in official street data, which was then implemented by geographers at IGN-TG. What started out as a small idea grew to engage the attention and collaboration of the institution's senior members, including the director of the technology department. The University of Panama's YouthMappers were engaged in the project's user workshops which led to collaboration on a VGI-to-official-dataset project framework. Participants expressed interest and capacity to further develop the project, proposing improvements and internalizing the ODK and Python framework through documentation and workshops. The incentives for officials included updating data at a reduced cost while integrating local knowledge into the geographic names. The incentives for volunteers included the opportunity to focus another category of efforts to respond directly to national official spatial data needs.

3   Challenges and Recommendations

Our case studies showcased a range of starting abilities, awareness, and capacity for incorporating open mapping with official spatial data infrastructures. Through the creativity of our YouthMappers interns, and the graciousness of our hosting institutions, windows of opportunity were discovered across various applications, from responding to health emergencies and natural hazards in Belize and Jamaica, to managing environmental conservation efforts in Colombia; from leveraging tourism assets in Costa Rica, to staging a new national spatial data infrastructure with youth in the Dominican Republic; from promoting gender equity in Mexico, to improving geographic names in Panama. These experiences sum up to a better understanding of both the criti-

cal challenges and the significant opportunities for incorporating VGI geospatial data into official geographic information.

The main recommendations we can derive from the collective experience of these YouthMappers-led case studies, as well as from the results of the larger research project (Wintemute et al. 2021), include the following from the lens of student experiences:

- Successful examples from the region help raise awareness of opportunities for youth and for open mapping, mitigating concerns of quality.
- Place-based knowledge from the field that students can supply can amplify goals for data relevance or completeness coverage that public agencies may need.
- Empowering local champions for official cartographic and VGI collaboration is key, and youth can serve in some of these leadership roles, too.
- Given the diversity of experience and capacity, efforts must be tailored to the individual

Fig. 26.3 An OpenMapKit tool tests how to update or add official geographic names of streets in Panama

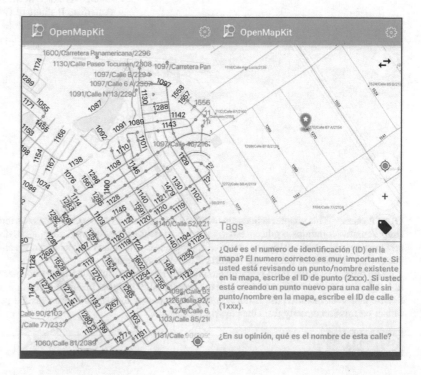

needs and opportunities present in each national context.

- Exchanges across countries can accelerate adoption in the region, as well as offer exciting learning opportunities for students.

It is our hope that this work spawns new champions and partnerships between the local OSM communities and government institutions. YouthMappers chapters can provide a supportive framework for efforts to build more effective and accountable institutions via fundamental spatial data, advancing SDG 16. At the same time, these opportunities can represent significant educational innovation for SDG 4, which in turn can build the spatial talent for the future of these same public institutions, advancing SDG 16 again, and indirectly all of the UN Sustainable Development Goals.

Acknowledgments Funding support for this project was provided by National Science Foundation Award #1907123, 5/19-03/22 (PIs Solís, Wintemute, Aguirre) and by USAID Award #AID-OAA-G-15-00007 and Cooperative Agreement Number: 7200AA18CA00015. Institutional support provided by the Pan American Institute of Geography and History, www.ipgh.org.

References

Brovelli MA, Ponte M, Shade S, Solís P (2019) Citizen science in support of digital earth. In: Guo H, Goodchild M, Annoni A (eds) Manual of digital earth. International Society for Digital Earth, Springer, Singapore, pp 593–622. https://doi.org/10.1007/978-981-32-9915-3

Norori M, Williams C, Torres E, Salazar L, Sáncho E, Flores E, Gómez EI, Benítez F, Menendez R, Núñez M, Martínez A, Wintemute J, Lugo, R., Zavala E, Ramírez N, Pérez J, Morales C, Samuels E, Mondragón R (2013) Proceso participativo de producir un mapa integrado de Centroamérica y sur de México. Revista Cartográfica del Instituto Panamericano de Geografía e Historia 89:47–60. Available from PAIGH. https://revistasipgh.org/index.php/rcar/article/view/485/503. Cited 5 Jan 2022

Stuart E, Samman E, Avis W, Berliner T (2015) The data revolution: finding the missing millions. ODI research report 03. Overseas Development Institute, London

United Nations Independent Expert Advisory Group (UN) (2014) A world that counts: mobilising the data revolution for Sustainable Development. A report to the UN Secretary General. United Nations, New York, p 28. https://www.undatarevolution.org/wp-content/uploads/2014/11/A-World-That-Counts.pdf. Cited 5 Jan 2022

Wintemute JP, Solís P, Aguirre N, Arriaga V, Birkenes A, Zhang C, Council D, Wulf E, Lay-Soler EK, Jones M, McCarley JS (2021) Volunteered mapping and data governance in the Americas. Report for the Pan-American Institute of Geography and History, Mexico City, Mexico. Available at: bit.ly/VGIamericas. Cited 10 Jan 2022

Stellamaris Nakacwa and Bert Manieson

Abstract

Along with increasing urban growth rates, especially in the global south, cities are becoming more fragile because of rapid climate change, insecurity, and increasing urban landscape challenges. With the limited budget sums, coupled with outdated and limited spatial and aspatial data, city planners, governors, and governments are left short of the optimal and efficient approaches to deploy and reckon just, smart, and sustainable cities across all populaces. This demands agile tools and applications for effective decision-making to maintain and sustainably improve quality of life with an assurance that no one is left behind. We demonstrate the potential utilization of OpenStreetMap datasets by urban planners and governing councils to enhance evidence-based planning and policy initiatives. Several projects have been pioneered and executed by youth to demonstrate their crucial role in the organization and collection of crowdsourced geospatial data as a manifestation of the broader theoretical underpinnings of urban governance encapsulated in SDG 16 – Peace, Justice, and Strong Institutions and SDG 11 – Sustainable Cities and

Communities. We argue youth are communicating through the collection of the data. We demonstrate practical approaches to the inclusion of OSM and the participation of local YouthMappers chapters towards objectively positive, just urban governance.

Keywords

Governance · Drones · Flooding · Smart cities · Justice · Uganda · Ghana

1 The Scope of the Challenge

In the current era of globalization and rapid urbanization, governance of cities remains a significant challenge of the 21st century and central to the achievement of sustainable development. As the share of the global population living in urban areas has surpassed 50%, cities have become a new historical phenomenon and have increasingly attracted attention as sites of the

S. Nakacwa (✉)
Makerere University, Kampala, Uganda

B. Manieson
University of Cape Coast, Cape Coast, Ghana

© The Author(s) 2023
P. Solís, M. Zeballos (eds.), *Open Mapping towards Sustainable Development Goals*, Sustainable Development Goals Series, https://doi.org/10.1007/978-3-031-05182-1_27

main threats to global sustainable development, but also as sites of the solutions (New Urban Agenda 2017). According to the UN Habitat (2010), rapid urbanization has taken place in Asia and Africa since 2007, with Sub-Saharan Africa estimated to be over 48% of the entire population by 2030. This inexorable urban transformation in Africa presents a pressing challenge to the landscape of urban governance concerning ideas of 'just', 'smart', and 'sustainable' cities (Gupta et al. 2015).

As a process by which local, regional, and national governments and stakeholders collectively decide how to plan, finance, and manage urban areas, urban governance influences whether the public benefits from economic growth, and also determines how they bring their influence to bear and whether political and institutional systems, processes, and mechanisms facilitate inclusive and pro-poor decisions and outcomes (William 2016). It involves a continuous process of negotiation and contestation over the allocation of social and material resources and political power.

With good urban governance, cities act as engines of growth and provide inhabitants with better opportunities and safe and innovative spaces, thus improving their global development pathways (Turok 2014). With distinctive targets and well-spelt-out responsibilities across all levels of governance, cities take a more central role in building national growth through increased revenue generation and political stability, as well as enabling shared and common lingual spaces for the global community. Without good governance, cities besieged by poor planning struggle to adapt to pressing political, social, and spatial challenges as a result of increased inequality, conflict, and environmental degradation (William 2016). Encapsulated in the 48% population growth predictions for sub-Saharan Africa, a burgeoning youth bulge currently estimated at a 20% for the youth demographic, between the ages of 15–24 (Eloundou-Enyegue 2021), will face insurmountable challenges with a very limited range of policy learning to mitigate the rapidly developing risk-taking behaviors that can be a catalyst of all forms of violence. These cities also generally have fewer resources and less governmental capacity to singlehandedly center and adopt policies to achieve sustainable development goals.

In this chapter, we reproduce OpenStreetMap (OSM) as a data platform that each country within the sub-Saharan region can capitalize as a data platform allowing cities to inform unique and fitting infrastructure and other service investment strategies in the efforts to build just and smart cities. By building and capitalizing on the data capabilities of web 2.0, we envisage data platforms at city scale in each country as concrete action that must be developed to allow for a well-informed, locally and nationally integrated targeted action that can be monitored and evaluated to ensure effective policy implementation (Smit and Pieterse 2014). The main proposition is that there have not been comprehensive, deliberate efforts by local and national governments, development agents, and urban systems to engage the value of data embedded within the volunteered database of OSM. Students are skilled, motivated, and strategically positioned to support this effort through data collection and data interpretation.

2 OSM for Urban Governance

Regarded as a web 2.0 development, OSM, a collaborative mapping project aimed at creating a database or digital map of the world (Quinn 2017), renders comprehensive coverage of spaces, places, and locations with free explicit geospatial data and information. Built as a crowd-sourced campaign, OSM proves an incorruptible space promoting visibility of every individual or community and propagate peaceful and harmonious exchange amongst city planners, policymakers, and the public, at a very commendable temporal resolution and in a smart manner. Its growth has proven a powerful antidote in the reproduction of community through various mapping engagements that aggregate a mix of profiles, editing habits, and robust styles across locations. It capitalizes 'citizens as sensors' thereby launching a domain that propagates unencumbered youth participation in building knowledge on emerging environmental, political, and socio-economic issues. Here, young people have the space to reconstruct and build pride within their cities through improving data and map quality. Accessibility to this newly collected data that is open and readily available, can pave the way for quick alternative policy interventions and at the same time drive policy attention to the most critical and unique aspects of their own cities' urban revolution.

3 Role of Youth Towards Achieving Positive Urban Governance

Youth participation provides young people the opportunity to share their knowledge and acquired skills. To ensure peace in developing countries it is important that young people are engaged, and their capacities improved to enable them to partake in the decision-making process. Through the use of OpenStreetMap, young people have been engaged in community mapping projects while engaging community leaders to tackle real-world challenges in their communities. YouthMappers have also created a space for young people to build their skills, techniques, and capacity in the geospatial through peer training programs. The YouthMappers model through the OSM platform has proved to be an effective tool in ensuring youth participation and capacity building.

Encapsulated in a cluster of case studies in Uganda and Ghana, is OSM data generated to support the process of effective decision-making regarding cross-sectional challenges constraining the achievement of just and smart cities. Overall, our stories and reflection intend to provide a contemporary knowledge tenet, towards scientific policy learning and positive thinking among policymakers. As YouthMappers, we have set ourselves on the course of reviving practical policies and OSM as a geospatial dataset to change the participation in governance and at the same time prompt agile realization of SDGs 16 and 11 targets.

4 YouthMappers Smart, Just Activities in Uganda

Ggaba is one of the rapidly growing informal settlements within the administrative bounds of Kampala. Unlike others, it is located within the Murchison Bay area of the world's largest freshwater lake, Lake Victoria, intercepted by Uganda, Kenya, and Tanzania. Coupled with the increasingly unpredictable rainy seasons, drastic water upsurge in the lake is making the livelihoods of the residents of Ggaba unpredictable, leaving the community beyond a breaking point. Encumbered by the 3.7% population growth rate, the severe encroachment on this bay is defined by the long-term failure of governments to plan, monitor, and collectively enforce development controls, resulting in rapid wetland ecosystem destruction, increased livelihood vulnerability, and high risks of ill health and poverty (Isunju et al. 2016). It is also defined as a landing site, lucrative for the fishing community and many recreational spaces for activities that instigate a significant amount of change in land use.

4.1 Flood-Risk Mapping in Ggaba

The Uganda Opening mapping program modeled flood resilience mapping mechanisms that Kampala and other urbanizing areas can reproduce in order to improve decision-making. The goal of the program was to develop methodology that integrates local capacity and available technology to improve access to baseline data through spatial data crowdsourcing. Baseline geospatial data can be used for disaster risk-reducing policy, designing interventions, for efficient disaster response, and to quicken decision-making (Fig. 27.1).

Uganda Bureau of Statistics (UBOS), Kampala Capital City Authority (KCCA), Office of Prime Minister (OPM), GeoGecko, community members, and YouthMappers, under the guidance of Map Uganda as the controlling organization, developed, discussed, executed, and overall participated in the successful implementation of the mapping activities in

Fig. 27.1 The study area of Ggaba is outlined, where mapped as well as unmapped regions are identified

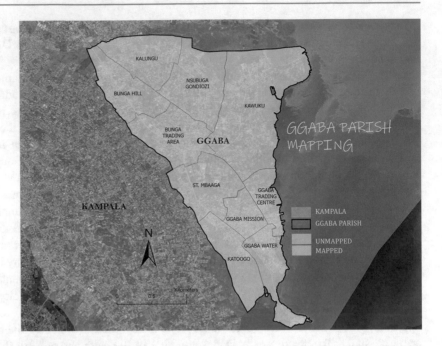

Ggaba. Chapter members of *GoodMappers*, *Geo YouthMappers*, and *Mappers For Life*, YouthMappers chapters in Uganda, were among those responsible for ensuring quality and field data collection based on the project model and general OSM quality standards. Overall, the project involved approximately 60% of youths to ensure robust rigorous and rapid data collection maximizing both time and cost. This was possible because of their signature advantages in terms of numbers but also their expertise in handling modern smartphone technology which consists of GPS hardware that can be exploited to derive spatial representation of their communities and other under-mapped societal issues (Figs. 27.2 and 27.3).

4.2 Project Model

A cost-effective methodical practice was developed by applying OSM to conduct flood-risk assessments and a disaster preparedness system with low administrative costs that provided geo data planning and data collection. The model was organized into three categories (1) remote and field mapping in OSM, (2) drone mapping, and (3) validation of collected remotely mapped fea-

Fig. 27.2 A data cleaning session unites team members from Uganda Bureau of Statistics, Map Uganda and YouthMappers

Fig. 27.3 The YouthMappers team coordinate remote mapping tasks using open tools

tures to reproduce a real-time spatial outlook of the livelihoods of the four villages that compose Ggaba. These would later be used to estimate the flood-risk maps and atlas for both community and authority engagements on the best decisions forwards towards a sustainable mitigation and management plan.

4.3   Remote Mapping

Open-source tools including the Humanitarian OpenStreetMap Team (HOT) Tasking Manager, Java OpenStreetMap (JOSM), and iD Editor were used to create mapping project tasks to support mass mapping process in OpenStreetMap. Buildings and roads data were generated by many people working on the same overall area of approximately 1.3 km² that included Ggaba Water, Katoogo, and Ggaba Trading Centre within Ggaba Parish. A total of 5657 buildings and 14 km of road data were created on OSM and made publicly available Visit https://tasks. hotosm.org/ and https://tasks.teachosm.org to learn about remote mapping.

4.4   Data Collection

A field mapping exercise was executed to create GPS spatial and aspatial properties and attributes of the mapped roads and buildings to derive the significant qualities of Ggaba infrastructure and flood-risk components. This was executed with open-source mobile applicationsOpenDataKit (ODK), OpenMappingKit (OMK), and OSMAnd for field performance monitoring and were downloaded onto individual smartphones. GPS tracks included the drainage footprint of the area and aspatial data included drains, roads, and buildings (Table 27.1).

These were integrated into a data model profiling detailed information on the unique features within the area that make a composition of the similar features that exist within other parts of Kampala and possibly other urbanizing areas of Uganda. This model was discussed and agreed upon by the community together with the governing authorities in the city under their participating capacity and henceforth handed over for mass extension in other flood-prone areas of the city.

Table 27.1 Categories of data features and attributes mapped

Drains/ditches	Buildings	Roads
Profiles	Purpose	Type of road
Connection	(Residential/commercial/social)	a. Surface
Width	Levels	b. Smoothness
Length	Building material	c. Width
	Roofing material	d. One-way
		d. Road Name/street Name

4.5 Drone Imagery and Validation

Drone imagery was collected for the 1.3 km^2 and was used within JOSM for the validation of the previously remotely mapped buildings and roads. This was necessary because it provided a desirable temporal and spatial resolution to clearly identify the actual number of buildings regarding the informal nature of the building styles within the area, especially in Ggaba Water. The drone mapping also complimented the derivation of topographic data that was later combined with the field-collected flood Z points to complete the derivation of a flood heatmap.

4.6 Results and Achievements

The development of a model to scale through other areas including a documented workflow combining OpenStreetMap online and offline open-mapping tools like the JOSM editor, iD Editor, the tasking manager, and field mapping (ODK and OMK) was a significant achievement designed during the planning phase of this activity. We succeeded in demonstrating that OSM is a low cost and integrative technology that can be used to map keys accurately and rapidly. Three villages were fully mapped, which created baseline data for the area establishing ingenuine methods for future accountability and performance monitoring and evaluation of the developments in the area. By October 2018, a total of 8945 buildings and 17 km of roads had been mapped, and through OSM, this data was made available for government and communal use.

Overall, all stakeholders from UBOS, KCCA, OPM, and communities became fully aware of the benefits and workflow processes necessary to create geospatial data. The greatest achievement is that all engaged authorities were exposed to tools and frameworks embedding unprecedented mechanism of democratic engagement of the communities, youths, and marginalized groups in critical issues regarding their communities hence, promoting inclusive and accountable citizens simultaneously propagating just and smart future cities.

This project annexed OSM for planning and policy realization including (a) National Free and Open-Source Software Policy (draft May 2016), (b) National Open Data Policy (draft May 2017), and (c) Uganda Spatial Data Infrastructure Policy (draft Feb 2018), towards inclusion, resource efficiency, mitigation, and adaptation to climate change.

5 YouthMappers Smart, Just Activities in Ghana

The city of Accra is estimated to have a population of about three million people with an annual population growth rate of about 4% between 2000 and 2010 (Ghana Statistical Service 2012). The size of the metropolis is approximately 300 km^2 (Amoako and Frimpong Boamah 2015; Møller-Jensen et al. 2005) and the city's vulnerability to floods has been an issue of great concern to the populace. Being one of the growing cities in Africa, the city is also faced with an uncontrolled increase in informal settlements in low-lying areas that are susceptible to floods. Residents living in informal settlements live in deprived conditions both physically and socio-economically. Most often these areas are situated along river catchments.

Because these settlements are informal, residents lack certain basic amenities and infrastructure such as proper waste disposal, sanitation, toilet facilities, drainage systems, etc. Over the last decade, incidents of flooding have reached an alarming level rendering some inhabitants homeless while leading to the loss of lives of others. According to the National Disaster Management Organization (2009), a total of six million dollars of assets are at risk each year. The causes of flooding range from the overflow of the Odaw River out of its catchments, insufficient channels for solid waste disposal, poor drainage, and uncontrolled growth.

5.1   Accra City Mapping (Open Cities Project)

The Accra City Mapping is an initiative of the Open Cities Project which is a collaborative effort by the Global Facility for Disaster Reduction and Recovery's (GFDDR) Open Data for Resilience Initiative (OpenDRI) and the World Bank Africa Urban, Resilience and Land unit with funding from the European Union's Africa Disaster Risk Financing (ADRF) program. This project was carried out in four low-income communities in Accra. These communities over the years have been affected by floods during the rainy season and have serious issues with solid waste management. The communities included Alajo, Nima, Agbogbloshie, and Akweteyman within the catchment area of the Odaw river basin. The purpose of the project was to provide assistance and support to the Greater Accra Resilience Integrated Development (GARID) project in the management of solid waste and floods. Dealing with the problem, therefore, required an evidence-based approach to ensure decision-making at the national, regional, and local levels of governments is supported by data.

5.2   OSM Approach

The OpenStreetMap (OSM) Approach provides a digital approach to solving challenges spatially. The project brought together OpenStreetMap communities in Ghana including the Humanitarian OpenStreetMap Team (HOT), OSM Ghana, members of YouthMappers Chapters in Ghana, and Mobile Web Ghana. The team created geospatial data to support future decision-making by providing updated data on affected communities. The use of OSM allows for effective participation of all stakeholders from several ministries such as the Ministry of Inner Cities and Zongo's as well as the Ghana Statistical Services and Metropolitan and Municipal District Assemblies (MMDAs).

This approach enables young people to bring their skills in mapping and digital tools to the forefront of the decision-making process. Over fifty community mappers were trained for the mapping activity. Attention was given to representation to ensure gender balance among mappers. Series of engagements were done to ensure participation of leaders even at mapping events as they also tried to understand how the geospatial data is being gathered.

5.3   Results and Achievements

Over 7279 buildings were mapped, 315 km of highway and 418 km of waterways were digitized. These maps are hosted online and are made available to stakeholders and the general public for use. The team from Ghana Web Ghana and HOT created a web application which will allow stakeholders make data-driven decisions on solid flood and solid waste management (Fig. 27.4).

The project is one of a kind that ensured a high level of participation among different stakeholders, youth, and gender balance. The high turnout of young people to volunteer in the mapping activities showed that they are ready and willing to come to the decision table if they are called upon. The project also allowed YouthMappers chapter members to showcase their skills in the use of digital tools as well as providing training

Fig. 27.4 YouthMappers in Ghana support data in the four communities (Alajo, Nima, Agbogbloshie and Akweteyman) for Open Cities Accra. Credit: Open Cities

Open Cities Accra - Alogboshie - Building Types or Use

for other young people to acquire very essential skills that could get them employed. The inclusion of girls in the mapping project proved as a step in the right direction to encourage girls in the use of digital tools.

6   Smart and Just Cities of the Future

The inclusion of young people makes us feel and know that our voices are being heard and that our contributions matter to the development of our communities. It gives us direct access to plan our communities in a way that addresses their needs, especially because we make up the majority of the population in developing countries. Youth participation is highly recommended in the planning of human settlements that are safe, inclusive, resilient, and sustainable for all. The projects described provided a platform for young people to design and provide solutions to the challenges in their communities.

Young people are not spared by the problems-facing developing countries. In most cases we are the most affected, therefore there is the need to ensure inclusion and participation at all levels of decision-making. The greatest benefits are that

OSM, open-source tools, and frameworks prevail as an unprecedented mechanism of democratic engagement of local communities, youths, and marginalized groups. The use of geospatial technologies not only advances global targets like SDG 16 – Peace, Justice, and Strong Institutions and SDG 11 – Sustainable Cities and Communities. But our direct involvement also allows for our participation in the search for solutions to critical issues regarding our own communities and overall, promoting inclusive and accountable citizens and institutions while propagating just and smart future cities. What a way to reproduce sustainable urban governance! Implicitly, our projects demonstrated the capacity of the young population as active responders to the governance priorities of their communities.

References

Amoako C, Frimpong Boamah E (2015) The three-dimensional causes of flooding in Accra, Ghana. Int J Urban Sustain Dev 7(1):109–129

Eloundou-Enyegue PM (2021) Development Sociology, College of Agriculture and Life Sciences. Available via Cornel Research. https://research.cornell.edu/research/exploding-youth-population-sub-saharan-africa. Cited 29 Jan 2022

Ghana Statistical Service (2012) 2010 population and housing census: summary report of final results. Accessed 8 June 2018

Gupta, J. et al. (eds) (2015) Geographies of urban governance. Springer, Cham. https://doi.org/10.1007/978-3-319-21272-2

Isunju JB, Orach CG, Kemp J (2016) Community-level adaptation to minimize vulnerability and exploit opportunities in Kampala's wetlands. Environ Urban 28(2):475–494. https://doi.org/10.1177/0956247816647342

Møller-Jensen L, Kofie RY, Yankson PW (2005) Large-area urban growth observations—a hierarchical kernel approach based on image texture. Geografisk Tidsskrift-Danish J Geogr 105(2):39–47

New Urban Agenda (2017) United Nations conference on housing and sustainable urban development. ISBN: 978-92-1-132731-1

Smit W, Pieterse E (2014) Decentralisation and institutional reconfiguration in urban Africa. In Parnell S, Pieterse E (eds) Africa's urban revolution. Bloomsbury Academic, London, pp 148–166. https://doi.org/10.5040/9781350218246.ch-008

Turok I (2014) Linking urbanisation and development in Africa's economic revival. https://doi.org/10.5040/9781350218246.ch-004

Supporting YouthMappers to Advance the SDGs Through Institutions and Partnerships

Mentoring Experiences in YouthMappers Chapters

Anthony Gidudu, María Adames de Newbill,
Jonathon Little, Maria Antonia Brovelli,
and Serena Coetzee

Abstract

YouthMappers brings together university students coalesced around institutional chapters carrying out collaborative mapping to address humanitarian and development concerns. Beginning at three universities, in six years, YouthMappers has grown into a global movement of nearly 300 chapters in more than 60 countries. This success, in part, is attributed to the strong emphasis on mentorship in the YouthMappers structure. In this book chapter, mentors from Italy in Europe, Panama in Central America, the United States of America in North America, South Africa, and Uganda in Africa share their reflections on the role of mentoring in their YouthMappers chapters. They share about the nature of their chapter activities, how mentoring is effected, what has been successful, and the challenges faced. As YouthMappers continues to grow, the role of mentors will remain central to attaining the YouthMappers mission.

Keywords

Education Partnerships · Mentoring · SDGs · Italy · Panama · South Africa · New York · Uganda

A. Gidudu (✉)
Department of Geomatics and Land Management, Makerere University, Kampala, Uganda
e-mail: Anthony.Gidudu@mak.ac.ug

M. A. de Newbill
Department of Geography, University of Panama, Panama City, Panama
e-mail: maria.newbill@up.ac.pa

J. Little
State University of New York, Monroe Community College, New York, USA
e-mail: jlittle@monroecc.edu

M. A. Brovelli
Department of Civil and Environmental Engineering, Politecnico di Milano, Milan, Italy
e-mail: maria.brovelli@polimi.it

S. Coetzee
Centre for Geoinformation Science, Department of Geography, Geoinformatics and Meteorology, University of Pretoria, Pretoria, South Africa
e-mail: serena.coetzee@up.ac.za

P. Solís, M. Zeballos (eds.), *Open Mapping towards Sustainable Development Goals*, Sustainable Development Goals Series, https://doi.org/10.1007/978-3-031-05182-1_28

1   The Role of Mentorship for a Student-Led Framework

YouthMappers is a global consortium of university-based students creating open data maps to address humanitarian and development concerns (Hite et al. 2018). Each student grouping is referred to as a chapter and is essentially student-led (Solís and Rajagopalan 2019) with oversight from faculty with support from senior students and alumni (Brovelli et al. 2020). To become a YouthMappers chapter, existing student organizations or new student led-groups apply for recognition by the YouthMappers steering committee (Brovelli et al. 2020).

YouthMappers formally began in 2015 as a brainchild of faculty from Texas State University, The George Washington University, and West Virginia University with support from the GeoCenter at the United States Agency for International Development (USAID) (Hite et al. 2018). Within two years of its launching YouthMappers had spread to 100 chapters in 38 countries (Solís and DeLucia 2019). By late 2018 this number had grown to 143 chapters in 41 countries (Brovelli et al. 2020). In 2019, this number grew to 157 chapters in 42 countries (Solís and Rajagopalan 2019), and as of the end of 2021, the number stood at 295 chapters in 64 Countries (YouthMappers 2021). Essentially, this growing global critical mass of mappers is leveraged through mapathons, which are collaborative efforts aimed at generating map data through remote mapping and field campaigns (Coetzee et al. 2018). Some of the users and beneficiaries of these maps are humanitarian and development actors such as the International Federation of Red Cross, Red Crescent, Doctors Without Borders, USAID, World Bank, and UN relief agencies, among others (Solís et al. 2018).

There are multifaceted reasons for the attraction to YouthMappers, and hence its exponential growth. For instance, YouthMappers provides students with the opportunity to collaborate either as individual chapters or in partnership with other chapters to address local, regional and global societal challenges. This consequently enables students to get practical experience in data capturing, open geospatial technologies, and how these can be harnessed to address humanitarian challenges around the world (Rautenbach et al. 2017). These chapters also provide for leadership development both locally within the chapters and through regional and international training opportunities; critical thinking through projects; networking and communication skills; awareness of humanitarian and development issues, all of which are essential soft skills for the job market (Solís and Rajagopalan 2019; Rautenbach et al. 2017; Hite et al. 2018).

The global relevance of YouthMappers is further appreciated in the context of the various Sustainable Development Goals (SDGs) that is addressed both directly and indirectly (Solís et al. 2018). For instance, YouthMappers' maps and projects have attempted to address poverty (Goal 1), hunger (Goal 2), gender equality (Goal 5), clean water and sanitation (Goal 6), energy (Goal 7), and especially partnerships (Goal 17) and quality education (Goal 4), given the engagement of students with community partners to achieve progress across all of the goals, as evidenced by the various examples in this current volume that illustrate this potential range, and the service-learning power of this model. Typical examples include improving food security in Bangladesh and Ghana, health and disease prevention in Mozambique and Kenya, enumerating households for malaria spray campaigns in Zambia, mapping for community resilience in Uganda, flooding preparations in the Philippines, earthquake response in Ecuador, and peace implementation in Colombia (Solís and DeLucia 2019).

But beyond the revolution that open geospatial data brought; beyond the broad applicability of spatial data for a range of problems; beyond the mere attraction of the intersection of youth, technology, and progress: there is an underlying dimension of support that is designed to ensure the success of YouthMappers as a student-led effort. That is the role of mentorship to the chapters' framework, gently structuring and enabling the experiences of students within the open mapping community for humanitarian and development needs, while balancing the creation of a space for leadership and ownership of truly youth-centered action. We assert that one of the significant contributions to the success and sustained growth of YouthMappers chapters has been the role played by its mentors. In this chapter, local chapter faculty mentors from some of the longest-running and inaugural chapters in Italy, Panama, South Africa, United States of America, and Uganda share experiences about their YouthMappers chapters and reflect on the role that mentorship has had on their respective students. In the conclusion, we consider the design element of mentorship in general and encourage a forward-looking stance toward continuous renewal of the movement through a mentorship frame.

2 Experiences from the PoliMappers, Italy: Advancing Curriculum and Interdisciplinary Innovation to Benefit Students

PoliMappers is the YouthMappers Chapter at the Politecnico di Milano, which is the top-ranked and largest technical university in Italy. The chapter started in 2016 with the inaugural group of universities, representing the first wave of expansion beyond co-founding institutions.

The precursor to the formation of PoliMappers was inspired by the first Italian conference on open-source GIS in 2000 and the European Free and Open Source Software for Geospatial (FOSS4G) conference in Italy in 2015. This led to the formation of an informal group of students interested in promoting collaborative mapping using open source GIS. It is this informal group that coalesced to form a more structured and organized grouping now called the PoliMappers. The chapter was structured such that its governance was vested in the students, with oversight from a university professor and a doctoral student. The mentors' roles have been focused on an advisory capacity, as well as providing an inter-

face with the university administration. It is through these official channels that the chapter has won leverage to introduce its activities during the official delivery of GIS lessons by showcasing the role of GIS in addressing humanitarian concerns.

As a student chapter within YouthMappers, PoliMappers has provided a platform for its members to participate in various projects, some of which have gone on to win international acclaim. Thus, the YouthMappers network has both leveraged the expertise of the institutional participation of the university in open mapping pioneership and, in turn, reinforced and provided a reinforcing network to validate and amplify the work of local students. These accolades have been a source of pride to the chapter and consequently have gone on to raise awareness about PoliMappers. The spin-off benefit of this has of course been the growth in chapter numbers. Furthermore, the student leaders have gone on to assume leadership roles in humanitarian mapping organizations, and with international entities like the United Nations.

One of the main benefits of PoliMappers to the university has been the opportunity for interdisciplinary learning. Members have had the opportunity to learn about geospatial applications on the effects of war, deforestation, informal settlements, and accessibility for fragile people. Members have been introduced to new technologies, critical thinking, and how to address societal challenges. At the onset of the pandemic, and as a result of the lockdown, the PoliMappers spearheaded the development and presentation of a 20-hour course unit on Human Collaborative Mapping. This course unit was well received within and outside the University. The benefit was not only in the offering of this course unit but also in ensuring camaraderie during a particularly difficult time. It goes without saying that PoliMappers has been a resounding success, not only to current and past students whose professional and social skills have been enhanced by leaps and bounds but also to the professors who have been inspired to be better educators and mentors.

3 Experiences from the YouthMappers UP, Panama: Local Sector and Regional Partnerships Supporting Student Success

The YouthMappers UP chapter in Panama, Central America was started in August 2016 in the Department of Geography at the University of Panama (*Capítulo de Jóvenes Mapeadores de la Universidad de Panamá*). This chapter was the second university that joined after the co-founding universities invited global membership (the first being University of Cape Coast, Ghana).

The success of YouthMappers UP is largely attributed to the support and mentorship of various professionals giving their time and knowledge consistently, year after year. The chapter has collaborated with YouthMappers partner organizations such as Esri, Digital Globe (today Maxar technologies), Mapillary, and national institutions such as the Panama Canal Authority and Tommy Guardia National Geographic Institute. These collaborations have greatly contributed to the development of students' professional skills besides the technical ones they have learned from using OSM. In this regard, the YouthMappers Humanitarian Mapping network has been very helpful in supporting projects and creating learning opportunities for YouthMappers UP members as individuals.

The spin-off benefits of these collaborations have been reflected in the achievements of the chapter. For instance, one of the chapter members was selected as a beneficiary of the leadership fellowship and was appointed regional ambassador in 2019. Furthermore, two chapter members won the Young Scholar award in 2019 and 2020, while representing Panama at the Esri Users Conference. On a regional level, the YouthMappers UP have collaborated with other YouthMappers chapters, such as a road trip from Panama to Costa Rica (and vice versa). This was organized through a friendship of the chapter mentors, to mutually share mapping priorities.

Such activities build an enduring framework for networking and sharing experiences that the university can celebrate and advance. In effect, this contributes to Objective 17 of the Sustainable Development Goals (SDG) regarding strengthening global partnerships with the purpose of working toward helping communities and achieving sustainable development.

In terms of activities, YouthMappers UP has been involved in mapathons to help collect information needed for communities in Panama and other countries. For instance, most recently, the chapter has worked with an NGO and the community to develop *The Turtle Eco-Route,* aimed at preserving turtles and promoting sustainable tourism. Another chapter project is the Smart University Campus app, which represents one of the first attempts to digitalize the university's buildings and assets.

One of the challenges, from a mentor's perspective, has been the diversity and fast-changing pace of technology being used by the chapter members. Although this seems to be an obstacle, it has also been an opportunity to learn and grow as a chapter. Another challenge has been the fact that students have their school and personal obligations. Despite these demands on their time, the students have been committed to the mission of YouthMappers. There has also been the challenge of growing numbers of students getting involved in the chapter activities. In spite of these challenges, students and professors from outside the chapter have been generous enough to participate and support the chapter activities.

Looking forward, the anticipation is that YouthMappers UP alumni members will be able to pursue careers in academia and contribute as mentors. The chapter also hopes to leverage the support of professors in multiple departments to contribute to inspiring new generations of YouthMappers UP. The chapter also hopes to expand activities to other regional centers of the University (in other cities) and support other public and private universities to start their chapters.

YouthMappers UP is more than just a network of students. Members have created lifelong friendships and had the opportunity to share their knowledge with their peers both inside and outside the university. It is hoped that more students and mentors will be inspired to join the chapter.

The ultimate intention of YouthMappers UP is to provide chapter members the opportunity to develop their leaderships skills, and by using geospatial technologies, help communities solve problems, address resilience, and improve their quality of life.

4   Experiences from the YouthMappers Chapter in Pretoria, South Africa: Generating a Social Circle of Peers and Pipelines

In 2016, the YouthMappers chapter of the University of Pretoria in South Africa was one of the inaugural chapters and only the second chapter in Africa at that stage. Events are arranged by postgraduate and senior undergraduate students under the auspices of a staff member, Dr. Victoria Rautenbach. There are two kinds of activities (Coetzee et al. 2018).

Firstly, mapathons are used to raise awareness of Geoinformatics study opportunities among high school students. Only a very small part of the school geography curriculum is dedicated to geographic information systems - and it is often difficult or even impossible to teach since it requires computer labs at schools. Therefore, many classrooms may have a negative connotation with the subject or are not aware of career opportunities in Geoinformatics. Prior to the pandemic, at least once a year, a group of local high school students visited the UP campus and mapped in OpenStreetMap for several hours. During their visit, they also received information about the university and the Geoinformatics study programs.

The second kind of event involves Geoinformatics students in the final semester of their bachelor's studies. Students work in project teams, simulating a future work environment, and implement a Geoinformatics solution that can benefit the local community (Coetzee and Rautenbach 2016). Data such as building footprints, community water points, and footpaths are mapped via OpenStreetMap, and integrated into a solution built with open-source tools. In some years, mapathons have been locally arranged, inviting a wider group of students so that more mapping can be done, or in other cases, the mapathons were linked to global humanitarian initiatives. In the final year, students take responsibility for validating the contributed data, creating an effective step-wise advancement for peer-to-peer mentorship.

Involving the students in mapathons via the YouthMappers chapter has inspired several students to pursue postgraduate projects related to topics advanced through the YouthMappers experience. For example, one student investigated why students participated in mapathons and found that they felt a sense of humanitarianism by contributing to communities in need.

The social aspect of mapathons was also evident: participants indicated that mapathons are fun and that they learned something new (Green et al. 2019). Another analyzed and characterized OpenStreetMap contributors in Mozambique, specifically after the 2019 cyclones, Idai and Kenneth, caused significant flood damage and a subsequent humanitarian crisis in Mozambique (Madubedube et al. 2021); and a third student is conducting a similar study for the whole of South Africa. Each of these students enjoyed the mentorship of the chapter faculty and more advanced peers. Being involved in the YouthMappers chapters has also opened opportunities for students to participate in their workshops, which connected them to a global community of young humanitarian mappers. Students have taken jobs in local organizations, national-level agencies, and beyond.

5   Experiences from a YouthMappers Chapter in a Two-Year Institution in the USA: Turnover Mentoring and Alumni Leverage

Monroe Community College of the State University of New York's YouthMappers chapter began in 2016 as a direct result of hosting a YouthMappers mapathon. The chapter's mapping events directly engage many of the United Nations Sustainability goals from reducing inequality, partnerships, and quality education, to innovation. A key component of this chapter is hosting at least one YouthMappers mapathon per semester. The chapter has been fortunate to collaborate with many organizations to support its mapathons from college librarians to the National GeoTech Center of Excellence in Kentucky, USA. With the world going virtual due to the pandemic, the club recently hosted over 400 participants in a mapathon during the fall of 2020 in English and one in Spanish with participants from over 10 Spanish-speaking countries.

Typically, students learn about the YouthMappers chapter through their Geospatial classes and often from the mapathons. Interested students are invited to one of the weekly meetings to learn more about the club and its goals. Interested students are mentored by faculty and soon to be graduates. The president of the club takes on an especially important role, both leading the club's meetings and mapathons, and mentoring new students. Depending on the interest of the president and the unique qualities of the officers, this may vary; however, the student president is very important to the mentoring and continued success of the club. Rather than getting right down to business at the start of each meeting, as the liaison for the chapter, the faculty mentor focuses on making connections with the club officers, fostering a positive attitude, and laughing!

One challenge is the need for new officers and presidents nearly every year. Often, as a two-year community college, students graduate within those years, requiring the need for a new president and many officers. Fortunately, a few recent Associate of Applied Science graduates have become mentors for new officers. The alumni mentors have provided countless hours of mentoring to the current students and are advocates for the YouthMappers chapter. The Alumni mentors have also co-led YouthMappers mapathons virtually with current officers. They are usually available at designated times in the evenings and on weekends to augment the availability of support to students and club officers. These virtual meetings again focus on fostering a personal connection. Alumni mentors also lead an online "Ask Me Anything (AMA)" recorded session to answer student questions about anything geospatial, career-related, or about the YouthMappers chapter. The alumni mentoring program began due in large part to a National Science Foundation Advanced Technological Education grant awarded to our college entitled "Meeting Workforce Needs for Skilled Geospatial Technicians".

Looking forward, the college is fortunate to have a new Geospatial lab, which will allow alumni mentors, club officers, and faculty to commingle face-to-face and online, for club meetings, geospatial classwork, and mapathons. It is hoped that this will become a new home to continue the chapter's tradition of mapathons, mentoring, meeting new people across the globe, and supporting the United Nations sustainability goals.

6 Experiences from YouthMappers in Uganda: A Multifaceted, Multinational, and Multidisciplinary Mentor Approach

Geo YouthMappers is the YouthMappers Chapter at Makerere University in Kampala Uganda. It was established in 2016 in the Department of Geomatics and Land Management. Chapter membership is open to students in the Department, but it is also open to students pursuing other courses interested in mapping and related opportunities. The location of the chapter in the department makes it possible to induct new members during their freshman/freshwoman years during their university orientation.

Student chapter alumni also play an active role in the chapter operations. Chapter activities essentially include various mapping activities organized under different themes, for example, mapathons and field data collection exercises. The data from these mapping activities is created using open source geospatial tools and is uploaded to the OpenStreetMap platform. The benefits of associating with the Geo YouthMappers chapter have included: the opportunity to relate what the students are studying in class with the real world; opportunity to apply what students are studying to enhance community resilience and address global challenges; building of professional networks across different academic years, programs, institutions, and countries; exposure to open source technologies; research; travel opportunities; leadership development; mentorship and more.

One of the enduring reasons for the success of Geo YouthMappers has been the role played by mentorship. Mentorship in Geo YouthMappers has been multifaceted and multilayered ranging from faculty mentorship to peer-to-peer mentorship. Faculty mentorship in Geo YouthMappers has been mostly advisory, as well as ensuring that the chapter has the necessary Faculty level support to ensure smooth operations. The main thrust of mentorship in Geo YouthMappers, however, has been articulated through peer-to-peer mentorship. Geo YouthMappers is privileged to have two regional ambassadors who have actively been involved in popularizing YouthMappers in the East African region. They are a reflection of the growth they have enjoyed through Geo YouthMappers, which also serves as a base and case study of the dividends of being part of this chapter.

The student chapter leadership also provides mentorship by designing and providing hands-on training, organizing mapping activities, sharing leadership development opportunities, encouraging participation in regional and international conferences, writing blogs, etc. Geo YouthMappers by design is student-focused and student-led, and so it is incumbent on the current leadership to ensure that there is a crop of incoming students excited to take up leadership roles in subsequent years.

Geo YouthMappers has been successful in attracting students which has ensured its vibrancy. All freshmen and freshwomen to the department have an opportunity to join and the previous success stories have been incentive enough. Geo YouthMappers are privileged to have had several successful applicants to the various YouthMappers leadership development trainings through fellow-

ships and conferences (e.g., in Nepal, South Africa, Germany, USA, Senegal), and these students have gone on to serve as chapter champions. The support of the Department has been immeasurable as this has provided access to University facilities like labs and internet. Geo YouthMappers have also successfully partnered with the local OSM community in Uganda and worked with organizations such as the Humanitarian OpenStreetMap Team (HOT Uganda) and MapUganda.

7   Conclusion

In the narrative from the different chapters, it is evident that YouthMappers is a force for good, with potential for far more good. In spite of the fact that the conglomeration of chapters is in different locations, with members of different academic backgrounds, gender, and other axes of difference in a world of difference, YouthMappers provides a platform for collaborative social engagement in a way that appeals to students' creativity, curiosity, innovation, and passion. It draws upon real-world community partnerships and redefines the way that education unfolds. Whether the approach is hyperlocal to multinational, peer-driven to turnover-ready, curriculum-focused, or research-driven, and everything in between, the mentors are the backstage managers of a strategic institutional leverage on behalf of the YouthMappers vision. An essential and required element of any chapter, attentive mentorship has perhaps been the key enabling design component that has helped realize the promise of youth engagement at the university level. In nearly every case, the mentorship requires a champion, but in an apparent oxymoron, the YouthMappers mentorship model is predicated upon ensuring students assuming ownership and leadership for flagship activities, including how to pass the baton.

The task for any mentor, be they faculty or alumni or peer, is to first and foremost ensure that student members are inspired to cultivate generations of young leaders to create resilient communities and to define their world by mapping it (YouthMappers 2021).

References

Brovelli M, Ponti M, Schade S, Solís P (2020) Citizen science in support of digital earth. In: Guo H, Goodchild MF, Annoni A (eds) Manual of digital earth. Springer, Singapore. https://doi.org/10.1007/978-981-32-9915-3_18

Coetzee S, Rautenbach V (2016) Reflections on a community-based service learning approach in a geoinformatics project module. In: Gruner S (ed) ICT Education. SACLA 2016, Commun Comput Inf Sci: 642. Springer, Cham

Coetzee S, Minghini M, Solís P, Rautenbach V, Green C (2018) Towards understanding the impact of mapathons – reflecting on YouthMappers experiences. FOSS4G2018, Dar es Salaam, Tanzania. Int Arch Photogramm Remote Sens Spat Inf Sci - ISPRS Arch XLII-4/W8:35–42. https://doi.org/10.5194/isprs-archives-XLII-4-W8-35-2018

Green C, Rautenbach V, Coetzee S (2019) Evaluating student motivation and productivity during mapathons. Int Arch Photogramm Remote Sens Spat Inf Sci – ISPRS Arch XLII-4/W14:85–91. https://doi.org/10.5194/isprs-archives-XLII-4-W14-85-2019

Hite R, Solís P, Wargo L, Larsen TB (2018) Exploring affective dimensions of authentic geographic education using a qualitative document analysis of students' YouthMappers blogs. Educ Sci 8:173. https://doi.org/10.3390/educsci8040173

Madubedube A, Coetzee S, Rautenbach V (2021) Contributor-focused intrinsic quality assessment of OpenStreetMap in Mozambique using unsupervised machine learning. ISPRS Int J Geo-Inf 10(3):156. https://doi.org/10.3390/ijgi10030156

Rautenbach V, Minghini M, Coetzee S, Solís P (2017) YouthMappers: a global network to empower students. In: Mapping the world, collection of open conferences in research transport 2017(35). https://doi.org/10.5281/zenodo.820529

Solís P, DeLucia P (2019) Exploring the impact of contextual information on student performance and interest in open humanitarian mapping. Prof Geogr 71(3):523–535. https://doi.org/10.1080/00330124.2018.1559655

Solís P, Rajagopalan S (2019) Workforce development and YouthMappers: understanding perceptions of students in humanitarian mapping. In: Proceedings of the academic track, state of the map 2019 September 21–23, 2019. Germany, Heidelberg

Solís P, McCusker B, Menkiti N, Cowan N, Blevins C (2018) Engaging global youth in participatory spatial data creation for the UN sustainable development goals: the case of open mapping for Malaria prevention. Appl Geogr 98:143–155

YouthMappers (2021) https://www.youthmappers.org/. Cited 20 Jul 2021

The Ecosystem Where YouthMappers Live and Thrive

Dara Carney-Nedelman and Courtney Clark

Abstract

YouthMappers live and thrive in an ecosystem of university chapters, organizers, sponsors, ambassadors, and partners. This system places youth at the center and is designed as an empowering network, which ultimately advances partnerships for the goals, in line with SDG 17. But because we don't just build maps, we build mappers- the result of this ecosystem also expands the capacity for students to advance to decent work and contribute to the economic growth in their careers and of their countries. We present here some of the voices of students, alumni, staff, and partners to describe this enabling ecosystem from their own perspectives.

Keywords

Work and Internships · Partnerships · OSM ecosystem · Youth Economic Development · SDGs

1 The Structure of a Partnership Ecosystem

Partnerships and collaborations are not always the solution to enhance an organization. Sometimes they can be inefficient, logistically complex, and cause more of a burden (Harold 2017; Ostrower 2005). This could especially be the case for a small staff team like YouthMappers. However, when organization leaders focus on the mission and put in place strategic pillars to guide an organization, partnerships can instead lead to impactful rewards. The YouthMappers network does this and has since its ideation design phase in 2014 and its formal creation in 2015, because of our intentional efforts to be student-led and community-based.

1.1 University Chapter Network and Regional Ambassador System

Currently, YouthMappers is not a non-profit organization, a corporation, or a governmental orga-

D. Carney-Nedelman (✉)
YouthMappers, and Texas Tech University, Lubbock, Texas, USA
e-mail: communications@youthmappers.org

C. Clark
YouthMappers and American Geographical Society, New York, NY, USA
e-mail: cclark@americangeo.org

P. Solís, M. Zeballos (eds.), *Open Mapping towards Sustainable Development Goals*, Sustainable Development Goals Series, https://doi.org/10.1007/978-3-031-05182-1_29

nization. Instead, YouthMappers is an international university network. In essence, this means that the individual chapters established on campuses form non-binding ties under a banner of a shared vision, by-laws, and ethics statement (Cowan et al. 2017). The name YouthMappers® is trademarked (USPTO 2017), and digital assets like URLs are secured using this unique handle, which serves as a focus for the identity of our consortium of members, chapters, organizers, sponsors, and partners. YouthMappers has been clearly defined as a community within the "community of communities" that make up OpenStreetMap (Brovelli et al. 2020; Solís 2017). However, the network does not fit neatly only within those boundaries, promoting youth capacity in ways that go beyond the orbit of OSM. We assert that YouthMappers has transformed into a youth movement due to the passion and drive of students (Solís et al. 2020).

Four founding partners have led the network and promoted its growth, and resources to support it since 2015 through a steering committee format. The three founding universities are Texas Tech University, the George Washington University, and West Virginia University. The United States Agency for International Development generously supports this program through a grant from the USAID GeoCenter. Today, a fifth organizing partner is the lead fiscal administrative organization, Arizona State University (Fig. 29.1). Fifteen other official orga-

nizational partners support YouthMappers in various ways that are formalized through memorandums of understanding (MOUs) (Fig. 29.2).

The network has been able to thrive in a grassroots capacity, thanks in no small part to a regional ambassador structure introduced in 2019. There are two types of regional ambassadors: outreach ambassadors and Everywhere She Maps (ESM) ambassadors. The outreach ambassadors provide local support to chapters and recruit new chapters, while the ESM ambassadors focus on building local capacity by and for women mappers. A key component of both types is to support students to be embedded and connected to the local OSM community. While YouthMappers has many formal partnerships, the regional ambassadors cultivate and manage multiple partnerships and connections that offer students opportunities beyond the consortium. Being open and encouraging informal collaborations – with guidance from highly motivated and qualified regional ambassadors – allows the network members to achieve goals that are locally sustainable and more effective than through some traditional means (Wei-Skillern and Marciano 2011).

Organizing Institutions

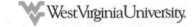

Fig. 29.1 Co-founding institutions of YouthMappers organize network-wide activities to link YouthMappers chapters around the world

Fig. 29.2 Partners of YouthMappers provide specific support to YouthMappers in alignment with their capacities and missions

1.2 Student, Alumni, and Staff Perspectives

Many YouthMappers students will describe being a part of the network as having friends in 67+ countries (and growing) around the world. Erneste Ntakobangize of Rwanda (2021) shared this sentiment in his blog:

I have seen that when you are connected to YouthMappers, you have the world in your hands. Living in foreign countries is not easy; most of the time, you feel lonely as it takes time to be familiar with the new conditions. Since I joined YouthMappers, I got the opportunity to be connected to many people from different countries … Nothing is better than having a network of people for whom you have the same feelings and commitments. Currently, I have at least friends in more than 60 countries all over the world. This is a privilege that I couldn't have if I had never joined YouthMappers.

The students, faculty, staff, and professionals who are connected to this movement feel a similarly strong connection across the globe because of their common cause. Innovation and collaboration have flourished for YouthMappers because there is a social and technical infrastructure for an innovative partnership model supporting and encouraging it.

From the perspective of a staff member, Dara Carney-Nedelman, who serves as communications coordinator, one of only three paid core staff:

I'm constantly amazed by the numerous conversations I have with individuals who have informal and formal partnerships with YouthMappers. Some organizations may be concerned about competitors, which limits their network and reach. However, YouthMappers embraces these kinds of collaborations. I recall a virtual meeting with an organization doing similar work in a part of the world YouthMappers is not very active. Instead of shying away from partnering with this organization, I remember my YouthMappers network colleague focusing on how we could collaborate, seeing this as an opportunity for YouthMappers students. I've seen this mentality of inclusiveness and openness time and again and know that an organization like ours that focuses on achieving its goals in a similar manner will succeed. YouthMappers has given me an opportunity I never expected I would have in an entry-level position, much less reach by 22 years of age. Youth's voices matter here! My ideas and perspective are always respected, which leads to successful teamwork and alliances internally and externally.

The core staff also appreciate the global diversity represented by those who join virtual meetings. We regularly speak with groups where all the representatives are based in different countries and regions of the world. Our favorite virtual calls are always those with the regional ambassadors because there are so many various perspectives and lived experiences that contribute to our conversations.

YouthMappers alumni Candan Eylül Kilsedar (2021) of Turkey and Italy shares this similar sentiment, as one can read from her blog where she wrote,

YouthMappers strives to address these [environmental, social, or economical] challenges through establishing collaborations for the use of and generation of open geospatial data. The collaborations I had helped me understand the peoples of the world and the challenges faced, whether locally or globally, better, thus enabling me to have a more solid ground and understanding in the projects I am involved with ... YouthMappers network has been a facilitator for me in this sense, as it reminds its members that young people have the capacity to make a positive change in the world; they already have what it takes to follow their vision.

Eylül explains well here that YouthMappers' collaborations led to her building skills for her future. There are many more examples from hundreds of blogs written by YouthMappers (Hite et al. 2018). These amount to a strong case of how YouthMappers as a partnership network clearly advances SDG 17. Because of engaging youth in administrative compensated roles of ambassadors and staff, YouthMappers is also building career networks, workforce talent, and capacity that contribute to work and economic development (SDG 8).

1.3 Connective Tissue for YouthMappers

The infrastructure of the network is bolstered by the health of our partnerships, both formal and informal. The network benefits from being largely decentralized, meaning the steering committee and staff members provide structure, leadership, and day-to-day support, but the students ultimately take ownership of the network and drive it forward, being supported regionally by the ambassadors, and locally by their own chapter mentors (See also Chap. 28). The network can have the impact it does because it has been purposely designed to enable students to have the power to shape and lead their chapters beyond anything a centralized organization could accomplish.

The student movement is phenomenal at developing flexible and timely partnerships with a vast array of local, national, and international organizations and government actors. Challenges do occur due to some disadvantages of decentralization. Some of these challenges include difficulties in capturing and recognizing all activities happening across the network, ensuring communications reach every student member (in a context of turnover like graduations), and unintended misunderstandings of YouthMappers' purpose or goals among new participants or supporters. However, the benefits of this enabling design far outweigh these challenges when it comes to the impact of the movement and students' ability to implement their own visions.

2   Components of the Ecosystem: Sponsors and Partners

As noted, YouthMappers organizers also engage with formal partnerships to support the network through MOUs to provide opportunities and resources. It is crucial to highlight that since the partnerships are dynamic, involvement fluctuates over the years. All partners that have collaborated with YouthMappers are not recognized in this book. However, the YouthMappers network is grateful for all partnerships, past and present- and recognizes that without the support of partner organizations accomplishing our motto of "building mappers" would be much more difficult if not impossible.

At the present time, twenty organizations are considered sponsors, organizers, or partners within the YouthMappers network, as follows: Texas Tech University, George Washington University, West Virginia University, The United States Agency for International Development, Arizona State University, the Humanitarian OpenStreetMap Team (HOT), the University Consortium for Geographic Information Science, The American Geographical Society (AGS), Mapillary, TeachOSM, Maxar, Missing Maps, GIS Certification Institute, OSGeo, Radiant Earth Foundation, Esri, MapBox, Hexagon Geospatial, Locana, and Crowd2Map. A few representatives of these organizations shared below their perspectives on partnerships and what that means for advancing the capacity of building mappers, which have the added benefit of growing the capacity of local mapping talent.

In other chapters, we consider the perspective of organizers and funding sponsors. This section focuses on some of the perspectives of other partners from private and nonprofit sectors that inform the mutually beneficial ecosystem that supports YouthMappers to thrive, while building mapping talent that ultimately will also in turn benefit these partners now and in the future (Fig. 29.2).

2.1   The American Geographical Society

Chief Executive Officer John Konarski and Manager of Special Projects Courtney Clark

The American Geographical Society (AGS) is deeply supportive of YouthMappers' mission. We recognize that the success of YouthMappers and the Everywhere She Maps program is critical to our vision of a world where geographic science and technology are used to address society's challenges and opportunities. Both YouthMappers and the AGS have a fundamental interest in promoting and supporting geographic thinking, education, and careers, particularly among youth. Like YouthMappers, we invest significantly to enable experiential geographic education among young people. Our approach complements that of YouthMappers – we focus on supporting geography teachers and students at the secondary level, while YouthMappers concentrates on postsecondary education. It is our intention to encourage high school students to consider participating in YouthMappers when they go to college.

Since YouthMappers' founding, the AGS has been proud to include members of the network in our annual Geography 2050 symposia and other special events we host to advance geographic and geospatial thinking. During our 2021 symposium, "Geography 2050: Towards a More Equitable Future," we were especially honored to feature YouthMappers' Everywhere She Maps Regional Ambassadors. These eight rising stars of geography from around the world shared their experiences of using geography to address the Sustainable Development Goals (SDGs), particularly SDG 5: Gender Equality, with our community and helped us more deeply understand barriers to gender equality within our field. YouthMappers helps us remain committed to a more equal discipline and world.

The AGS significantly expanded our partnership with YouthMappers through the addition of a shared staff resource to both our teams in 2021. We joined forces to form a creative resource-sharing model that enabled YouthMappers to hire a director for its Everywhere She Maps program and us to bring on board a manager for our Geography Educator Initiative.

2.2   Mapillary

Program Manager Edorado Neerhut and Data Analyst Christopher Beddow

Mapillary has had a long history of association with YouthMappers since its early days. During this time, we've been able to collaborate with YouthMappers chapters across the world in places like Bangladesh, Italy, Sierra Leone, Turkey, and Uganda. Much of this work has centered around the role street-level imagery can play as a tool to collect map data. Students have learned how to collect street-level imagery, how computer vision can be applied to derive map data automatically from images, and how to use derived data with OpenStreetMap editing tools and QGIS. Working with YouthMappers has given us the opportunity to connect with future leaders in the geospatial community. YouthMappers have regularly shown us new applications of street-level imagery that opened our eyes to the value it can provide in different contexts.

Mapillary is a critical tool for collecting and analyzing geospatial data using photos. It produces openly licensed data and is integrated with OpenStreetMap, which is the most critical infrastructure for YouthMappers. Having a direct relationship with YouthMappers allows us to think of our product in terms of how it serves the goals of YouthMappers, and helps YouthMappers to have open communications with us about their needs and activities. In the long term, it means we can build excellent tools together that enable the mission of YouthMappers in improving maps and map data around the world.

In terms of successes, we have learned about new applications for street-level imagery such as mapping waste disposal sites, mapping electrical infrastructure, and mapping slums. Also, imagery is now available for the OpenStreetMap community in new areas that would often get ignored by organizations mapping with strictly commercial incentives. In addition, many of the students we've worked with have developed into prominent leaders in their fields. It has been incredibly satisfying to see some of them undertaking postgraduate studies in places like the United States and Germany, joining local organizations as advocates for open map data, or international organizations such as HOT where they provide important local context to HOT's work. A YouthMappers alumni from Italy and Turkey is now working at Facebook as a critical part of Mapillary's community team. Finally, through our Map2020 project, YouthMappers were able to connect with the global geospatial community at State of the Map 2019 in Heidelberg, Germany.

With respect to some challenges, we see that access to high-speed internet can vary greatly between YouthMappers chapters and even students within the same chapter. As a high bandwidth application, students sometimes had to wait days for their street-level imagery to upload. Their persistence through these challenges was admirable and many worked with local organizations or their university to find ways to upload the imagery as quickly as possible. Similarly, socioeconomic conditions also vary between students. Data collection can involve long journeys and not all students were able to access vehicles easily or afford the fuel required for these trips (See also Chap. 11). Training participants to use Mapillary effectively can take time, since hands-on collaboration and open learning sessions are more effective than just reading instructions in order to ensure the best data quality and understanding of the tools available.

Mapillary's partnership with YouthMappers empowers organization members to use images to document and index what exists on the earth, in a given location at a specific time. This helps to map infrastructure, and allows reference to current and past geospatial data to carry out planning for sustainable future industrialization as well as innovation all based on maintaining or improving upon what exists. The photographic component of Mapillary gives a visual element to geospatial projects in the sustainable development realm, removing the abstraction between map data and the physical world. The machine learning capabilities of Mapillary help to scale the amount of impact any one person or organization can have, creating map data across the globe from community-contributed images that give a better understanding of infrastructure on the map than any manual data collection can achieve in a limited time period.

2.3   TeachOSM

Organizer Steven Johnson

Our partnership is structurally informal but TeachOSM and YouthMappers complement each other's capabilities and services quite nicely. TeachOSM is founded on the belief that OpenStreetMap makes an excellent educational tool not only for teaching cartography and geography but as an example of applied geography in action. YouthMappers is an example of an organized group applying geography to social needs. Where our organizations have a common cause is in fostering equitable and inclusive leadership across the open mapping community. For our audience of educators, YouthMappers is an organization we can point to as an entrée into open mapping and valuable community service.

Operationally, we find our organizations partnering in and around the educational domain greater than the OpenStreetMap ecosystem. TeachOSM and AGS are active collaborators on programming for educators. TeachOSM has also had YouthMappers members co-facilitate our events, and enlisted them to present open mapping to educators and learners. YouthMappers often uses TeachOSM infrastructure (Tasking Manager) to stage mapping projects. Increasingly, YouthMappers are stepping up to help validate TeachOSM projects, sometimes providing important feedback to educators and learners.

There are three reasons our partnership is important. First, leadership development, as explained above, is enhanced through our com-

mon commitment to making the open mapping community responsive to issues of equity and inclusion. For example, TeachOSM currently has an internship opportunity co-facilitating events. Through this internship, YouthMappers support young mappers through technical support (i.e. validation), and by co-facilitating TeachOSM events (i.e. Map-Alongs).

Second, in terms of spatial citizenship, both YouthMappers and TeachOSM are citizen science projects writ large. Also, both of our organizations are developing organizational capacity and social capital. In that sense, the work of both TeachOSM and YouthMappers is implicitly promoting mapping as an act of spatial citizenship. Both of our organizations orient our contributors around an ethical framework for the use of mapping technologies, such as protecting individual privacy, while sharing public information. Both YouthMappers and TeachOSM educate our contributors how to map, and by extension, the reflexive use of geospatial technologies. Combined, our organizational activities draw a line between mapping, social benefits, and engagement as global citizens.

Third, regarding map maintenance and community management, OpenStreetMap is not only a large map but a large community and operational ecosystem. Both YouthMappers and TeachOSM are addressing the need to fill the pipeline with new mappers and new contributors. Not just to edit the map, but also to manage and administer the operating infrastructure of the project. The next generation will be responsible for maintaining the world's only publicly available open geographic data set, which is an incredibly valuable public resource.

At least two emerging successes of the TeachOSM-YouthMappers partnership are worth noting. First, validation of TeachOSM projects is important where YouthMappers are increasingly lending support to new mappers by validating TeachOSM projects. Contributors to these projects are frequently drawn from educators, and many of them find it encouraging to get a message that their work has been checked. Moreover, some YouthMappers validators provide valuable feedback to new mappers, encouraging them and helping retain mappers.

Secondly, several YouthMappers have co-facilitated a TeachOSM "Map-Along," our weekly open mapping program for educators. Like validation, having practitioners map with them and demonstrate why mapping is important is an invaluable experience for educators. (Bear in mind, many educators have had minimal formal training in geography.) Also, the Map-Along events offer YouthMappers an opportunity to lead events and showcase their geographic talents and abilities.

Our partnership probably best addresses SDG17.7 and SDG17.9, relating to diffusion of technology and capacity-building. Both YouthMappers and TeachOSM fill a void by making mapping accessible and available. TeachOSM addresses this void through education and YouthMappers through 300+ chapters who are contributing data and developing organizational and community infrastructure. Through open-source methods we foster a broad cadre of spatially-informed citizens among teachers and their students, who connect applied learning to social issues.

2.4 Locana

Senior Consultant Chad Blevins

Partnering with YouthMappers is important as we share interest of promoting open-source geospatial data and software. We have a shared interest in creating, maintaining, and utilizing high-quality geospatial data. This partnership also exposes students to professional projects and provides a chance for them to discover potential job opportunities. For example, Locana has been supporting the OSM community for over a decade and has worked on several projects where background knowledge about OSM is essential.

Locana and YouthMappers signed an MOU in September 2020 which was right in the middle of the COVID-19 pandemic which inhibited many new opportunities from taking place. That said, Locana was able to partner with YouthMappers stu-dents and alumni over the past year to create data in support of municipal waste mapping projects under our wastesites.io campaign (See also Chap. 20). We worked together with YouthMappers from Zambia, Kenya, and the USA to pilot projects and develop concepts to educate and improve waste management on a local scale.

In addition, Locana has subcontracted with YouthMappers alumni as local data experts in projects taking place in developing countries. Locana's goal is to create and provide more opportunities for YouthMappers students and alumni throughout the world.

Locana believes youth are an invaluable resource in helping countries achieve their development objectives and we consider the UN's Sustainable Development Goals (SDGs) in our management processes and when selecting and delivering projects. Our projects integrate environmental, social, and governance considerations as we seek to work with environmentally and socially conscious clients who share the same beliefs as us.

2.5 Mapbox

Community Team Lead Mikel Maron

Our partnership with YouthMappers has had good notes of engagement, and we look forward to strengthening our relationship for the partnership to reach its full potential. In 2016, Mapbox team members assisted with mapping in a collaboration between YouthMappers and USAID to create data on OpenStreetMap for the benefit of the President's Malaria Initiative. We also lent our expertise by evaluating imagery and validating data for a YouthMappers project in Kenya. We were pleased to host the YouthMappers Research Fellows for a session at our DC office in 2018. During the session, we introduced the fellows to Mapbox and how our services and tools are used in disaster response and resilience projects, and we also provided hands-on training in Mapbox studio. Mapbox has also provided

YouthMappers with some support in using our products and donates the cost of hosting maps on the YouthMappers website.

We are ready to engage in a closer mentorship relationship with YouthMappers' Research Fellows and other implementers, and we would like to provide support related to the data the fellows collect and analyze. Our team members have a lot of experience developing data-driven projects that have an impact, and we would like to support a greater variety and kinds of training and experience. We highly value our partnership with YouthMappers, because it's an unparalleled opportunity in the mapping space to have the kind of engagement with young people that YouthMappers has created. It's exciting to see YouthMappers students directly involved in mapping for their local causes, community, and regions. After they graduate, many YouthMappers students will set forth in a professional career related to geography, and it's the purpose of the Community Team at Mapbox to make long-term investments in the mapping space – which includes supporting the next generation of mapping professionals.

3 Partners Ecosystem for the Implementation of SDGs 17 and 8

YouthMappers, as a student-led chapter-based university network, capitalizes on a unique position and creates meaningful partnerships to further support the role of cross-sectoral implementation. The network leverages collaborations to invest in student initiatives, which has resulted in the creation of a movement of Digital Humanitarians within this academic space (Solís et al. 2020). "Universities are uniquely placed to lead the cross-sectoral implementation of the SDGs, providing an invaluable source of expertise in research and education on all sectors of the SDGs, in addition to being widely considered as neutral and influential players," according to El-Jardali et al. (2018).

The university center of partnerships itself thus advances SDG 17 directly. But as we have demonstrated, these connections to private and nonprofit sectors also serve to advance SDG 8 in terms of enabling the knowledge talent and capacity for economic development across all goals. A tangible instance of our student investment includes the internship programming, such as the match program through Everywhere She Maps, focused in increasing gender equity in the network and providing professional development and career opportunities to women and nonbinary YouthMappers students. The program pairs qualified YouthMappers students with partner organizations, who host the students as interns. The students gain invaluable professional experience and build their networks, while the host organizations benefit from the students' expertise, time, and energy. For example, YouthMappers' partner Crowd2Map Tanzania, which mobilizes volunteers to add key areas of Tanzania to OSM to assist with female genital mutilation (FGM) prevention and response efforts, hosted nine women YouthMappers students from across Africa as interns in 2021. The interns improved their mapping skills as they learned how to validate edits made to OpenStreetMap and implement data quality control measures, and they learned about the issue of FGM and the role of geographic data in preventing and responding to its instances. Crowd2Map Tanzania, benefited from the interns' consistent, dedicated efforts to improve the quality of its data.

The internship match program certainly faces challenges. For many of our partners, the prospect of starting an internship program or expanding an existing one is prohibitively expensive. While we aim to continue finding internship opportunities for YouthMappers students, we are also working with our partners to create additional professional development programs focused on mentorship and skill building. These activities align with SDG 8 because through them, we are focusing on workforce development, not just building maps, but also building mappers.

References

Brovelli M, Ponti M, Schade S, Solís P (2020) In: Guo H, Goodchild MF, Annoni A (eds) Citizen science in support of digital earth. Manual of digital earth, Springer, Singapore. https://doi.org/10.1007/978-981-32-9915-3_18

Cowan N, McCusker B, Solís P (2017) Code of ethics for YouthMappers Chapters. Washington, DC. Available from YouthMappers. http://bit.ly/2uwifGf. Cited 5 Jan 2022

El-Jardali F, Ataya N, Fadlallah R (2018) Changing roles of universities in the era of SDGs: rising up to the global challenge through institutionalising partnerships with governments and communities. Health Res Policy Syst 16:38. https://doi.org/10.1186/s12961-018-0318-9

Harold J (2017) The collaboration game: solving the puzzle of nonprofit partnership. Stanf Soc Innov Rev. https://doi.org/10.48558/Z0QV-MQ13

Hite R, Solís P, Wargo L, Larsen TB (2018) Exploring affective dimensions of authentic geographic education using a qualitative document analysis of students' YouthMappers Blogs. Educ Sci 8:173. https://doi.org/10.3390/educsci8040173

Kilsedar E (2021) Alumni Reflections: Eylül Kilsedar. YouthMappers Blog. Available via YouthMappers. https://www.youthmappers.org/post/choose-to-grow-up-with-youthmappers-to-become-a-mapper-you-dreamed-to-be. Cited 3 Sept 2021

Ntakobangize E (2021) Choose to grow up with YouthMappers to become a Mapper you dreamed to be. YouthMappers Blog. Available via YouthMappers. https://www.youthmappers.org/post/choose-to-grow-up-with-youthmappers-to-become-a-mapper-you-dreamed-to-be. Cited 2 Sept 2021

Ostrower F (2005) The Reality Underneath the Buzz of Partnerships. Stanf Soc Innov Rev 3(1):34–41. https://doi.org/10.48558/33S9-4637

Solís P (2017) Building mappers not just maps: challenges and opportunities from YouthMappers on scaling up the crowd in crowd-sourced open mapping for development, AAG Annual Meeting. American Association of Geographers, Boston

Solís P, Rajagopalan S, Villa L, Mohiuddin MB, Boateng E, Wavamunno Nakacwa S, Peña Valencia MF (2020) Digital humanitarians for the sustainable development goals: YouthMappers as a hybrid movement. J Geogr High Educ 46:1–21. https://doi.org/10.1080/03098265.2020.1849067

United States Patent and Trademark Office (USPTO) (2017) Registered Trademark, YouthMappers®, Reg. No. 5.203.611, Class 41, Ser. No. 87–165,163

Wei-Skillern J, Marciano S (2011) The networked nonprofit. Stanf Soc Innov Rev 6(2):38–43. https://doi.org/10.48558/GCY2-RN71

A Free and Open Map of the Entire World: Opportunities for YouthMappers Within the Unusual Partnership Model of OpenStreetMap

Mikel Maron and Heather Leson

Abstract

OpenStreetMap (OSM) is a very unusual kind of partnership, not only in the context of the United Nations Sustainable Development Goals (SDGs) but also frankly everywhere. YouthMappers has expanded OSM through localized engagement on a global scale. Examining this unique approach is instructive to learn not only about partnerships (SDG 17) but also about innovation in the open technology industry (SDG 9). We consider this joint journey so far and ponder on how to amplify our collective impact in the future.

Keywords

Geospatial industry partnerships · OpenStreetMap community · Communication · Innovation · SDGs

M. Maron (✉)
Mapbox, Washington, DC, USA
e-mail: mikel@mapbox.com

H. Leson
Solferino Academy, International Federation of Red Cross and Red Crescent Societies, Geneva, Switzerland
e-mail: heather.leson@ifrc.org

1 The Unusual Model of OpenStreetMap

To join the more than 8 million registered users of OpenStreetMap (OSM), there is no agreement to negotiate. Simply start to contribute to the map in your part of the world where you are. The ultimate aim is a clear, shared, and measurable mission to accomplish one thing: Map the entire world in the open. This simple but provocative invitation leaves room for creativity in the technical methodologies of mapping, and particularly in the models for organizing people together to build the map.

That said, getting involved in the OSM "partnership" has many of its own complications. There are almost two decades of implicit understanding increating the map within OSM, and it's not always obvious or transparent what kinds of best practices have been loosely adopted over time. It takes extensive engagement, experimentation, and patience to understand how OSM works in practice. There's no official front door, so for an individual, group, or organization to enter OSM requires jumping in and participating and building alliances across existing communities.

OpenStreetMap is a community of communities (Hagen 2019), and is not a static thing frozen in time. It adapts and expands. OSM as a whole has a lot to learn from the success of YouthMappers in engaging youth and rapidly expanding across the world. Participation in the project as a whole is still heavily weighted in the advanced economies where it started, and in some ways has saturated awareness in the areas of open-source software and geography. To continue to grow and map an ever-changing world requires expanding engagement.

2   The OSM and YouthMappers Journey

Growth is not always easy, and the OSM+YouthMappers journey has seen its pain points. These have centered on different online communication expectations in geographies across the world; the different contexts of learning environments vs. the development of map data in an accessible, production database; the tension of scale of entire groups editing *en masse* relative to the ability of others in the OSM community to monitor quality; and the swirl of global cultures clashing and cooperating.

For students, OSM represents a platform to realize opportunities to do something real and contribute to something that will make a difference right now. This is among the few opportunities to have an impact from within a learning environment. Further, YouthMappers are working right alongside professional humanitarians, university researchers, open-source volunteers, and paid corporate mappers. This innately leads to accelerated learning.

Other approaches to partnership can learn a lot from the OSM approach, as it allows widely divergent interests – from individual enthusiasts mapping their local parks to the largest multilateral institutions working on disaster risk reduction – to collaborate in the same shared database. OSM involves nontraditional partners and addresses power and inclusion in new ways.

YouthMappers demonstrates a decentralized network with open principles and "starfish and spider" approach (Brafman and Beckstrom 2008). We are in a time when the humanitarian and development industry needs a flipped model with locals at the forefront. The opportunity of YouthMappers for OSM – and vice versa – centers on the fact that students are already a part of many local communities and a global network. By taking a more strategic approach to local partnerships from hubs and labs and universities to local NGOs and governments, this alliance could become stronger and have more impact. And, in that impact, we can build OSM together, via sustained leadership while maintaining a youth-led approach. This kind of partnership innovation adds a new way of working and collaborating which can and should modulate traditional modes.

2.1   The Origins of OpenStreetMap

OpenStreetMap was started by a student. OSM began in the United Kingdom in 2004 to build freely accessible geographic data where none existed before. Specifically, OSM was a youthful rebellion against the expensive data licensing of the UK national mapping agency, the Ordnance Survey. As a student, Steve Coast wanted to program mapping applications, but could not do so without free to use road network data (Coast 2015). This was difficult to obtain, so he decided to create the map data himself.

Taking advantage of the new availability of consumer-grade GPS devices, decent internet connectivity, and open-source software, OSM was modeled after the contribution model of Wikipedia. It was not the first "wiki map," but was perhaps the first with the audacious goal to create a free and open map of the whole entire world. OSM rapidly gathered adherents among open-source software user groups, frequently hosted by computer science departments of universities. Interest and contribution from many others quickly followed, from artistic communities, startups, and established companies in the mapping industry. The first State of the Map conference was held in 2007 – hosted at a university, no less – and featured talks on everything from commercial mapping for cycling to disaster response, highlighting the growing breadth of applications enabled by open geographic data (The State of the Map 2007).

2.2 Expansion of OSM User-Creator Communities

OSM is essentially a platform for anyone who wants to produce or use open data of any kind. No permission is necessary. Although sometimes newcomers experience gatekeeping attitudes or barriers as noted above, as long as data is collected that accurately represents the world, and can be freely shared and attributed, contribution and use are welcome. This has allowed for completely unexpected developments and innovation.

Among the most notable is the Humanitarian OpenStreetMap Team (HOT; See also Chap. 31). Many areas of the world, particularly the most vulnerable to disaster, lack available data at all. A disaster can significantly alter the landscape, so it must be easy to make and distribute updates to the map. OSM was a natural fit in some ways for this need, though a very odd fit organizationally. Disaster response agencies must operate with regimented structure, yet in OSM essentially no one is in charge. HOT began as a community of interest within OSM, and quickly took on the challenge to build partnerships and respectfully negotiate with very different kinds of entities. The 2010 earthquake in Haiti was the first event when the approach of HOT was tested, and the ready availability and quality of OSM made it the base map for the response among humanitarian entities. The Haiti earthquake remains one of the most well-known uses of OSM for real-world collective impact.

2.3 A Stage Set for the Emergence of YouthMappers

The devastation in Haiti and the key role of open data caught the attention of even more unusual partners. In Washington DC, the State Department identified an opportunity to support humanitarian response by making satellite imagery available post-disasters to HOT and others through its program, MapGive. The World Bank began work to improve OSM in preparation for future disasters, most notably in Nepal under the OpenCities program where data preparation works greatly helped the response years later after the 2015 earthquake there. The entity that would become the main sponsor of YouthMappers, USAID GeoCenter, had also been working with the OSM community in Haiti as part of recovery efforts there. The people in these institutions involved in these efforts were to some extent already connected in professional settings, but they set out to purposefully connect with each other, as the OSM community was expanding, and began to form a community of their own. At George Washington University, a forward-thinking geography department incorporated OSM mapping for HOT tasks directly into the curriculum. Researchers at West Virginia University were transforming open and acquired geospatial data into decision-ready support for USAID Missions. And since 2003, the scholars who at that time were in residence at Texas Tech had been designing innovative programming to directly engage students in spatial data efforts for sustainable development research at universities with USAID in scores of countries. Together, this group of academics – with varying degrees of engagement to OSM – began to coalesce their ideas formally, directly connecting the learning experience to a real contribution to the world. This fertile ground of open collaboration that had emerged from OSM user-creators contributed to what would become the core design of YouthMappers.

3 Present Innovative Patterns of Collaboration

In many ways, YouthMappers as a distributed and youth-led program is a natural fit for the unusual partnership and community model of OpenStreetMap. With "no permission necessary," YouthMappers aligns with OpenStreetMap's inclusivity, to allow anyone to get involved in their community and their world via a map. Students don't need to ask faculty. Traditionally marginalized communities don't need to ask those who are marginalizing them. Likewise, the experience of the traditional OSM community expanding and encountering new ways "to do OSM" has potential for the OSM movement to more explicitly identify ways to nurture the necessary growth of the mapping community.

In other ways, it is a challenging collaboration to engage relatively inexperienced newcomers within an established community with many implicit cultural practices. Here conflicts can and did arise. Sometimes newcomers experience gatekeeping attitudes, especially toward "remote" volunteers, who sometimes have trouble interpreting satellite imagery of far way terrains. Within OSM there have been naysayers doubting the need for "another" program, because of the historical presence of students within OSM from the very beginning. But this posture lacks understanding of the implicit gap to joining OSM that YouthMappers helps to bridge.

3.1 The Innovative Model from Individuals to Groups

While it originally may have been envisioned as a space for individuals to contribute, OSM has emerged as a "community of communities," which align on both local geography and applications of interest (Solís 2017; Brovelli et al. 2020). In addition to purpose, these also can align with personal or social identities, such as in the case of students (Brovelli et al. 2020:600). Universities have an inherent spatial distribution, and students are practiced in forming clubs and interest groups within their schools. While starting a YouthMappers chapter is more formal than starting up a local meetup in OSM, the requirements to form a chapter are minimal: a faculty sponsor and agreement to the terms of participation in the program. The first recom-

mended activity for a chapter is to map their local place on OSM – their campus. Then groups are encouraged to organize around HOT mapping activities on the tasking manager and other YouthMappers organized campaigns.

Collaboration follows this design. YouthMappers is a global network, and groups in very different places will face common challenges as they map their schools, organize events, and engage in global mapping activities. OSM has spread and grown through individual enthusiasm, cooperation, and willingness to help others. OSM is incredibly generous by definition, in that mappers are sharing their time and knowledge to build the map, and by and large that extends working with others around the world. Local OSM communities will share what they have tried, failed, and succeeded at in welcoming new members, growing their community, and figuring out tricky mapping challenges. The fact that a YouthMappers chapter in South Asia can reach out and directly learn both from local mappers outside of their university or from colleagues in South America is a reflection of the core of OpenStreetMap's potential to serve as a learning network spanning the globe.

3.2 Disruption as a Component of Innovation

Nevertheless, with a variety of backgrounds and disciplines, time zones, languages, and cultures, opportunities for misunderstanding abound. The communication styles of, say, American and German software developers have so many subtleties of meaning that miscommunication is inevitable. Like many global efforts, English is the primary language of the global OSM community, and those with English as a second or even third language are not able to participate as quickly and fully. Text-based communication lacks the nuance of in-person communication, where a joke can easily be interpreted as an insult. This is the backdrop of so much of the strife on the internet as a whole, and global OSM communication channels are not immune either. For YouthMappers, the tendency has been to not engage, with the result that the story of YouthMappers within the OSM community is often told by others.

Some of the criticism is fair, even if unfairly delivered. HOT tasks are designed to be completed quickly, by large numbers of people, and not always involving people mapping locally in those places (if there are any at all). This can result in lower quality data that is not consistently reviewed by others with more experience. OSM is a wiki, and a learning journey for all, so it is expected that newcomers may need more help. However, the potential scale of contributes within YouthMappers engagements means that the existing communities' ability to give feedback and help mappers develop skills can be overwhelmed. YouthMappers has worked to address this through training and dedicated validation teams.

Also working against groups like HOT and YouthMappers is the common tendency within OSM to focus on editing problem areas, even when the majority of efforts might be high quality. A similar dynamic with companies employing large teams of editors in OSM led to the adoption of the Organized Editing Guidelines by the OpenStreetMap Foundation, which provide specific instruction on how to make organized efforts visible to, and part of a dialogue with, other mappers (OpenStreetMap Foundation 2018). Ensuring these guidelines are helpful to build healthy relationships between organized efforts and the broader community is a work in progress. There is a need for more genuine investment in the process by companies and HOT, as well as many others in local OSM communities. Guiding rules of partnership are not meant to be an enforcement mechanism, but a means to work together toward a shared goal. In this way, it can be concluded that the emergence of these communities has led to the kinds of innovation for editing that OSM needs as it grows and evolves.

3.3   Efforts Off the Map

Organized efforts in OSM have sometimes overwhelmed community efforts away from making the map. For many years, the State of the Map conference has held a scholarship to cover the costs of traveling to and attending. This has opened up participation in the global community to people in parts of the world that would have otherwise struggled to access the conference and enriched the global community through sharing of new perspectives on the craft. Scholars are one of the best parts of the conference, and many excellent YouthMappers have been scholars at the conference. What has been overwhelming is the application process.

Understandably SOTM events are attractive opportunities for many YouthMappers, but in recent years, more than half of the applications have come from students in YouthMappers chapters, and unfortunately, the typical quality of applications is low. That is not necessarily a reflection on the applicants themselves. The process requires writing about their contributions to OSM, and how the conference attendance will benefit their community back home. Without experience and skill in writing for an international audience, an understanding of what the scholarship program is looking for, and a means to self-assess their own likelihood of selection, the result is a large number of applications which recapitulate the texts of the YouthMappers program itself. For the application review team, the work of selecting applicants is that much more arduous.

In response, YouthMappers organizers have since put into place a screening and recommendation process to help reviewers and now offer writing workshops to improve the professional development of students seeking to apply. In this way, the experience of OSM, in turn, is also helping YouthMappers innovate and improve how to support and build the capacity of students off the map, too.

3.4 Communication Channels for a Multicultural Movement

The relative lack of explicit partnership guidelines within OSM is a strength and a weakness. There are many expectations embedded in the cultural practice of OSM that are only understood after effort, and a few bruises. While some initiatives like the Organized Editing Guidelines have tried to make more explicit how to engage, the general message of this open community is to jump in, get involved, and do it. Direction given to newcomers may be as little as to "just" post on the mailing list, or edit the wiki.

The YouthMappers chapters have perhaps been confused by this messaging, and some have felt reluctant to get directly involved in the community and communication dynamics chapters sometimes find themselves in. Social participation and partnership are not as straightforward as editing the map. The opportunity to work in the same space as professions is one reason why the YouthMappers network has gone viral, however students often feel intimidated by more experienced people. But it is more than newcomers feeling shy. This hesitation also stems from the fact that there are so many unspoken assumptions implicit in a fast-growing, global, complex community of individuals and groups from countries around the world and different sectors. They may not even realize they are not following convention.

Another concern is that because YouthMappers are far more inclusive in terms of gender, people of color, and people from majoritarian nations than OSM at large represents, the existing community may also not realize the barriers to inclusion. The dynamic makes for a culture of "fear of mailing lists" and "fear of engagement" due to some deep systematic issues, which are not entirely of OSM's doing given the nature of technology communities. There are now working groups within OSM established for grappling with digital, cultural, and diversity and inclusion

divides. The OSM culture shifts to tackle safe and healthy communities is ongoing. This groundwork should precipitate a shift to also engage more deeply across the project rather than on separate tracks.

Intentional institutional support is necessary to retain the power of the loose partnership model of OSM, yet bridge the strengths of traditional institutions. This was the early approach of HOT, where members of HOT spent extensive time and energy to build networks and understand the dynamics and structures of the disaster response field, while likewise helping to educate them about how to work with OSM. YouthMappers members have a lot to understand about the history of OSM, how the OSM Foundation functions, what channels various communities use, and what topics are part of age-old debates. Focused training on how to collaborate and partner within OSM, like through the YouthMappers Academy, has been put in place to better prepare YouthMappers to succeed as OSM mappers.

The benefit to OSM from the YouthMappers movement is potentially huge. A framework to understand how to work with the OSM community will be very useful to YouthMappers but also to other newcomers as well. Similarly, making sure the OSM community understands how to engage with one specific community among the many in OSM would help to ensure its continued growth and impact. Perhaps the OSM community is sometimes simply not self-aware of its own dynamics. We understand the OSM community not as a separate thing, but made up of everyone who participates – whether they come as individuals or groups of individuals. A different viewpoint like the one students bring can help this culture develop and grow. And OSM must develop and grow. Indeed, the mission is to map the entire world, and the world is massive and ever-changing. The energy of global youth today, and their investment in OSM as they go into their careers, are clearly essential for OSM to grow in the decades ahead.

4   Recommendations for New Innovations Within the OSM + YouthMappers Communities

Considering these ideas, we see a significant opportunity for YouthMappers to shine. Be it civic engagement or global movements - the need for a skilled and impact-driven network remains a challenge, but a necessity. YouthMappers' design as a chapter network represents the epitome of the intersection of "partnerships" and "distributed network" approaches. In the urban civic space, as well as development and humanitarian spaces, the need for more quality data and youth engagement proves to have an impact on achieving the Sustainable Development Goals (SDGs) for nations – but also for organizational goals for the institutions that may be participating in seeking the SDGs. There are a plethora of existing examples of YouthMappers engaging across their civic institutions (city, local and national governments), NGOs/INGOS (HOT, World Bank, UN, etc.), and universities. We hope to continue to see that grow.

Reflecting on all of this trajectory, we emphasize that SDG 17 indicates that partnerships are key to the success of SDGs. It is a marriage of purpose and skill. Having a rich and thoughtful partnership strategy can lay a strong platform and pipeline for YouthMappers to have an impact in their own lives and in their communities. Add to this the benefit of mentorship and future employment to build on the important lessons of civic engagement that OSM brings and that YouthMappers as a peer network tailors to that experience. YouthMappers gain valuable experience in social impact, negotiation, and project management while the suite of OSM partners receive an influx of experienced youth to inspire and support the necessary digital transformation and information management needs of multiple sectors. This in turn sows the seeds for true innovation, which is something that also advances SDG 9.

Still, we believe that more innovation is needed to continue to evolve this unique model with impact, and to effectively address such a dynamic, fast-moving context like OSM. Fortunately, YouthMappers will not need to create a new framework to pursue this innovation, but there are some important lessons learned and recommendations from the perspective of OSM to test and apply. We consider the three most important ones here, starting with a first step that will enable all of them:

4.1   Stronger Connections to Civic and Open Communities

In communities around the world, there are open social entrepreneurs, innovation-driving entities, civic technology organizations, hubs, and labs. Alignment in partnership with these groups is already part of some YouthMappers chapters' local engagement. By sharing best practices from civic engagement within the YouthMappers chapter network and building alliances with these networks, especially at local and regional levels, there are endless possibilities to support the local YouthMappers' SDG journeys. A shared mission can build on the lessons from YouthMappers collaborating in universities, and enrich students and these communities in mutually beneficial ways.

4.2   OSM Contributions to Locally Defined Priorities of the SDGs

Certainly, YouthMappers is part of the global OSM community. With millions of contributors on every type of geospatial content, the stories and OSM data activities can align with SDGs in every possible way. How this unfolds in each community is clearly unique. While HOT has been doing this often with their work, they might not tackle some of the important non-urgent non-disaster types of community needs like accessibility of buildings, infrastructure gaps, and climate change issues, that are more long term or even mundane. OSM's working groups, specifically the Local Communities and Chapters Working Group, already have YouthMappers engaged to contribute to this kind of action that can inform any of the SDGs that matter most in a particular local context.

YouthMappers participants and OSM contributors at large can learn much from each other. What if there were an OSM mentor from a "sister city" or their very own city, supporting the YouthMappers OSM journey in more explicit and strategic ways? What if OSM had more YouthMappers supporting working groups and supporting the project's evolution to learn from YouthMappers' lessons? Two such examples are the work of Map Kibera and OSM Philippines. These offer models that could be replicated or adapted to local circumstances, if willing and capable youth leaders could be so directed.

4.3 Deeper Links to Corporate Social Impact

Many businesses have expanded their focus to include corporate social responsibility, foundations, and social impact offices and programs. These organizations seek people to mentor and to have a sense of purpose in their communities. By aligning more with global and local businesses on the SDGs, the opportunities for immediate and long-term impact on their lives and the lives within their communities are large.

Recognizing that implementing this recommendation could potentially grow overly complex, and given the fact that YouthMappers students do have university priorities, we recommend a practical starting tactic: align with organizations that have open source program offices or social impact teams. This represents one way to have OSM and YouthMappers adapt together due to shared open values. Businesses are surely keen to mentor YouthMappers chapters thereby helping the network and the individuals on their learning and leadership journeys.

5 Looking Ahead

The next stage of partnerships across the OSM community will intersect with YouthMappers everywhere. We should keep an eye on enabling youth to have an impact by building these partnerships themselves. Overall, there are existing network challenges and ample opportunities to shift how YouthMappers might support the SDGs via partnerships. And there remains a need to better engage students within the existing, evolving OSM context while still maintaining a productive, inspiring youth-led space for them to thrive.

A final, overarching recommendation is that YouthMappers (both as a global network and as individual chapters) work to map out a partnership strategy that suits their strategic goals, keeping in mind that fundamental engagement with and through OSM and the OSM Foundation serves both the youth that are mobilized around a free open map of the world and that very free open map community itself. Doing so means that the skills to develop and sustain partnerships must be nourished.

YouthMappers of the future could rise to become a model of "citizenship" in open spaces, while also evolving to support the SDGs in the use of those open spaces. Reciprocally, we hope for members of the OSM community that are not already a part of the youth orbit to embrace this innovation – even this disruption. Work to understand, engage, welcome, and support youth in their local mapping spaces within the incredible, unique movement that is OpenStreetMap.

References

Brafman O, Beckstrom RA (2008) The Starfish and the Spider: the unstoppable power of leaderless organizations. Decentralized revolution. Penguin Group, London, paperback 240 pp

Brovelli M, Ponti M, Schade S, Solís P (2020) Citizen science in support of digital earth. In: Guo H, Goodchild MF, Annoni A (eds) Manual of digital earth. Springer, Singapore. https://doi.org/10.1007/978-981-32-9915-3_18

Coast S (2015) The Book of OSM. CreateSpace Independent Publishing Platform. ISBN 151423274X, 290 pp

Hagen E (2019) Sustainability in OpenStreetMap: Building a more stable ecosystem in OSM for Development and Humanitarianism. White Paper, Open Data for Resilience Initiative, GFDRR Labs, 10 December. Available from World Bank. https://opendri.org/resource/sustainability-in-openstreetmap/. Cited 9 Jan 2022

OpenStreetMap Foundation (2018) Organized Editing Guidelines. Data Working Group, 15 November. Available from OSM Wiki. https://wiki.openstreetmap.org/wiki/Organised_Editing_Guidelines. Cited 9 Jan 2022

Solís P (2017) Building mappers not just maps: challenges and opportunities from YouthMappers on scaling up the crowd in crowd-sourced open mapping for development, AAG Annual Meeting. American Association of Geographers, Boston

The State of the Map (2007) Conference Schedule. Manchester, United Kingdom, 14–15 July. Available at The State of the Map. https://2007.stateofthemap.org/programme-overview/. Cited 9 Jan 2022

Youth and Humanitarian Action: Open Mapping Partnerships for Disaster Response and the SDGs

31

Tyler Radford, Geoffrey Kateregga,
Harry Machmud, Carly Redhead,
and Immaculata Mwanja

Abstract

Use of data and digital technologies has dramatically changed the look, feel, and efficacy of humanitarian action over the past decade. Partnerships among data producers and data users in all sectors have increased access to and sharing of data across agencies and responders, and thus amplified the impact of youth in this domain. At the heart of those partnerships are young people, often university students, especially those who make up the YouthMappers network. Our close partnership between YouthMappers and the Humanitarian OpenStreetMap Team has not only advanced the global goals for disaster response and for development in their future workplaces and innovations, but also for global goals to enable better alliances across public, private, and nonprofit partnerships.

Keywords

Humanitarian response partnerships · HOT · Data · Disaster vulnerability · Indonesia · Tanzania · Peru

1 The Need for Digital Mapping in Humanitarian Response

In disasters, crises, and other complex emergencies, humanitarian responders often need to answer a basic set of questions: where did the event occur, which people were affected and where, and what critical needs are emerging? Since the early years of international humanitarian action, answering these questions has required

T. Radford (✉)
Humanitarian OpenStreetMap Team,
Washington, DC, USA
e-mail: tyler.radford@hotosm.org

G. Kateregga
Humanitarian OpenStreetMap Team,
Kampala, Uganda
e-mail: geoffrey.kateregga@hotosm.org

H. Machmud
Humanitarian OpenStreetMap Team,
Jakarta, Indonesia
e-mail: harry.mahardhika@hotosm.org

C. Redhead
Humanitarian OpenStreetMap Team, London, UK
e-mail: carly.redhead@hotosm.org

I. Mwanja
Humanitarian OpenStreetMap Team,
Dar es Salaam, Tanzania
e-mail: immaculate.mwanja@hotosm.org

© The Author(s) 2023
P. Solís, M. Zeballos (eds.), *Open Mapping towards Sustainable Development Goals*, Sustainable Development Goals Series, https://doi.org/10.1007/978-3-031-05182-1_31

basic information on people and places to respond effectively and prioritize aid. This information often arrives in the form of a map.

Since the international response to the 2010 Haiti earthquake, these maps are increasingly derived from digital map data, based on the world's free and open map source, OpenStreetMap (OSM). The organization providing tools that enable the creation of this base map data, and provide it to the humanitarian community is the Humanitarian OpenStreetMap Team (HOT).

2 Early Precedents for Student Engagement in "Crisis" Mapping

In January 2010, Patrick Meier launched the Ushahidi Crisis Map for Haiti. In February that same year, the Ushahidi Crisis Map for Chile was launched after a severe earthquake off the coast. University students at the Tufts University Fletcher School and Columbia's School of International and Public Affairs formed the backbone of what was then called "crisis mapping." At the time, Meier (2010) imagined a network of students at 12 universities around the world. While this vision for mobilization advocated for honoring the incredible energy of youth around the world, the reality of systematically tapping this energy for the "near real time" needs of humanitarian action required investment and strategic design, not only within university systems but also across the landscape of agencies working together on disaster responses to effectively utilize that energy in real-world response contexts. Students work at an academic pace with class loads and exam schedules that may be at odds with the real-time rush of data needs during disasters. They need specialized training and interface support, especially in the face of consequential humanitarian decisions in the fast-moving professional space of humanitarian agencies. Still, the experiences of the early generation of Crisis Mappers made the first case for how universities could be key to the future of humanitarian action.

Over the past decade, use of data and digital technologies has dramatically changed the look, feel, and efficacy of humanitarian action – in ways that have benefited not only the communities served but also in ways that enable youth engagement. Partnerships between data producers and data users have made this possible, especially in increasing access to and sharing of data across agencies. The most successful partnerships that HOT has developed have nearly always involved students as data producers. At the heart of those partnerships with young people are often university students. Since 2014, when the network was launched, this has included students that make up the YouthMappers network.

3 HOT Partnerships Engaging YouthMappers Students

From the early days of OpenStreetMap it was anticipated that open, free map data would be a tremendous benefit for humanitarian aid and economic development. The idea was proven during the Haiti earthquake in 2010. HOT was incorporated in the immediate aftermath, in August 2010, as a U.S. nonprofit, and then became a registered 501(c)3 charitable organization in 2013. HOT is the main connecting point between humanitarian actors and open mapping communities, welcoming anyone to contribute to its mission in line with its code of conduct. The HOT community has grown exponentially and is made up of volunteers, local community leaders, and professionals committed to the mission of helping reach those in need through maps. This diverse group includes dedicated voting members, staff, contractors, volunteers, interns, and board members. And of course, students.

Often, a government agency working in public health, disaster or emergency management, or statistics will identify a data gap. Working in OpenStreetMap, HOT volunteers, including YouthMappers, use satellite imagery and their knowledge of specific locations to add knowledge to OpenStreetMap to fill that gap. During disasters and crises, data produced by

YouthMappers and other volunteers in OpenStreetMap in the morning is exported by HOT to the United Nations Office for the Coordination of Humanitarian Affairs (OCHA) Humanitarian Data Exchange later that same day. This partnership makes young people a critical component of disaster and emergency response.

3.1   Data to Inform Both Humanitarian Action and the SDGs

Data created for immediate disaster needs are also used long beyond an immediate response effort into early recovery and towards the achievement of the SDGs. Today, open mapping has become a global movement to create free and open geographic data for impact. The movement relies on a complex ecosystem of partnerships across government, civil society, academia, and the private sector, which come together to collaboratively create this critical data resource for the SDGs – involving many of the same organizations as response teams.

Data created through open mapping supports monitoring SDG progress by making available fundamental, detailed, and timely information on the natural and built environment, including facilities, services, infrastructure, human settlements, and land use. The process through which data is produced, which often includes partnering with young people, involving women and girls in acquiring marketable technology skills, and directly engaging disaster survivors and those experiencing multidimensional poverty helps, in itself, to advance SDG targets such as those related to employment, gender equality, and disaster resilience.

Furthermore, open mapping is transforming how governments and citizens – often youth – can collaborate by allowing them to co-create and support critical government functions through maps and data. Free and open digital tools with low barriers to entry enable almost anyone with geographic knowledge, whether an individual enthusiast or a professional from a public or private sector institution, to contribute map data. The community openly designs schemas for mapping features as they are encountered in the world, develops new workflows and applications to contribute as technology advances, applies spatial data for an ever-growing array of uses, and supports and advocates for a growing, open collaboration across sectors.

3.2   Relationships to Enhance Citizen Generated Data in Vulnerable Places

The open mapping community is one that cares deeply about the quality and use of geographic data and includes people from all parts of society and across the globe, from experts in geospatial technologies to those in their own neighborhoods. It is hugely powerful for people from around the globe to work in the same database. To date, more than 8 million people are registered on OpenStreetMap, and approximately 50,000–60,000 make contributions – in the form of edits – during any given month.

This collaborative approach has been phenomenally successful at creating maps in under-mapped and under-served places, addressing critical needs, as seen in responses ranging from the Nepal earthquake and flooding in Bangladesh to mapping local schools and health care services in informal settlements. Through excellence in (open) data, many OpenStreetMap participants have gone on to become more engaged citizens. As confirmed by Lämmerhirt et al. (2018), as valuable as data itself may be, how open data seeds new relationships and becomes the center of partnerships is critical for joint action: "Citizen Generated Data initiatives open up new types of relationships between individuals, civil society and public institutions. This includes local development and educational programmes, community outreach, and collaborative strategies for monitoring, auditing, planning and decision-making."

4 Cases of HOT Partnerships Engaging YouthMappers Students

This chapter explores real-world cases in which partnerships facilitated by HOT among government, humanitarian agencies, and YouthMappers were not only beneficial to meet the immediate purpose of the data but also resulted in supporting a foundational data framework necessary for long-term systemic change that will ultimately enable achievement of the SDGs.

In these examples, we contemplate the unique barriers, challenges, and struggles that humanitarian action entails – especially such action that engages youth. We reflect upon the unique commitments that both HOT and YouthMappers innovated in order to overcome or simply acknowledge them as learning opportunities. In many ways, this is an ongoing experiment in order to tangibly realize the vision that many within the open mapping community have long held about tapping into the talents and energy of students – doing so in ways that meet the humanitarian needs of local communities and the agencies that serve them in disaster contexts. As such, the idea that youth already comprise part of the communities where they live, where humanitarian action is needed, needs to constantly be balanced with the idea that youth can be responders and partners within the humanitarian realm, alongside agencies. This remains an ongoing point for reflection and future innovation around both disaster response and the SDGs. The following cases share our experiences together with mapping for critical infrastructure in Indonesia, urban flooding in Tanzania, malaria elimination across sub-Saharan Africa, and disaster preparedness in Peru.

4.1 Indonesia: Supporting Disaster Risk Management and Contingency Planning

Humanitarian OpenStreetMap Team (HOT) began activities in Indonesia in 2012 in collaboration with the local Disaster Management Agency (BPBD) DKI Jakarta under the Jakarta Mapping Program. This program aimed to map critical infrastructure to support flood contingency planning, and was funded by Australian Aid (AusAID) through the Australia-Indonesia Facility for Disaster Reduction (AIFDR). As part of that program, HOT started its first university engagement with the Geography Department of the University of Indonesia. Students were taught about OpenStreetMap and how to contribute to it through a program where they worked with village representatives to add critical infrastructure location and information as part of a flood contingency planning assessment in Jakarta. In three months, they completely mapped the area. Along with newer OSM users registered from the university students, contributions from universities have steadily been increasing through today.

Since the successful collaboration with the university in mapping Jakarta, HOT and Perkumpulan OpenStreetMap Indonesia (POI), a national partner of HOT in the country, have been involving local universities in most activities. Moreover, academic institutions have become one of the main partners of HOT and POI in establishing and building OSM communities in the country. The involvement of universities in HOT and POI programs has covered various activities from data collection, capacity building, data quality assurance, community engagement, and technology innovation.

In the latter part of the decade, additional disaster management projects in Indonesia continued to demonstrate the importance of university and student participation. Starting in 2016, the United States Agency for International Development (USAID) funded the Pacific Disaster Center (PDC) and HOT under the InAWARE project which focused on collecting critical lifeline infrastructure in three of Indonesia's major metropolitan areas.

The first phase of the project ran from 2016 to 2018 to map three cities: Surabaya, Jakarta, and Semarang. The data from the mapping activity was uploaded into OSM and also integrated into InAWARE, a disaster monitoring platform created by PDC, and used by the National Disaster Management Agency (BNPB) to monitor and create the disaster management policy in priority areas. When conducting the mapping activity, HOT and POI collaborated with local universities in each city to conduct mapathons, mapping events where the participants remotely map the priority areas together at the same period of time, with the aim to complement initial data on building footprints and existing roads. Moreover, when conducting field surveys, HOT Indonesia worked together with some of the students to collect information from priority facilities and critical infrastructure. Their local knowledge and a better understanding of the mapping area contributed significantly to the accuracy and detail of the collected data.

Furthermore, students from the University of Gadjah Mada exchanged knowledge and methods with visiting student YouthMappers from co-founding university Texas Tech University around the HOT tasks to map areas around Mount Sinabung together, an active stratovolcano in Indonesia's North Sumatra Province. USAID GeoCenter and the Office of Foreign Disaster Assistance were working to improve existing infrastructure data in the area; particularly build-

ings, roads, and inland water features (Ituarte-Arreola and Thomas 2017). These features were important as they depict areas subject to volcanic and seismic activity which can be catastrophic in neighboring villages.

After mapping each city, the project aimed to transfer the knowledge from HOT and POI to the local community including universities not only in the priority cities that HOT mapped but also in other areas in the country. Several activities were conducted such as mapping and data collection workshops, public webinars, and mapping design competition. These activities aimed not only to develop the capacity of the participants but also to make sure they could contribute to wider society (Figs. 31.1 and 31.2).

In 2021, HOT, through POI, extended their collaboration with universities through the University Roadshow Program, a collaboration with Amazon Web Services (AWS), to promote digital information technologies including open digital mapping to university students, as well as to map poorly mapped areas in Indonesia. This initiative emerged from the fact that there are

Fig. 31.1 Geography students at Semarang State University work with HOT Indonesia to build capacity and engagement

Fig. 31.2 Students in Yogyakarta participate in a mapathon to trace building data for tasks around Mount Sinabung, an active stratovolcano in Indonesia's North Sumatra Province, for USAID GeoCenter and the Office of Foreign Disaster Assistance. (Credit: J. Ituarte-Arreola and R. Thomas, Texas Tech University)

many areas in the country that had still not been mapped, particularly those whose location falls outside Java and Bali islands. The program has two main activities which are training and a mapping competition. Students map their local area, while POI promotes digital information technologies including OSM to the academic institutions and builds relationships for future collaboration in disaster management. These activities are focused on mapping the highest disaster risk areas, particularly in Sumatra and Kalimantan islands where the availability of exposure data is still low.

There are several ways in which universities – and YouthMappers students – have become centered in disaster management projects in Indonesia with the support of HOT. First, effective disaster management relies upon timely data and technology innovations to make that data available to decision-makers. Universities are frequently compelled towards innovation and enlist relatively high engagement of students and faculty in activities that are urgent and cutting edge. Moreover, university students can be seen as flexible agents of change in communities that may otherwise be resistant to new tools. Students are perceived to be keen to learn new technologies which not only could help them thrive but also make a contribution to society on large scale. While in reality, they may need guidance, their eagerness to learn and apply new technology bodes well for introducing tools such as open mapping.

The retention rate of university students is relatively high in the context of contribution to mapping. Students participate not because they are paid to do so but because they want to do so as a voluntary act – they seek meaningful applications of what they are learning. In HOT projects, students are likely to continue to map in OSM even when the projects they were involved with are finished. Thus, even if not all students continue, more student mappers lead to more volunteer contributors in the long run. Their contributions on the whole are relatively sustainable for long-term partnerships. Although individuals may leave when they graduate, the chapter-based structure of YouthMappers as a group activity encourages the long-term engagement of students in HOT projects.

Lastly, data generated from the collaboration can produce practical products in various forms such as digital maps, infographics, pilot project learnings, and research to analyze and solve many development challenges and problems. These often end up becoming an inspiration for student theses or dissertations – serving both to expand formal knowledge but also apply to the needs for knowledge of local organizations that are created from the data.

The more that HOT and POI collaborate with universities, the more we realize that there are some things that need to be improved in the future such as creating long-lasting partnerships. Most HOT and POI partnerships with universities rely on particular donor-funded projects to sustain mapping efforts in a geographic area. Thus, after the project is completed, frequently the partnership struggles to find resources to keep going. Therefore, an emerging practice is to follow up collaboration by establishing Memoranda of Understanding (MoU) with universities instead of particular individuals (or the chapter groups) to formalize and hopefully solidify long term partnerships to make programs more sustainable and regularized as well as provide more opportunities for university students to be involved and contribute in future cohorts.

Collaboration with university students and YouthMappers chapters has been expanding to support the government, particularly the National Disaster Management Agency (BNPB) and the local disaster management agency (BPBD), in several areas from remote mapping, to field survey and technical support. YouthMappers sponsored a Regional Ambassador to support this and other work with HOT in Indonesia to build upon these experiences. This collaboration can reduce the dependency of the government on HOT and POI and instead build and activate the resources of the community youth in local areas.

4.2   Tanzania: Flood Resilience Through Open Map Data

Using OpenStreetMap to enable community mapping in Dar es Salaam has its roots in a 2011 pilot project in the Tandale ward, known as "Ramani Tandale." The project demonstrated that a few dedicated individuals – including university students and Tandale community members – could generate and manage data about their community well beyond what was previously available to any "expert" inside or outside Tanzania. It was unclear at the time what the future would bring. With Africa being the fastest urbanizing continent on earth, and Dar es Salaam its fastest-growing city, new and innovative partnerships would be needed for urban planning and disaster management data to keep pace.

The kick-off of the Dar Ramani Huria (meaning "open mapping" in Swahili) project took place in early 2015. Over the years, Ramani Huria has been groundbreaking in many ways. It has expanded on the Tandale pilot to include more than 1000 dedicated university students and over 3000 community members not at a cursory level, but truly relying upon their skills to carry out the mapping process. The project has moved from mapping one ward to comprehensively mapping 78 out of 95 wards by 2020 and covering an area home to an estimated 4 million people. It has incorporated Unmanned Aerial Vehicles (UAVs), a street-view mapping bajaj (tricycle / tuk tuk), and Real-Time Kinematics (RTK) to increase mapping precision. It is the largest and most complex urban mapping project in East Africa while being driven almost entirely by students including from YouthMappers chapters, and community members serving as HOT staff and volunteers (Dar Ramani Huria 2018).

Each team member's desire to put the city or neighborhood they call home "on the map" was multifaceted. One important, yet little-publicized motivating factor was that several HOT team members and YouthMappers students in Tanzania are disaster survivors themselves, having been severely affected by flooding in their own homes. Working to map their city was a powerful way to make the invisible visible and contribute real data for government-led urban planning and disaster management. Dar Ramani Huria not only generated near real-time, comprehensive, and detailed flood data and maps. It also transformed residents from victims of disaster to survivors having the capacity and ability to advocate for and effect positive social change.

Moving forward, an incredible wealth of data now exists in some of the previously least-mapped areas in the country. This data has incredible potential to be used not only for improved urban planning and disaster risk reduction but also to contribute to achieving some of the Sustainable Development Goals. Each day, Dar Ramani Huria team members work to update OSM with water, sanitation, and hygiene-related map features and add data for more sectors, such as bicycle routes and other sustainable transportation.

To date, HOT, in collaboration with its local partner OpenMap Development Tanzania (OMDTZ), has engaged a total of six universities in projects and mapping activities that have contributed greatly to OSM. Through these collaborations, YouthMappers chapters have been formed and some that were inactive have been revived to rejuvenate the continual coming and going of students. Most of these partnerships, however, rely on projects that are being implemented in a specific region and the consent or readiness of a university to work with HOT. They wrestle with the natural cycle of student life – graduating and renewing participation – but as alumni leave the university, they assume at best a place for ongoing support, and a least a knowledgeable citizen about the power of mapping.

Through microgrant programs, more YouthMappers chapters have been provided with funds to conduct projects that are solving issues in their communities while also contributing to OSM. This has proven to be an effective way to keep chapters churning and act as a catalyst to have more contributors on the map – during and beyond their studies. The Ramani Huria project not only engaged university students as part of their student internship and industrial training, but it also increased students' knowledge of open-source technology and increased the students' marketability for employment.

4.3   Malaria Elimination Mapping Campaign

Malaria is a preventable and treatable infectious disease transmitted by mosquitoes that kills more than one million people each year, most of them in sub-Saharan Africa, where malaria is the leading cause of death for children under five. According to the World Malaria Report (WHO 2020), in 2019 alone, there were 229 million new cases of malaria and 409,000 deaths. One child dies from malaria every two minutes. Because malaria is a global emergency that affects mostly poor women and children, malaria perpetuates a vicious cycle of poverty in the developing world.

Malaria related-illnesses and mortality cost Africa's economy USD 12 billion per year.

To combat the spread and occurrence of malaria, one needs to disrupt the transmission cycle of malaria parasites. This can be done by preventing mosquito bites, reducing the number of mosquitos, and by decreasing the prevalence rate of malaria parasites in both mosquitos and people. Aside from trying not to get bitten, this can be accomplished by interventions such as Indoor Residual Spraying (IRS), bed net distribution, and enabling better and faster diagnosis and treatment. All of these interventions imply knowing *where* transmissions are happening.

In 2016 and 2017, the HOT community, in partnership with Maxar (then DigitalGlobe), mapped an area of interest covering over 560,000 square kilometers in Southern Africa (Botswana, Zambia, Zimbabwe), Southeast Asia (Cambodia, Laos), and Central America (Guatemala, Honduras) to support the Clinton Health Access Initiative (CHAI) Malaria Program and Program for Appropriate Technology in Health (PATH) Visualize No Malaria campaign. In this partnership, Maxar provided satellite imagery, HOT coordinated volunteer contributors including YouthMappers, and CHAI supported government agencies in actually using the data in each country for IRS campaigns.

The first part of the mapping was remote mapping, where over 5000 mappers, mostly YouthMappers from different countries, were engaged in tracing over 3.6 million buildings from satellite imagery, resulting in information that is useful to know where people live. Volunteers from private sector organizations joined in by mapping remotely from offices in multiple countries. To engage the university students, online mapping competitions were organized, where the best mappers won individual prizes. There were also university prizes where the best universities for each month won prizes, which encouraged teamwork among the participants.

There were some challenges in engaging YouthMappers. Not everyone who wanted to participate had access to a computer and internet. Some of the participants had to share computers and use them in turns. Also, most of them were full-time students and had to find time for the mapping activities in between attending to their studies. However, mapping was also a learning opportunity for many of them, as it was a new skill they were learning, and opened up new doors for them to explore the world of Geographic Information Systems (GIS) and open mapping. In some universities, HOT conducted training for students on GIS and mapping skills, to make sure students benefited more from the program, beyond just carrying out mapping.

To complement the remote mapping, and to make OSM data more useful, HOT worked with the Ministry of Health and Wellness staff in Botswana and the National Malaria Program (NMP) to enhance and expand on the available data in Botswana for use in Malaria interventions by conducting field surveys collecting information such as types/use of buildings, building and roofing materials and road conditions in six malaria-endemic regions in Botswana. This data was used in supporting malaria elimination interventions including IRS, bed net distribution, and enabling better (faster) diagnosis and treatment.

Using this data, spraying teams were able to get a full overview of the buildings they needed to cover, especially those away from the known population and built-up areas, and plan for the needed supplies to cover all the residential buildings in the targeted areas. Though the primary use of the data was for the Malaria elimination campaign, the data is still openly available in OSM and can be used by other organizations including governments in different countries. In some cases, students also used the data for their own research as well. The YouthMappers program was able to leverage this joint HOT partnership as a model for future multi-country campaigns (Solís et al. 2018) (Fig. 31.3).

Through programs like this, and partnerships between governments, international organizations, the private sector, and volunteer groups like YouthMappers and HOT, youth in different countries are able to become active agents of change by providing data that is critical in solving big challenges like malaria that affect growth and development in their countries. They can share these experiences through global HOT teams and other chapters in the YouthMappers network. Using technology to decentralize humanitarian and development work ensures local ownership and impact but also promises to extend those impacts to become global in reach.

Fig. 31.3 Health workers spray for mosquitos outside a building in Botswana to prevent the spread of malaria, utilizing OpenStreetMap data on buildings and roads. (Credit: N. de Gier, PMI AIRS)

4.4 Peru: Collaboration Among Government, Academia, and the Regional Group of Earth Observations (AmeriGEO)

Engaging local students in community events and cross-sector collaborations has been critical in capturing local knowledge and ensuring participants understand local challenges for disaster preparedness. For example, in June 2021, HOT participated in an engagement with AmeriGEO (a consortium of Earth Observation government agencies on the continent), where CONIDA and CENEPRED organized a mapathon event to celebrate the bicentennial of Peru's independence. This focused on showcasing and democratizing local capabilities for Earth Observation through Peru's National Satellite. By convening a diverse group of over 150 students, local government representatives, and GIS professionals, and providing access to recent satellite imagery, valuable open data was created and used to generate products such as risk management tools and maps that could help in disaster preparedness and policymaking.

Education programs in the Americas region have seen a recent uptake of GIS courses and constant updates on new available technologies. As student mappers graduate, they integrate as professionals into the workforce of their countries and positions within public administration that will allow them to access state-of-the-art technologies to improve their decision-making processes. Early introduction to open data projects like OpenStreetMap through HOT events and YouthMappers chapters have enabled them to have a wider set of tools at their disposal to support their agencies in meeting and monitoring the SDGs.

With all volunteering initiatives, a leader or group of leaders to take ownership and drive an initiative forward and develop a strong value proposition or reason for volunteers to engage is critical. In 2019, HOT's Women Connect project (in partnership with YouthMappers and supported by USAID) founded the first formal official Peruvian YouthMappers chapter at UNSAAC university in Cusco. During the COVID-19 pandemic, HOT and GAL Group supported public health officials at all levels of government to plan how to support communities by mapping rural and vulnerable populations in the Cusco region (Firth 2021). Students supported the mapping of large projects to generate critical geospatial data at scale, and also used OSM data in their Masters projects for mapping investigations into challenges including solid waste management, sustainable transport (bicycles), agriculture and irrigation, and access to labor markets for people with physical disabilities.

5 Looking Forward: Youth Leading on Sustainable Action in Their Communities Towards the SDGs

Looking ahead, we will continue to innovate around the unique role of youth for the SDGs that the arena for humanitarian mapping presents. The unique, urgent character of disaster response urges innovation, which in turn attracts and benefits from student involvement as seen by the examples of HOT and YouthMappers joint experiences to date. Beyond continuing to advance such areas of fruitful action across the current established stakeholders and processes, we envision that in the future, we will amplify these partnerships in ways that help us respond better. To this end, private sector corporations that prioritize youth action have become an important ally in partnering for the SDGs. HOT has supported more than 25 corporate organizations to participate in mapping as a virtual volunteering opportunity, many supported the remote mapping of places most at risk during the Covid-19 response activation.

Youth skills development is a priority for many private sector organizations and a focus of their Corporate Social Responsibility strategies, given their interest in cultivating talent for future workforce pipelines. We are excited to go deeper and further than virtual volunteering with these corporate partners, for example, by leveraging the YouthMappers network and connecting local youth with cross-sector organizations so organizations can contribute their skills and expertise to support youth-led community projects. This opportunity includes the professional capacity of HOT as an organization itself. Since the beginning of the HOT-YouthMappers partnership, a number of alumni have emerged as leaders to join HOT as voting members and employees.

Institutions across sectors should seek partnerships for the SDGs that are community-driven, that is, community needs and priorities drive who and how we partner for impact. There are a plethora of ways young people are leading, collaborating, and contributing to local initiatives that are important to them and support the SDGs. This is because they are already part of that community. One way to enable this vision is through the opportunity that YouthMappers creates for students to use spatial data to explore and create solutions to social problems – and their partnership with HOT, which enables deeper local engagement while strengthening global relationships. These students may become future leaders in various sectors, from public and private to non-profit, which means an understanding of open tools, technical know-how, and the power of open community projects experienced together can be taken forward to future workplaces and innovations.

References

Dar Ramani Huria (2018) YouthMappers and Dar Ramani Huria in Tanzania. Available via YouthMappers Channel. https://www.youtube.com/watch?v=kB5APZ2u82Q. 27 Dec 2021

Firth R (2021) Covid-19 pandemic in Peru: Mapping Health Implications. News from Humanitarian OpenStreetMap Team, 24 August. Available from HOT. https://www.hotosm.org/updates/covid-19-pandemic-in-peru-mapping-health-implications/. Cited 27 Dec 2021

Ituarte-Arreola J, Thomas R (2017) From Texas, USA to Yogyakarta, Indonesia. YouthMappers Blog, 18 April. https://www.youthmappers.org/post/2017/04/18/from-texas-usa-to-yogyakarta-indonesia. Cited 27 Dec 2021

Lämmerhirt D, Gray J, Venturini T, Meunier A (2018) Advancing sustainability together? Citizen generated data and the SDGs. Report from the global partnership for sustainable development data, Open Knowledge International and Public Data Lab. Available from Data4SDGs. https://www.data4sdgs.org/sites/default/files/services_files/Advancing%20Sustainability%20Together%20CGD%20Report_1.pdf. Cited 27 Dec 2021

Meier P (2010) Why Universities are key for the future of crisis mapping. Blog on iRevolutions, 1 April. https://irevolutions.org/2010/04/01/universities-crisis-mapping/. Cited 27 Dec 2021

Solís P, McCusker B, Menkiti N, Cowan N, Blevins C (2018) Engaging global youth in participatory spatial data creation for the UN sustainable development goals: the case of open mapping for malaria prevention. Appl Geogr 98:143–155. https://doi.org/10.1016/j.apgeog.2018.07.013

WHO (2020) World malaria report 2020. World Health Organization, 30 November. Available from WHO. https://www.who.int/publications-detail-redirect/9789240015791. Accessed 10 Aug. 2021. Cited 27 Dec 2021

Part V

The Paths Ahead

Generation 2030: The Strategic Imperative of Youth Civic and Political Engagement

Michael McCabe and Steven Gale

Abstract

The young are deeply concerned about the world they will inherit, yet trends indicate that youth globally experience barriers to opportunities for civic engagement, lack of participation access, distrust, and voicelessness on the issues they care about most. The youth want to be more engaged in meeting the development needs of their communities and want to help lead democracy and social justice efforts despite contexts that discourage them. The authors argue for a renewed sense of engagement that is meaningful and puts youth at the center, and in the lead, in ways that capture the energy of a new generation.

Keywords

Youth engagement · Global partnerships · Leadership · Decision-making

M. McCabe (✉)
United States Agency for International Development (USAID), Bureau for Development, Democracy, and Innovation, Inclusive Development Hub Youth Unit, Washington, DC, USA
e-mail: mimccabe@usaid.gov

S. Gale
United States Agency for International Development (USAID), Bureau for Policy, Planning and Learning, Washington, DC, USA
e-mail: sgale@usaid.gov

1 The State of Youth Engagement

According to a recent poll, young people are deeply concerned about the world they will inherit, want to be more engaged in meeting the development needs of their communities, and are helping to lead democracy or social justice protests in their countries. At the same time, new research (Perry 2021) shows a large decline in trust and admiration for democratic governance. According to Freedom House (Repucci and Slipowitz 2021), for the first time in decades, authoritarian-leaning regimes outnumber democratic-leaning ones, with a majority of the world's population now living in authoritarian-leaning countries.

Engaging "Generation 2030"—the 2.4 billion youth globally—is essential to effectively support our common development priorities, such as the Sustainable Development Goals, while also stemming the tide of global disillusionment. Policies that are fair from an intergenerational standpoint allow people of all ages to meet their needs in a way that does not shortchange or undercut the ability of future

P. Solís, M. Zeballos (eds.), *Open Mapping towards Sustainable Development Goals*, Sustainable Development Goals Series, https://doi.org/10.1007/978-3-031-05182-1_32

generations to meet their own needs. This "intergenerational fairness" is emerging as a defining theme of our time, yet there is no generally accepted way to measure it. The Framework for Intergenerational Fairness (SOIF 2021) is a promising methodology in development that can be used to assess whether a policy decision might be considered "fair" to different generations, now and into the future. It is an appealing first step in systematically examining the extent to which a public policy can be rated on a scale from "completely unfair" at one end to "clearly fair" at the other. Ultimately, addressing intergenerational fairness will require comprehensive efforts to facilitate sustained and meaningful partnerships with youth.

2   Six Troubling Trends in Youth Engagement

Despite some glimmers of hope from youth engagement at the recent COP26 (Conference of Parties to the United Nations Global Climate Change Conference, UN Framework Convention), and their determination to be heard, there are six troubling trends we see in youth engagement.

2.1   Lost Confidence in Democracy

Recent data show that across the globe, youth satisfaction with democracy is declining not only in absolute terms but also relative to how older generations felt at the same stages in life (Fao et al. 2020). Much of the youth discontent stems from the perception of economic exclusion and being left out of decision-making that impacts their future. That perception is shared across emerging democracies, from Latin America to sub-Saharan Africa, priority regions for USAID and UN development objectives.

2.2   Stepping Back from Political Engagement

According to the recent Global Youth Development Index, which measures progress in 181 countries, youth participation in politics—a key driver in strengthening democracy and civil society—continues to deteriorate (Commonwealth 2021). In addition, an analysis from the Pew Research Center on global attitudes towards the United States and democracy revealed that the most worrisome indicator in their survey was young people's record low level of confidence in United States democracy (Wike et al. 2019).

2.3   Influences Shifting on Trust and Opinions

USAID's new survey and dashboard of youth civic engagement in ten countries found that relatives are the most likely source to influence youths' opinions on civic and political engagement (USAID 2021). It also found trust in government institutions and elections by youth is very low.

2.4   Absent from Decision-Making

A 2017 Global Shapers Annual Survey of almost 25,000 people aged between 18 and 35 from more than 180 countries and territories showed that 55.9 percent of respondents believe their views are not taken seriously into account before important decisions are made (Saouter et al. 2017).

2.5   COVID-19 Makes Matters Worse

Economic uncertainty has risen for almost all age groups as a result of the global pandemic, but today's youth have been hit especially hard. Data collected by the OECD (2020) reveals that today's youth have less disposable income than previous youth and are more than twice as likely to be unemployed than their middle-aged counterparts. These economic uncertainties are taking a toll on career prospects, as well as social well-being, and point to the need to accelerate youth engagement.

2.6   Voiceless on Key Issues Like Climate Change

The effects of climate change on future generations are much more debilitating to youth, who will inherit the consequences of decisions made today by older generations (Thiery et al. 2021). Youth are more likely to inherit a hotter, more extreme, and more uncertain and punishing climate. Yet, today's youth feel locked out of the very decisions on climate change that will impact them the most. This has significant implications for youth mental health outcomes, according to a recent 10-country study, which shows a sharp rise in climate anxiety among youth aged 16–25 (Marks et al. 2021).

Some of these concerns may reflect a deeper frustration with political engagement rather than democracy *per se* but neither is a desirable outcome.

At a recent conference with young change-makers from around the world, led by the School of International Futures, local youth from more than 25 countries raised an alarming but familiar set of concerns. They are frustrated that time after time, they feel stranded on the periphery of decision-making that will have profound influences on their future. On the positive ledger, they are already taking on poverty-reducing grassroots projects in their own communities, anxious to build and expand these projects into regional networks with like-minded youth, and eager to serve as a "Sensing Network" to bring forth issues *well before* they become established trends. These youth changemakers clamor to help shape their future and to improve lives through innovative approaches like strategic foresight (Fig. 32.1).

3   Building a Meaningful Compact with Young People

The pieces for meaningful engagement and partnerships with youth exist but need to be better articulated through comprehensive efforts. Some of these efforts are underway. USAID's new Global LEAD initiative seeks to support one million young changemakers over the next four years through increased investments in educa-

Fig. 32.1 Youth around the world are eager to be changemakers. (Credit: USAID.gov)

tion, civic and political engagement, and leadership development programming across all sectors. Its YouthLead.org platform now has 14,000 young changemakers sharing resources and conducting peer-to-peer learning. USAID and the State Department's support of young leaders through the Young African Leadership Initiative and other regional programs has strengthened the skills and opportunities of over 25,000 youth in the past 11 years.

USAID is also partnering with UNDP 16×16 and with UNICEF's Generation Unlimited partnership to foster youth civic engagement. The Generation Unlimited Youth Challenge funds youth-led innovations in education, employment, and civic engagement.

These efforts, however, are just a small part of the investment needed to build trust and a real partnership with Generation 2030—a partnership that provides them with the networks, skills, opportunities, and resources to revitalize democracy, tackle climate change, lead pandemic and humanitarian prevention and response efforts in their communities, innovate on skills for entrepreneurship and employment, and much more.

> Donors and governments need to create new and invest more in existing models of youth civic and political participation, along with recognizing youth as equal partners in development. That alone will inspire a generation of young leaders to work with other senior leaders to build their communities back better through collective action, public service and advocacy. – Darya Onyshko, a youth democracy leader from Poland who helps support the Community of Democracies #YouthLead effort and the European Youth Democracy Network

As USAID turns 60 years old this year, and as the UN completes 75 years of work, our collective vision for the next decade should align with a new generation of dynamic, diverse, innovative young leaders and citizens who champion democratic values and are motivated and empowered

to organize for a more peaceful and just world. A high-level, coordinated commitment to this new compact with Generation 2030 will be the key to building long-term partners in development and democracy. And, the really good news is that young people in the developing world, according to a recent telephone survey of 21,000 people from across the globe, are the most optimistic age group of all (Cain Miller and Parlapiano 2021).

YouthMappers is a powerful example of global youth who are bucking these trends. They are one model of possible interventions for youth-focused – indeed youth-led programs, creating opportunities for young people in ways that align with the Global LEAD framework of education, civic and political engagement, youth leadership, and youth organizing (Elisberg 2021). Students create and use their own mapping data within efforts with the purpose of development while learning and gaining experience. This model offers a framework that is highly interdisciplinary and highly adaptable, making it possible for youth to plan for a range of development objectives while also increasing civic participation and leadership.

Let's all work together to seize this engagement opportunity.

Acknowledgments The views expressed are those of the authors and do not necessarily reflect those of USAID or the U.S. Government. This article originally appeared on New Security Beat, the blog of the Wilson Center's Environmental Change and Security Program. Reprinted with permission. Available via Wilson Center. https://bit.ly/3qvZT1k.

References

Cain Miller C, Parlapiano A (2021) Where are young people most optimistic? In Poorer Nations. The New York Times. 17 November. https://www.nytimes.com/2021/11/17/upshot/global-survey-optimism.html

Commonwealth Secretariat (2021) Global youth development index and report 2020. Available via APS Group. https://thecommonwealth.org/sites/default/files/inline/5023_V3_Book_lores_smaller.pdf. Cited 30 Nov 2021

Elisberg J (2021) Global LEAD Toolkit: Resources to support opportunities for young people to contribute and lead community development. Available via

United States Agency for International Development, YouthPower. https://www.youthpower.org/sites/default/files/YouthPower/files/resources/Global-LEAD-toolkit_110421.pdf. Cited 27 Dec 2021

Fao R, Klassen A, Wenger D, Rand A, Slade M (2020) Youth and satisfaction with democracy. Centre for the Future of Democracy, Bennett Institute for Public Policy, October. Available via Analysis & Policy Observatory. https://apo.org.au/node/308978. Cited 30 Nov 2021

Marks E, Hickman C, Pihkala P, Clayton S, Lewandowski ER, Mayall EE, Wray B, Mellor C, van Susteren L (2021) Young people's voices on climate anxiety, Government Betrayal and Moral injury: a global phenomenon. Available at SSRN. https://doi.org/10.2139/ssrn.3918955

OECD (2020) Governance for youth, trust and intergenerational justice: fit for all generations? Organisation for Economic Cooperation and Development Public Governance Reviews. Available via OECD Publishing, Paris. https://doi.org/10.1787/c3e5cb8a-en. Cited 30 Nov 2021

Perry J (2021) Trust in public institutions: trends and implications for economic security. United Nations Department of Economic and Social Affairs Policy Brief #104, June. Available via UN/DESA. https://www.un.org/development/desa/dspd/2021/07/trust-public-institutions/. Cited 30 Nov 2021

Repucci S, Slipowitz A (2021) Freedom in the World 2021: democracy under Siege. Freedom House, February. Available via Freedom House. https://freedomhouse.org/sites/default/files/2021-02/FIW2021_World_02252021_FINAL-web-upload.pdf. Cited 30 Nov 2021

Saouter P, Jaffar HM, Babington-Ashaye Y, Andeleji K, Forsyth J, Berehe M, Popper N, Tyrakowski J (2017) Global Shapers Survey, World Economic Forum. Available via Amnesty International. https://www.es.amnesty.org/fileadmin/noticias/ShapersSurvey2017_Full_Report_24Aug__002__01.pdf. Cited 30 Nov 2021

SOIF (2021) Framework for intergenerational fairness. Specialist report by the by the Calouste Gulbenkian Foundation and School of International Futures. Available via School of International Futures. https://gulbenkian.pt/de-hoje-para-amanha/wp-content/uploads/sites/46/2021/07/IGF_Framework_SpecialistReport_EN.pdf. Cited 30 Nov 2021

Thiery W, Lange S, Rogel J et al (2021) Intergenerational inequities in exposure to climate extremes. Science 374(6564):158–160. https://doi.org/10.1126/science.abi7339

USAID (2021) 2021 youth civic engagement country snapshots, study brief. Available via United States Agency for International Development, YouthPower. https://shar.es/aWiGzb. Cited 27 Dec 2021

Wike R, Silver L, Castillo A (2019) Many across the globe are dissatisfied with how democracy is working. Pew Research Center Report, April. Available via Pew Research Center. https://pewrsr.ch/2IRgoRw. Cited 27 Dec 2021

Reflecting on the YouthMappers Movement

33

Jennings Anderson, Chad Blevins, Nuala Cowan,
Dara Carney-Nedelman, Courtney Clark,
Michael Crino, Ryan Engstrom, Richard Hinton,
Michael Mann, Brent McCusker, Rory Nealon,
Patricia Solís, and Marcela Zeballos

Abstract

YouthMappers co-founders and organizers reflect on the contributions narrated in this volume, and in a larger sense, the contributions of the more than 300 university campus chapters in more than 60 countries. These reflections provide a point of departure to imagine what's next and "where to go from here." We call for greater investment in youth, especially in majoritarian countries, as these contributions represent the dividend of previous investment in local capacity for the SDGs.

Keywords

Youth · Open Data as a Public Good · Movement · Innovation · Action · Sustainable Development Goals

J. Anderson (✉)
YetiGeoLabs, LLC, and YouthMappers,
Helena, MT, USA

C. Blevins
Locana, Alexandria, VA, USA
e-mail: chad.blevins@locana.co

N. Cowan
The World Bank, Washington, DC, USA
e-mail: ncowan@worldbank.org

D. Carney-Nedelman · M. Zeballos
YouthMappers, and Texas Tech University,
Lubbock, TX, USA
e-mail: communications@youthmappers.org;
marcela.zeballos@ttu.edu

C. Clark
Everywhere She Maps, YouthMappers and American
Geographical Society, New York, NY, USA
e-mail: cclark@americangeo.org

M. Crino · R. Nealon
USAID GeoCenter, Washington, DC, USA
e-mail: mcrino@usaid.gov; rnealon@usaid.gov

R. Engstrom · R. Hinton · M. Mann
The George Washington University,
Washington, DC, USA
e-mail: rengstro@gwu.edu; rhinton@email.gwu.edu;
mmann1123@email.gwu.edu

B. McCusker
West Virginia University, Morgantown, WV, USA
e-mail: brent.mccusker@mail.wvu.edu

P. Solís
Knowledge Exchange for Resilience, Global Futures
Laboratory, and School of Geographical Sciences and
Urban Planning, Arizona State University,
Tempe, AZ, USA
e-mail: patricia.solis@asu.edu

P. Solís, M. Zeballos (eds.), *Open Mapping towards Sustainable Development Goals*, Sustainable
Development Goals Series, https://doi.org/10.1007/978-3-031-05182-1_33

1 Marveling at the Movement as a Digital Public Good

One of the major trends we see in the sustainable development discourse, and one that is emerging globally, is the idea that open data is a digital public good. This means that the open data movement is democratizing information, and actions to create open data can be thought of as contributing to democracy. YouthMappers have emerged as a major contributor to making data free, open, and available at a local level, in many corners of the planet. This data is always contributed with a purpose, often advancing local, regional, and national advances for the United Nations Sustainable Development Goals, as the many stories in this book have narrated. Embedded within these stories are the many ways youth are growing a geospatial mapping movement in itself but we also find in the open sharing of data, ideas, time, and talent, the many ways they also are contributing to this larger picture of a digital economy and creating digital public goods. We believe it would be hard to find a better example of sustainable development on the ground. The bottom-up approach, which typically gets a lot of lip service, is rarely seen in the same light as these youth authors have illuminated for us here.

In the sections that follow, we reflect on a few questions that this movement and the youth that we engage with have compelled us to consider.

1.1 What Has Been the Main Contribution of YouthMappers to the Potential for Reaching SDGs?

As noted, we see that YouthMappers is at the forefront of the democratization of data, and data is a foundational element for free and open societies. The data contributed by YouthMappers serves a global digital public good that underpins modern digital economies. It underpins deci-sions, by policymakers, actors, and the behavior of everyday people. All of this informs the way that development unfolds, be it sustainable or not, and preferably the former.

As a network, perhaps a space for action, YouthMappers has provided these young people with a platform for self-organization, and self-expression. In order to attain collective goals like the Sustainable Development Goals (SDGs), the next generation will need the ability to self-organize, support one's peers, and strategize around helping one's local community, let alone beyond their borders. Reading the pages of this book, we find evidence that the program has not only built mappers, it has also built leaders.

1.2 How Does YouthMappers Strengthen Youth Links to the UN SDGs?

YouthMappers provides pathways to find the tools and resources to use for students to address and work towards the sustainable development goals in their communities. It enables discovery of what the SDGs mean in different local contexts, where students work collaboratively across chapters, nations, and continents. We have seen the projects and the work YouthMappers chapters do, which can spread to motivate others to take on similar projects. Ideas and methods often will spread throughout the network addressing a particular SDG. Waste mapping is one example.

More importantly, related to this exchange it is important to remember that mapping had traditionally been the domain of wealthier countries with established mapping agencies. YouthMappers helps to bring mapping back to a local level and enables contributions from people and communities living at locations of interest and need.

This is the digital generation. Connecting the digital with sustainable development is influencing how youth are interpreting what it means to be a digital humanitarian.

1.3 In What Ways Do You Believe Leadership Assumed by YouthMappers Has Been Able to Contribute, Galvanize, or Mobilize Their Communities?

Giving an identity platform for youth to break open their careers and be heard is something that inspires us and inspires each other, as peers. We have seen many of the members and alumni thrive and move forward onto the professional scene and further mentor and support others. This is what digital leadership in an open data space can look like.

YouthMappers students inspire and engage local community leaders to address challenges faced locally. And then they provide solutions to resolve those challenges through public geospatial data that is created through grassroots initiatives. When these lead to tangible impacts, like those we read about in many of this book's stories, it gives youth a voice in the development in their own communities where they otherwise have limited opportunities to be involved in decision-making. Technology is often associated with youth and leveraging it to improve their community's development makes them more respected and valued. These are essential early experiences for leaders, who are not just leaders of the future, but leaders today.

2 Reflecting on Innovation, a Spirit of Overcoming, and Action Towards the SDGs

A common thread visible in the research and audible in the stories of many YouthMappers is that of a common struggle and challenge in resource and opportunity poor environments; overcome with spirit and action; resulting in an instance that could be only described as *innovation*.

2.1 What Are Some of the Most Unique Methodologies and Technologies Used by YouthMappers That Relate to the UN SDGs? And What Is Unique About Them?

Not only is leveraging a suite of Open Source geospatial technology and tools to create open data something we might call an innovation, but also the ways that youth have been stepping into the OpenStreetMap community and emerged as a recognized valuable human resource. We have seen multiple times students introduced to OSM through YouthMappers grow to take on technical challenges, and share solutions with local OSM communities, and the global OSM as well.

But because YouthMappers chapters are present in majoritarian nations, and because they are comprised more representative of the gender balance of humanity, they are at the forefront of leading gender equality in the technology and geospatial world. They are pushing the boundaries of Whose Geospatial we mean, when we call OSM the "people's map of the world." Still, connectivity and access to technology issues are real and the much-touted digital divide is a barrier. Nevertheless, YouthMappers are meeting local issues head-on using technology and community involvement to solve these problems such as those related to SDGs.

2.2 What New Innovations or Methods Do YouthMappers Have with the Potential to Advance?

There are many cases where GeoAI and machine learningare a part of the methodologies, when augmented feature extraction, or tag verification is used. How to do that right, ethically, and build capacity while doing it is something that the network is still wrestling with, balancing the need for training, quality, accuracy, and all of the innovative questions that leading the use case landscape entails. But it is clear. Students are not just sensors. Responsible innovation in GeoAI for disasters, in particular, and purpose-driven mapping in general needs to be trustworthy, and also inclusive. This is a space where YouthMappers themselves may have much to say in tandem with the increased utility of AI and ML.

These developments aside, YouthMappers are poised to advance the revamping of analog technologies. Demonstrated in the cases throughout this book, YouthMappers have tremendous capacity to confirm, validate, create and contribute local information, all the more important as the knowledgeable, (and ethical!) humans-in-the-loop while mapping trends towards GeoAI.

Going back to ideas about democratization of data and of tools, we furthermore see this digital/analog presence of youth as holding the keys to the potential to advance geospatial literacy and educational trends towards Open Source/Data and tools in the first place. Because youth and young people are in the learning mode, this leverage is significant. Just merely knowing what OSM is and the existence of QGIS lets them enter the global geospatial workforce a rung above the rest. They will take that leverage to innovate.

2.3 What Conceptual Ideas Do YouthMappers Have the Potential to Advance?

Beyond the practical and the technical, there are ideas in motion that are of a more conceptual or academic character, and YouthMappers are anticipated to become a part of these advancements in the field. Two immediately come to mind. First, the creation of Digital Twins(Digital Cities or other terms) is coming. This means creating base data that can be used and applied to urban planning, advancing technology, and even the metaverse. Second, a mature subfield of open mapping with roots in geography includes quite a few examples of community mapping and/or participatorymapping. This continues to evolve. Looking forward, if we could progress this to

having an influential seat at the table when it comes to actual spatial planning for our own communities, YouthMappers will be a part of that process. Taking participatory mapping all the way through to participatory planning – and having data bridge that gap represents a new old approach that holds a lot of promise for SDGs.

3 Identifying Where We Still Have a Lot of Work to Do

Despite this good news, the YouthMappers movement continues to face barriers, challenges, and room to grow and improve. A few of these areas where much work is needed, and which youth who assume roles as future leaders should embrace, include:

- *Data informing decisions* – This area is fraught with challenges, and connecting the dots to the actual problem-solving moments can be problematic (at large, as well as with geospatial); trade-offs and unforeseen consequences are but the start of efforts that seek change. Careful engagement and wisdom are needed here.

- *Inclusivity* – while many strides have been made with respect to the acceptance of students, we anticipate that critiques about whether they are generating the quality that we know they produce will likely continue; commitment to training, validation, and quality control will remain critical to future success. Similarly, bringing greater diversity in terms of location, ethnicity, gender, and other axes of difference has been an important contribution of YouthMappers, but we believe a struggle that will continue.

- *Resource allocations* – On some campuses, students continue to wrestle with gaining the space and support from their home institutions, given that this is an informal (non-classroom) activity that they are leading. While mentoring attenuates this in many places, there are still locations that fall short of recognizing the powerful energy they can bring now to research and the reputation of their universities. We predict ongoing resource battles, but explicit attention and tracking of the value that youth brings to their institutions could attenuate this challenge.

- *Connectivity* – The digital divide and thegender digital divide will likely remain an issue for many of our chapters going forward. There are not yet any cost-effective solutions for this in the short or medium term. If we could partner with, and potentially get the support of new actors, for instance, telecom firms, particularly in Africa, and/or other private sector support for open mapping by youth, we may be able to cover this cost and amplify our ability to give connectivity grants which would mean so much to so many chapters.
- *Privilege* – Students with the privilege of attending university are already privileged in some ways, and still sometimes need guidance on connecting to those who are the most vulnerable in their communities. We need to continue to be vigilant and work across chapters between high-income and lower-income nations, as well, to guard against reproducing post-colonial relationships that would erode the solidarity of the movement built. This includes avoiding activities that consider mappers only as "sensors," or being (unwitting) extractive tools for development that is not ethical or truly sustainable.

4 Where Are YouthMappers Going Next

Given the vast involvement with YouthMappers addressing every SDG, we can definitely conclude that this action is impacting the world in a positive way, using open geospatial data and the engagement of youth. Looking forward we should encourage more and more work that more closely informs sector experts in helping map and apply geodata to sustainable development action that has potential to transform decision-making for the public good.

YouthMappers is much more than a data or geospatial program. It has turned into a well-rounded network program that supports the youth, digital inclusion, gender, and many other sectors serving international development. The future will retain its geospatial roots but the community is on a trajectory to mature into a program that responds to development and SDG matters. It is not an exaggeration to say that YouthMappers has transformed into a movement. Because current students and alumni alike are so dedicated to the purpose and to each other – since they have grown it from the start – it has the potential to grow and thrive beyond its initial vision.

Whatever version of sustainable development goals that comes after the SDGs should find fertile ground in this movement. Youth are plugged in and maybe could help the international community refine/redefine / revolutionize what as a global community we need to measure, track, and aim for. They will be the generation to live this reality.

5 How Are We to Support This Journey from Here?

Reflecting on our experiences as individuals and a group of collaborators who co-founded and organize YouthMappers, we rightly ponder on how to support the journey from here. The answer needs to emphasize the importance of doing behind-the-scenes work with the right mindset when it comes to how we choose to engage with youth. In short:

Be a good ancestor.

We conclude with three ways that we can be good ancestors – as organizers, mentors, and the alumni who are joining in the leadership of this movement. First, we must be sure the youth understand the trajectory of the past in getting to this moment, but then let go as they **choose their own paths forward.** As Kahlil Gibran wrote, "their souls dwell in the house of tomorrow, which you cannot visit, not even in your dreams." Make sure that strong youth leadership is in place in your chapters, give them access to the tools they need, and let them lead the way. You can also read into this, "then get out of their way!"

Second, **connect young people to each other.** This principle is one of the design aspirations of YouthMappers being a network that works collaboratively around global campaigns for mapping tasks and meanwhile encourages youth to develop local project mapping that can be shared with others. We are deeply inspired by the stories and research results in this book, which have leveraged lasting connections that can be forged by meeting around the map.

Finally, **raise visibility of their ideas and amplify their accomplishments**. Young people typically have not yet built up the means, mechanisms, or reach that we have cultivated over time as older advisors, and indeed often inherited from our own ancestral professional communities. We need to continue to open up safe spaces for youth to participate, even rebuild those mechanisms. The world is changing at an accelerated pace, so we need to accelerate the transitions for youth to step up. Then give them recognition, and make sure credit is distributed fairly. This allows all of us to not only be inspired by their thoughts, innovations, and energy, but it also keeps us accountable for a sustainable future.

Index

A

Ableism, 186
Aburrá Valley Metropolitan Area (AMVA), 222, 226
Academia, 347
Academic institutions, 348
Access to electricity, 125, 136, 138
Accessibility, 182, 186, 187, 307
    in Brazil, 186–187
Accra City Mapping, 311
Action cameras, 138
Aerial Unmanned Vehicle Systems International
            (AUVSI), 167
Africa RISING program, 32, 37–39
African Drone and Data Academy (ADDA), 162–164
Agriculture, 30
Agriculture-related activities, 266
Airin Akter, 105
Akure, Nigeria, 235
Amazon Web Services (AWS), 349
Amazonian framework, 289
Amazonian landscape, 288
American Geographical Society (AGS), 329
AmeriGEO, 354
Anthromes, 282, 283
Anthropocene, 282, 284
Anthropocene Epoch, 278
Anthropocene, YouthMappers
    Coastmap project, 279
    participatory approach to mapping coastal
            communities, 279
    phased methodological processes with community
            participation, 280, 282
ArcGIS 10.1, 53
ArcGIS Online platform, 60, 62
Architecture for Smart City, 18, 19
Arizona
    extreme heat deaths and mobile homes, 241–248
    pattern beyond indicators, 243
    unseen climate vulnerable locations, 243–244
Artificial intelligence (AI), 126, 127, 132
Asia, YouthMappers
    climate change, 89, 91
    collaboration and engagement, 91

COVID-19 global pandemic
    digital innovation, 91
    mapping essential services, 88
    mapping health facilities, 87
    mapping residential households, 89
open spatial data, 86
resilience and sustainable development goals, 87
"Ask Me Anything (AMA)", 322
Atrato River, 282
Australia-Indonesia Facility for Disaster Reduction
            (AIFDR), 348
Australian Aid (AusAID), 348
Authentic community development, 1
Awareness programs, 252, 256, 257

B

Bagsakan/drop-off center, 273
Bangladesh, 94, 95
Bangladesh Challenge, 87
Bangladesh Red Crescent Society, 90
Batteries, 169
Bed net distribution, 352
Belize, 297–299, 303
Bicycle Master Plan, 223
Bicycle paths, 226
Biodiversity, 190, 191, 195–196, 198, 202
Biomphalaria or *Bulinus*, 70
Body mass index (BMI), 51
Brazil, 291
Brazilian Institute of Geography and Statistics (IBGE),
            186
Building mappers, 328

C

Capacity-Building Service Group in Bangladesh, 105
Capacity-building training, 23
Cape Coast to Tamale, 32
Car battery chargers, 169
Cartographic field, 102
CENEPRED, 354
Central business district (CBD), 15

© The Editor(s) (if applicable) and The Author(s) 2023
P. Solís, M. Zeballos (eds.), *Open Mapping towards Sustainable Development Goals*, Sustainable
Development Goals Series, https://doi.org/10.1007/978-3-031-05182-1

Child malnutrition, 51, 53, 54
Chi-square test, 53
Citizen generated data
  in vulnerable places, 347
Citizen science projects, 81
Civic communities, 341
Civic engagement, 341
Civic-tech organization, 103
Civil society, 347
Climate change, 89, 279, 280, 361
Climate change in Bangladesh
  climate change-induced disasters on the coastal area,
      252–253
  gendered impacts, 253
  and gender equality, 259
  "hot spots", 252
  SDG 13, 252
Climate heat extremes, 242
Climate-induced disasters, 251–252
Clinton Health Access Initiative (CHAI) Malaria
      Program, 352
Coastal anthropocene, 283, 285
Coastal communities, 283
  participatory approach to mapping, 279, 280
  SDGs, 277, 278
Coastmap project, 279
COBIRRIS-Paez, 115–117, 121, 122
Collaboration, 156, 158
Collaborative and Humanitarian Mapping, 20
Collaborative cartography, 288–290
Collaborative favela mapping, 184, 186
Collaborative geospatial technologies, 26
Collaborative mapping, 14
  on OSM, 289, 291
  of riparian communities, 289
Collaborative strategies, 347
Colombia, 222, 297, 299, 303, SDG 11 (Sustainable
      cities and communities)
Communication, 336, 338
  channels of multicultural movement, 340
  coordinator, 327
Communities, 37, 354, 355
Community-based flood vulnerability mapping, 218
Community disaster response team (CDRT) members, 90
Community-engaged learning (CEL), 59
Community flood mapping, 218
Community flood mapping, informal settlements
  Bwaise and Kalerwe communities, 210–211
  community mapping results, 213–214
  data collection exercise, 212–213
  pre-fieldwork preparations, 211–212
Community mapping, 368
Community of communities, 326, 338
Community outreach, 347
Community projects, 26
Community resource mapping projects, 58
Complex metropolitan system, 222
CONIDA, 354
Connectivity, 283, 369

Conservation of the Mangrove Ecosystem, 281
Consumer-grade GPS devices, 336
Contingency planning, 348–351
Coping strategies index (CSI), 51
Corporate Social Responsibility strategies, 354
Costa Rica, 297, 299, 300, 303
Costa Rican System of Juristic Information (SCIJ), 118
COVID-19, 225, 297, 354, 360
COVID-19 community resources in Dutchess County
  challenges and opportunities, 62, 63
  community input and dedication, 58
  Hudson Valley Mappers, 58, 59
  humanitarian and disaster relief efforts, 58
  immediate GIS response, 59–61
  longer-term GIS support, 61, 62
  open mapping challenges, 63, 64
  Poughkeepsie, 58, 59
COVID-19 Community Resources Map, 61–63
COVID-19 mapping efforts, 58
COVID-19 pandemic, 6, 19, 122, 184, 332, 354
  county residents, 60
  crisis scenario, 64
  local mapping effort, 65
  mapping process, 62
  mapping projects, 59
  maps, 63
  reliable healthcare location, 57
  restrictions, 63
  testing sites, 58
  YouthMappers students, 68
COVID-19 restrictions, 63, 138
Crisis mapping, 346
Critigen, 332
Cross-continental YouthMapathon, 76
Cross-sectional curvature (CSC) value, 119
Currulao, 283
Cycling community, 225–227
Cutting-edge AI tools, 138
Cycling infrastructure, 222, 223, 226–228
Cyclists, 223–228

D
Data, 22
  citizen generated in vulnerable places, 347
  and data producers, 346
  and digital technologies, 346
  humanitarian action, 347
  SDGs, 347
Data informing decisions, 368
Data producers, 346
Data quality, 330
Data users, 102
Decent internet connectivity, 336
Decentralization, 328
Decision-making, 360
Declaration on the Elimination of Discrimination Against
      Women, 272
Dental services, 265, 266

Department of Geomatics and Land Management, 322
Descriptive research method, 263
Digital assets, 326
Digital divide, 173, 174, 369
Digital elevation model (DEM), 117, 281
Digital humanitarians, 222, 279, 333
Digital inequality, 181–182 SDG 10 (Reduced
    inequalities)
Digital innovation, pandemic, 91
Digital invisibility, 184
Digital mapping, 345, 346
Digital technologies, 346
Digital Twins, 368
Disaster awareness programs, 252
Disaster contexts, 348
Disaster risk reduction (DRR), 257
Disaster vulnerability, 252, 253
Distributed network approaches, 341
Dominican Republic, 297, 300, 301, 303
Drainage system, 216
Drone-based orthophoto mosaics, 281
Drone imagery, 310
Drone mapping, 310
Drones, 308
    African Drone and Data Academy, 162
    and data technology, 162
    innovation in health prevention efforts
        challenge, 169
        data collection, 168, 169
        malaria prevalence, 166
        mosquito breeding sites in Kasungu, 166, 167
        pre-flight preparation, 167–168
    innovation region, 163
    in Kasungu, 162
    in Malawi, 162
    promoting development and humanitarian
        assistance, 161
    for SDGs, 169
    UAVs, 161
    UNICEF's deployment, 162
    and youth, 162–163
    youth in Malawi
        collaborations, youth and community
            involvement, 165–166
        Dzaleka Mapping Project, 164, 165
        Dzaleka Refugee Camp, 165
        UAV imagery, 165
Dzaleka Mapping Project, 165

E
Earth as a System course, 96, 97
Ecological zoning, 198, 203
Economic development, 126, 138, 156, 346
Ecosystem
    AGS, 329
    Critigen, 332
    Mapbox, 332, 333
    Mapillary, 329, 330
    partnerships, 325–328

    private and nonprofit sectors, 329
    TeachOSM, 330, 331
Eco-tourism, 196
Ecuador, 191, 193, 195
Education, 37, 268, 270, 273
    course, 97
    SDGs, 94, 99
    solutionaries, 96
Education for Sustainable Development (ESD), 96
Educational programmes, 347, 354
Electricity
    affordable, 141
    energy crisis, 142
    epileptic power supply, 142
    field mapping, 145
    mappers for the SDGs, 148
    mapping challenges, 146, 147
    remote mapping, 144
    resultant effect, 142
    SDGs, 143, 144
    YouthMappers in Nigeria, 142–143
Electricity access, 142, 143
Electricity agencies, 135
Emergency management system (EMS), 254
Emergency medical services, 266
Emergency medical transportation, 265, 266
Emergency response, 253, 258
Emotional cartography, 223
Employment models, 262
ENA software, 53
Energy, 114, 121, 126, 135, 139 Water–energy–land
    nexus methodology
Environmental education, 257
Environmental Protection Agency (EPA), 232
Environmental resilience in Latin America
    ecological zoning for conservation and sustainability,
        198, 203
    equitable resilient future, 205, 206
    Kuna Nega, 193, 195
    mapping a hidden paradise during a
        pandemic, 193
    mapping trees and palms around old Panama historic
        monument, 203
    open spatial data, 190, 191
    Smart Campus, 203
Epileptic power supply, 142
European Free and Open Source Software for Geospatial
    (FOSS4G), 319
Evacuation planning, 253, 254, 256
Everywhere She Maps (ESM) program
    aim, 103
    core activities, 103
    ESM ambassadors, 326
    mapping campaigns, gender equality, 104
    professional development and internship matching,
        103, 104
    regional ambassadors, 105, 106
    technical capacity building, 103
    WiT leadership fellowship, 104, 105
Extreme heat deaths, 242, 243

F

Fallen lands, 291
False stereotypes, 103
Farm management practices, 41
Favelas, 182, 184, 185, 187
Federal University of Technology in Akure (FUTA),
    142–144, 148, 236
Fellowship workshops, 105
Female genital mutilation (FGM), 178, 333
Flood disaster resilience planning, 218
Flood mapping, 217–219
Flood resilience
    mapping, 307
    Open Map Data, 351–352
Flood vulnerability, 210, 217, 218
Food and Agriculture Organization (FOA), 49
Food consumption score (FCS), 51
Food insecurity and malnutrition, northern Ghana
    measuring and mapping hunger
        current indicators, 50
        spatial patterns, 52
    research, 48
    YouthMappers
        baseline data, 48
        locating the regions, 48
Food security
    agricultural sector, 30
    economic and physical access, 29
    in Ghana, 30
    malnutrition and stunted growth, 30
    open data, 30, 31
    SDGs goal, 29
Foundational data framework, 348
Free and open-source software (FOSS), 222
Freedom House, 359

G

Gender-based violence (GBV), 24, 178, 275
Gender digital divide, 174, 369
Gender equality, 112, 329
Gender-equal maps, 103
Gender equity, 94, 275
Gender gap
    in mobility, 262, 263
Gender inclusion, 102
Gender-inclusive mapping communities, 95
Gender inequity, 93
Gender-relevant topics, 112
Generation 2030, 359, 362
GeoAI tools, 127, 367, 368
Geographic attributes, 103
Geographic data, 347
Geographic Information Systems (GIS), 58, 292, 353
Geographic knowledge, 102
Geography, 336
Geo-Inquiry Process, 97
GeoJSON-type file formats, 225
GeoLab, 222
Geomatics Engineering Students' Association of Nepal
    (GESAN), 111, 154–158

George Washington University, 318
Geospatial data, 181
Geospatial industry, 156
Geospatial Industry Partnerships, 337, 341
Geospatial literacy, 368
Geospatial mapping movement, 366
Geospatial skills, 172, 173, 177, 178
Geospatial technologies, 177–178, 347
Geoventurers YouthMappers chapter, 96
Geo YouthMappers, 322, 323
Ghana
    Accra City Mapping, 311
    agricultural sector, 30
    food security, 30, 42
    incidents of flooding, 311
    metropolis, 310
    Ministry of Food and Agriculture, 30
    OSM approach, 307, 311
Ghana Health and Demographic Survey (GHDS), 50
Ghana Statistical Service, 38
Global citizens, 331
Global community, 339
Global diversity, 328
Global map, 102
Global movements, 341
Global OSM community, 338
Global Positioning Station services, 138
Google maps, 87, 263
Governance, 116, 305, 306
Government institutions, 297
GPS tracks, 309
Grab Map operations, 88
Great Acceleration, 278
Greater Accra Resilience Integrated Development
    (GARID), 311

H

Habal-habal/modified motorcycle, 270
Habitat of humans, 76
Habitat of snails, 72, 76
Haiti earthquake in 2010, 346
Harvard Humanitarian Initiative, 94
Health, 242
Healthcare-related facilities, 59
Healthcare services, 273
    dental services, 265, 266
    emergency medical transport, 265
    emergency medical transportation/medical services, 266
    health clinic, 265
    mental health, 265
    pharmacy, 265
    travel time and travel cost, 265
    women *vs.* men, 264
Health clinic, 265
Health Facilities in Saint-Louis, Senegal
    data validation, user stories, 66
    Gaston Berger University YouthMappers, 65
    mobilization, 65
    organizations, 65
    UGB YouthMappers, 65

Healthsites OpenDataKit (ODK) collection tool, 58
Healthsites' open-source data collection tool, 65
Healthsites project manager, 65
Heat, 242
Heat-related deaths, 243
Heat resilience, 247
Height-for-age (HAZ), 53
High-quality geospatial data, 332
HOT's Women Connect project, 354
HOT Tasking Manager, 290, 291
Hotspots (unsafe places), 24
Household dietary diversity score (HDDS), 51–53
Household essentials, 273
    location, 264
    longer travel times, 264
    "men's work", 263, 264
    road condition, 264
    women travel longer distances, 264
Household food insecurity and access scale (HFIAS), 51
Household food security status
    anthropometric measurements, 52
    data analysis, 53
    degree of malnutrition, 54
    household-level categorical variables, 55
    malnutrition status, 53
    prevalence of food insecurity, 53
Household hunger scale (HHS), 51
Household survey, 262
Housing, 242–245
Hudson River Housing, 59
Hudson Valley Mappers
    community geography, 58, 67
    community partners, 63
    community priorities, 60
    COVID-19 response mapping projects, 63
    Esri software, 64
    history, 59
    long-term relationships, 59
    shareable online web map, 58
    Vassar students, 64
    web map, 59
    YouthMappers chapter, 58
Human Development Index (HDI), 14
Humanitarian action
    and SDGs, 347
Humanitarian agencies, 348
Humanitarian assistance community, 67
Humanitarian mapping, 97, 98, 354
Humanitarian OpenStreetMap Team (HOTOSM/HOT),
        89, 172, 173, 177, 184, 323, 330, 337, 340, 346
    citizen generated data in vulnerable places, 347
    data, 347
    and economic development, 346
    government agency, 346, 348
    Indonesia, 348–351
    Malaria, 352–353
    open mapping communities, 346
    Peru, 354
    Tanzania, 351–352
    Tasking Manager, 309, 311

Humanitarian response partnerships
    crisis mapping, 346
    digital mapping, 345, 346
    HOT Humanitarian OpenStreetMap Team
        (HOTOSM/HOT)

I
IBM SPSS version 21, 34
iD Editor, 309, 310
IITA Ghana office, 31
IMM proximity, 22
Impostor syndrome, 104
InAWARE project, 348
Inclusivity, 368
Individual youth activists, 4
Indonesia, 348–351
Indoor Residual Spraying (IRS), 352
Inequalities, 186
Informal settlements
    Bwaise and Kalerwe, 215–217
    data gathering and mapping, 22
    geographic contexts, 26
    Global South, 14
    Italy, 14
    in Kenya, 24
    life, 14
    living conditions, 14
    problems, 15
    Rwanda, 15
    spontaneous development, 14
    uneven distribution, 15
    urbanization, 14
Informed decision-making, 172
Infrastructure project, 272
Innovations, 162, 367–368, 370
    and collaboration, 327
    communication channels, 340
    component, 338, 339
    cultural practices, 338
    evolution, 161–162
    model from individuals to groups, 338
    organized efforts, 339
    remote volunteers, 338
Innovative partnership model, 327
In-service teacher education, 94
Integrated modification methodology (IMM), 18
Intentional institutional support, 340
Intergenerational fairness, 360
Intergovernmental Conference on Environmental
    Education, 94
International Cartographic Association (ICA), 182
International Federation of the Red Cross (IFRC), 90, 318
International Institute of Tropical Agriculture (IITA)
    activity, 31
    farm technologies, 31
    farmer profiling project, 31
    farmers' needs, 31
    interventions, 33
    national and international agencies, 31

International Institute of Tropical Agriculture
      (IITA) (*cont.*)
   objectives, 32
   OSMtracker, 34
   programs, 32
   R4D organization, 31
   research lab, 31
   support services, 41
International organizations, 330
International Society for Photogrammetry and Remote
      Sensing (ISPRS), 182
International university network, 326
Internet, 129, 138, 148
Internet connectivity problem, 146–147
Internet system, 291
Internship Match program, 104
Invisibility, 288
Invisible communities, 288–289
IoT devices, 145
Italian YouthMappers
   benchmark development plan, 18
   collaborative activities, 19, 21
   HDI, 18
   humanitarian activities, 18
   IMM methodology, 18
   results, 21, 22
   slums upgrading, 18
Italy, 76, 319

J
Jakarta Mapping Program, 348
Jamaica, 297, 298, 303
Java OpenStreetMap (JOSM), 309, 310
Joint youth action on SDGs, 81, 82
Jomo Kenyatta University of Agriculture and Technology
      (JKUAT), 24
JOSM editor, 20, 32, 34

K
Kampala Capital City Authority (KCCA), 218
Kenya, 233, 234
Kenyan YouthMappers
   joint initiatives, 23
   methodology, 24
   outcomes, 25
   partnership with map Kibera trust, 23
Kigali GeoDR, 15
Knowledge Exchange Resilience (KER), 243, 247
Knowledgeable students, 1
Kobo Toolbox software, 32–34, 42, 95, 122, 236
Kobo Toolbox v1.29, 117
Kuna Nega, 193, 195

L
Land Water-energy-land nexus methodology
Land grabbers, 289
Land tenure system, 37

Latin America
   youth and open data in, 189, 190
Leadership, 362
Leadership development, 331
Learning environment, 336
Light detection and ranging (LiDAR), 155
Livelihoods, 273, 274
   engaged in agriculture, 266
   location of agriculture, 268
   microenterprise and micro-services in agriculture, 266
   in non-agricultural activities, 268
   production support services, 267
   source of income, 266
   travel cost/time, 268
   travel time/cost, 268
   unclassified/tertiary roads, 268
   women's role in activities, 267
Livestock, 266
Local OSM community, 326
Lusaka, Zambia, 234

M
Machine learning, 330, 367
Maize cultivation, 39
Maize leaf stripping, 39
Majoritarian nations, 340
Malaria, 352–353
Malaria elimination campaign, 352–353
Malaria related-illnesses, 352
Malnutrition, 30
Map literacy programs, 111, 153, 156–158
Map Worlds: A History of Women in Cartography
      (book), 101
Mapagaling, 87
"Map-Along" events, 331
Mapathon, 21, 144, 148
MapBeks, 88
Mapbox, 332, 333
MapContrib, 87
MapGive, 337
Mapillary, 329, 330
MapMalawi, 164
Mappers4Med initiative, 64
*Mapping Amazonian riparian locations in the northern
      municipality of Tefé – AM*, 290
Mapping campaigns, 104
Mapping community, 178
Mapping electrical infrastructure, 330
Mapping features, 347
Mapping Modi Gaupalika, 107
   achievements, 109
   collaboration, 107
   description, 107
   impact on community, 109
   Kshitiz, 107
   mobile applications, 107
   overprotective Nepali parents, 107
   QGIS, 107
   security, 107

Mapping slums, 330
Mapping solid waste, 233
    YouthMappers in
        Akure, Nigeria, 235, 237
        Lusaka, Zambia, 234
        Nairobi, Kenya, 233
Mapping waste disposal sites, 330
Maps
    inclusiveness, 102
    OSM database, 102
    sectors essential, 102
    society development, 111
    symbols and instruments, 102
    YouthMappers, 102
MapUganda, 323
MASTR-SLS project, 76
Mayflies, 78
Medellín, 222
Memoranda of Understanding (MoU),
        328, 351
Mental health, 265
Mentoring, 322
Mentorship, 333
Mentorship, YouthMappers
    New York, 321
    Panama, 320, 321
    PoliMappers, Italy, 319
    Pretoria, South Africa, 321
    student led-groups, 318, 319
    Uganda, 322
Mexico, 297, 299, 301, 303
Micro, small and medium enterprises
        (MSMEs), 261
Microenterprise, 266–268
Microgrant programs, 352
Micro-services in agriculture, 266
Migration decision, 95
Millennium Development Goals (MDGs), 172
Mobile applications, 256
Mobile homes
    energy standards, 242
    heat resilience, 247
    light materials, 242
    in Maricopa County, 243
    Mesa mobile home parks data mapping, 247
    parks, 245
    parks in the US, 244
    as primary homes within OpenStreetMap, 244
    residents, 242
    and RVs, 243–245
    tags, 245
Mobility
    gender gap, 262, 263
    transportation
        education, 268, 270, 273
        gender-based violence, 275
        healthcare services, 264–266, 273
        household essentials, 263–264, 273
        livelihoods, 266–268, 273, 274
        microenterprise, 266–268

        safety, 270–272, 274
        security, 270–272, 274
        well-being, 272
Multicultural movement, 340

N
Nairobi, Kenya, 233
National Disaster Management Agency (BNPB), 348,
        351
National Malaria Program (NMP), 353
National monuments, 203
National NGO Program on Humanitarian Leadership
        (NNPHL), 94
National Plan for Disaster Management (NPDM), 256
National Science Foundation, 322
National spatial data infrastructure (NSDI), 296, 298,
        300–302
Near real time, 346
Neglected tropical diseases (NTDs), 70
Nepal earthquake, 347
New York, 58, 321
Newcomers, 78
NGOs/INGOS, 341
Nigeria, 235
Nigeria Sustainable Energy 4 All (SE4ALL) program,
        144, 148
Nigerian energy supply crisis, 141
Non-agricultural activities, 268
Non-cycling community, 225
Northern Ghana, 37 Food insecurity and malnutrition,
        northern Ghana
Northern Regions of Ghana, 31

O
Oceania
    sustainable development, 261, 262
Office of Community-Engaged Learning (OCEL), 59
Official cartographies, 296–298, 301, 303
Old Panama historic monument, 203
Online programs, 155
Open communities, 341
Open culture, 26
Open data, 30
    as a public good, 366
Open Data Kit (ODK), 302, 309, 310
Open data mapping, 254, 256, 259
Open map data, 330, 351–352
OpenMap Development Tanzania (OMDTZ), 352
Open map source, 346
Open mapping, 347, 351
Open mapping communities, 346, 348
OpenMappingKit (OMK), 309, 310
Open mapping methods and activities
    essential participation of community as researcher,
        225, 226
    remote data collection, 223
    socio-spatial survey of cyclists, 224, 225
    thematic mapping and data analysis, 225

Open principles, 336
Open-source businesses, 342
Open-source geospatial data, 178, 179, 216–217
    and software, 332
Open-Source Geospatial Foundation (OSGeo), 182
Open source geospatial tools, 127
Open-source mobile applications, 309
Open-source platform's six satellite imagery sources, 263
Open-Source Program Offices (OSPOs), 342
Open-source software, 336
Open-source tools, 309
Open-source volunteers, 336
Open spatial data opportunity, 233
Open spatial data for resilience, 190
OpenStreetMap (OSM), 15, 34, 64, 65, 118, 173, 174,
            182–184, 187, 217, 236, 254, 256, 259, 306,
            307, 310, 321, 323, 330, 346, 347
    attributes, 244–246
    community, 5, 330, 367
    editing tools, 329
    geography, 336
    individual, group or organization, 335
    innovations Innovations
    local meetup, 338
    open-source software, 336
    participation, 336, 347
    partnership, 335
    platform, 2, 3, 6, 223
    in Sierra Leone, 126, 132, 134, 139
    technical methodologies, 335
    unusual partnership and community model, 337
    for urban governanct, 306–307
    use, 311
    working groups, 342
    YouthMappers communities, 341–342
    and YouthMappers journey
        learning environment, 336
        monitor quality, 336
        nontraditional partners, 336
        online communication, 336
        open principles, 336
        origins of OSM, 336
        stage set for emergence, 337
        starfish and spider approach, 336
        user-creator communities, 337
OpenStreetMap Foundation, 102
OpenStreetMap free, 263
OpenStreetMap-Philippines, 88
Open tools, 185
Operational ecosystem, 331
Ordnance Survey, 336
Organizational capacity, 331
Organized Editing Guidelines, 339, 340
Origin-Destination Survey of 2017, 223
Orthomosaic, 165–167
OSMAnd, 309
OSM Colombia´s Tasking Manager platform, 223
OSM community, 337
OSM Data Model, 187
OSM ecosystem, 326, 332, 333

OSM mapping, 77, 81
Outreach ambassadors, 326

P
Pacific Disaster Center (PDC), 348
Pacific Region, 261, 262
Paid corporate mappers, 336
Palms, 203
Panama, 193, 195, 198, 202–204, 297, 302, 303, 319,
            320
Pan American Institute of Geography and History
            (PAIGH), 297
Participatory mapping, 95, 97, 98, 217, 368
Participatory planning, 228
Partnerships
    and collaborations, 325
    connective tissue for YouthMappers, 328
    and distributed network approaches, 341
    HOT Humanitarian OpenStreetMap Team (HOT)
    mapillary, 329, 330
    regional ambassador system, 325, 326
    SDGs 17 and 8, 333
    student, alumni and staff perspectives, 327–328
    student-led and community-based, 325
    university chapter network, 325, 326
Paternalistic perceptions, 103
Pathogens, 69
Pedagogical approaches, 96, 97
Pedagogy, 94
People with disabilities (PwD), 186
Perkumpulan OpenStreetMap Indonesia (POI), 348–350
Peru's National Satellite, 354
Peru's independence, 354
Pharmacy, 265
Philippines, 262, 263, 272, 275
Pink Shirt Day Panama Foundation, 192
PoliMappers, 319
    coordinated surveying, 19
    difficulties, 19
    IMMdesignlab, 19
    innovative teaching program, 20
    Italian YouthMappers chapter, 18
    participation, 18
    Twitch channel, 21
PoliMappersOSMCha tool, 21
Polisocial project MASTR-SLS, 72
Poultry, 266
Power grid mapping in West Africa
    COVID-19 pandemic, 127
    local fieldwork implementation, 127
        authorization and identification, 129–130
        capacity assessment and First-Pass Remote
            Mapping, 128
        field mapping and ground truthing, 128
        field mapping and monitoring, 131–132
        mapillary verification tasks, 132
        output metrics and mapping outcomes, 132–134
        setting up and deploying mapping teams per
            location, 128–129

open source geospatial tools, 127
power access, 125
for SDGs
    affordable and clean energy, 135–136
    challenges and recommendations, 138–140
    infrastructure and foster innovation, 136–138
    YouthMappers in Sierra Leone, 126
Power mappers, 78
Preservice teacher education, 94
Principal investigators (PI), 297
Private sector, 347, 354
Privilege, 369
Program for Appropriate Technology in Health
        (PATH), 352
Programs, 297
Project #1270, 20
Proof-of-concept methodology, 281
Protection and Restoration Zones, 203
Public health, 232, 235
Public institutions, 347
Python program, 302

Q
Quality education, 58, 93, 95
Quantitative changes, 291
Quantum GIS (QGIS), 107, 329

R
"Ramani Tandale", 351
Rapid urbanization, 305, 306
Real-Time Kinematics (RTK), 351
Real-world cases, 348
Real-world response, 346
Recreational vehicles (RVs), 242–245
"Reduce, Reuse, Recycle", 238
Regional ambassador system, 325, 326
Regional ambassadors, 94, 105, 326
Regular mappers, 78
Remote data collection, 223
Remote mapathons, 144, 148
Remote mapping, 148, 309
Remote volunteers, 338
Renewable energy, 126
Research collaborative mapping, 182
Research-for-development (R4D) organization, 31
Resource allocations, 368
Resource Conservation and Recovery Act (RCRA), 232
Resource management process, 115
Responsible consumption, 238
Riparian communities, 287, 289, 290, 292
Risk management, 354
Rlood-risk assessments, 308
Road-related accidents, 270
Rural agricultural communities, 262–263
Rural households, 274
Rural-urban migration, 209–210 SDG 11 (Sustainable
        cities and communities)
Rwanda YouthMappers, 15–17

S
Safe places, 24
Safety
    during commutes, 272
    and security, 270–272, 274
Sahara Desert, 35
San Humberto, 19
Schistosomiasis, 70–72, 74, 76, 81
School of International Futures, 361
Science education, 98
Science, technology, engineering, and mathematics
        (STEM), 103
SDG 1 (No poverty), 14–26, 29–44 Food security;
        Rwanda YouthMappers
SDG 2 (Zero hunger), 29–44, 47–56 Food security; Food
        insecurity and malnutrition, northern Ghana;
        Household food security status
SDG 3 (Good health and well-being), 57–83, 85–91,
        231–238, 241–248 Arizona; Asia,
        YouthMappers; COVID-19 pandemic;
        Senegal, schistosomiasis transmission
SDG 4 (Quality education), 57–68, 93–99, 295–304,
        317–323, 325–333 COVID-19 pandemic;
        Ecosystem; Education; Mentorship,
        YouthMappers; Official cartographies
SDG 5 (Gender equality), 93–99, 101–112, 251–259,
        261–275 Education; Everywhere She Maps
        (ESM) program; Transportation
SDG 6 (Clean water and sanitation), 69–83, 113–122,
        209–219 Community flood mapping, informal
        settlements; Senegal, schistosomiasis
        transmission; Water–energy–land nexus
        methodology
SDG 7 (Affordable and clean energy), 113–122, 125–148
        Electricity; Power grid mapping in West
        Africa; Water–energy–land nexus
        methodology
SDG 8 (Decent work and economic growth), 153–158,
        171–179 YouthMappers in Nepal
SDG 9 (Industry, innovation, and infrastructure),
        125–140, 153–158, 161–170, 221–228,
        335–342 Drones; OpenStreetMap (OSM);
        Power grid mapping in West Africa;
        YouthMappers in Nepal
SDG 10 (Reduced inequalities), 101–112, 171–179,
        181–206 Environmental resilience in Latin
        America; Everywhere She Maps (ESM)
        program
SDG 11 (Sustainable cities and communities), 14–26,
        141–148, 181–188, 209–219, 221–228,
        305–313, 345–355 Community flood
        mapping, informal settlements; Electricity;
        Rwanda YouthMappers; Uganda
SDG 12 (Responsible consumption and production),
        47–56, 161–170, 231–238 Drones; Food
        insecurity and malnutrition, northern Ghana;
        Household food security status
SDG 13 (Climate action), 85–91, 241–248, 251–259,
        277–285 Anthropocene, YouthMappers;
        Arizona; Asia, YouthMappers

SDG 14 (Life below water), 261–275, 277–285
        Anthropocene, YouthMappers; Transportation
SDG 15 (Life on land), 189–206, 287–292
        Environmental resilience in Latin America
SDG 16 (Peace and justice and strong institutions),
        287–292, 295–313 Official cartographies;
        Uganda
SDG 17 (Partnerships for the goals), 317–323, 325–333,
        335–342, 345–355, 359–362 Ecosystem;
        Mentorship, YouthMappers; OpenStreetMap
        (OSM)
Security and safety, 270–272, 274
Senegal, schistosomiasis transmission
    cross-continental effort
        drone mapping, 76
        OSM mapping, 77
        joint youth action on SDGs, 81
        mapping the habitat of humans, 76
        mapping the habitat of snails, 72, 76
Senegalese government, 67
Sensing Network, 361
Sensors, 139
Services, 103
Sexual harassment, 270, 271
SIGenBici, 222, 223, 225–228
Sister city, 342
Skill building, 333
Skills-focused courses, 263
Slovin's formula, 263
Smallholder farmers, 39, 41
Small-scale farmers, 32, 42
Smart cities, 306, 307
Smartphones, 138
Snails, 70, 72, 74
Social benefits, 331
Social capital, 331
Social participation and partnership, 340
Socioecological systems, 278
Socioeconomic conditions, 330
Socio-economic inequality, 59
Socio-economic realities, 6
Socio-spatial survey of cyclists, 224
Solar-powered mini-grids, 135, 139
Solid Waste Disposal Act (SWDA), 232
Solid waste management, 234, 235, 238
Solutionaries
    humanitarian mapping, 98
    interdisciplinarity, 99
    knowledge and skills, 94, 98
    students, 94
    SUNY, 95, 96
    University of Dhaka, 94, 95
SOTM events, 339
South Africa, 319, 321, 323
Spatial analysis, 48, 54
Starfish and spider approach, 336
Street-level imagery, 19, 330
Student engagement
    in crisis mapping, 346
Student-led faculty-mentored design, 94

SUNY Fredonia campus, 97
Supporting disaster risk management, 348–351
Sustainable action, 354, 355
Sustainable cities, 217
Sustainable development, 58, 172, 305
    gender gap in mobility, 262, 263
    mobility Mobility
    in Oceania, 261, 262
    transportation Transportation
Sustainable Development Goals (SDGs), 57, 318,
        329, 332, 341, 342, 347, 351, 354, 355,
        366–369
    building resilience, 6
    collective action, 7
    geospatial technologies, 3
    interlinked global goals, 2
    open mapping, 222
    primary, 5
    science and technology, 3
    secondary, 5
    socio-economic infrastructure, 6
    teachers, 94
    urbanization, 6
    world with no poverty, 6
    YouthMappers activities, 2
Sustainable mobility, 222
Sustainable transportation, 221
Systematized collaborative mapping, 183–184

T
Tanzania, 351–352
Tapagra Hydrological Reserve, 198
Teacher education, 94
TeachOSM, 20, 97, 330, 331
TeachOSM Tasking Manager, 76
Technical-Vocational schooling, 263
Technological connectivity, 278–279
Technology-based emergency management, 256
TECHO Colombia, 19
TECHO volunteers, 20
Texas State University, 318
Text-based communication, 338
Thematic mapping and data analysis, 225
Thematic maps, 225
Tourism, 190, 193, 196
Transportation
    mobility
        education, 268, 270, 273
        gender-based violence, 275
        healthcare services, 264–266, 273
        household essentials, 263–264, 273
        livelihoods, 266–268, 273, 274
        microenterprise, 266–268
        safety, 270–272, 274
        security, 270–272, 274
        well-being, 272
Trees, 203
Trust and opinion, 360
Typha, 72, 74, 76, 77

U

UCC YouthMappers, 32, 35
Uganda, 318, 319, 322–323
    flood-risk mapping in Ggaba, 307–308
      OSM data, 307
      YouthMappers Smart, 307
Uganda Opening mapping program, 307
UGB YouthMappers, 67
UK national mapping agency, 336
UN relief agencies, 318
Unclassified/tertiary roads, 268
UNESCO World Conference on Education for
        Sustainable Development, 96
Uninterrupted power supply (UPS), 169
United Nations Development Program, 222
United Nations General Assembly, 2–3
United Nations International Strategy for Disaster
        Reduction (UNISDR), 258
United Nations Office for the Coordination of
        Humanitarian Affairs (OCHA) Humanitarian
        Data Exchange, 347
United Nations Sustainable Development Goal 11, 210
United States Agency for International Development
        (USAID), 23, 318, 348
United States Department of Agriculture (USDA), 49
University chapter network, 325, 326
University of the Philippines Resilience Institute (UPRI),
        87, 89
University researchers, 336
Unmanned aerial vehicles (UAVs), 161, 165, 351
Unusual partnership model
    OSM OpenStreetMap (OSM)
UPNEPA (mobile application), 145
Upper East Region, 32
Urban governance, 306
Urban Waste Mapping project, 236
Urbanization, 14
URLs, 326
USAID GeoCenter, 337, 349
USAID Missions, 337
User-creator communities, 337

V

Vaccination providers, 63
Vassar students, 59
Verbal and physical abuse, 270, 271
VGI-to-official-dataset project framework, 302
Violence, 262, 275
Volunteered geographic information, 296
Volunteer mapping programs, 301

W

Waste disposal, 216–218
Waste management, 232–235, 237, 238
Wastesites.io, 237, 238, SDG 3 (Good health and
      well-being); SDG 12 (Responsible
      consumption and production)
Water-based disease, 70

Water–energy–land nexus methodology
    Birris and Paez basins in Costa Rica, 114
    challenges, 122
    COBIRRIS, 115
    COBIRRIS-Paez, 116
    geophysical and spatial modeling interpretation,
        118–121
    Kobo Toolbox v1.29, 117
    land-use distribution and resource management, 114–115
    local governance, 116–117
    local water associations, 116
    Nexus application, 121
    open-source and geodata applications, 117–118
    SDGs 6 and 7, 114
    stakeholders, 115
Weight-for-age (WAZ), 53
Weight-for-height (WHZ), 53
Well-being, 272
West Virginia University, 318
Wicked problems, 97, 99
Wiki map, 336
Women
    affected by climate-induced disasters, 251–252
    disaster information, 253
    vulnerabilities, 252
Women contributed fields, 101
Women cyclists, 227
Women in Technology (WiT), 105
Women's full participation barriers, 103
Work and internships, 331, 333
Workforce development, 333
World Bank, 318
World Malaria Report, 352

Y

Young African Leaders Initiative (YALI), 23, 362
Youth economic development, 333
Youth engagement
    meaningful compact with young people, 361
    six troubling trends in
        absent from decision-making, 360
        climate change, 361
        COVID-19, 360
        influences shifting on trust and opinions, 360
        lost confidence in democracy, 360
        stepping back from political engagement, 360
    state of, 359
YouthLead.org platform, 362
YouthMappers, 2, 211, 212, 217, 218, 307, 325
    ability, 2
    blog, 2
    capacity and experience, 2
    climate change education and emergency response
        awareness, 257
    community mapping projects, 67
    connective tissue, 328
    definition, 1
    educational goals, 5
    emergence of, 2

YouthMappers (*cont.*)
  establishment, 2
  at Federal University of Paraná, 182–183
  flexible inclusive design, 3
  at FUTA, 142–144, 148
  global university consortium, 14
  Leadership Fellowship program, 104
  network, 5, 14, 110, 328
  and OSM, 256–257
  SDGs, 2, 3, 5
  in Sierra Leone, 126, 132, 135, 139
YouthMappers in Africa
  challenges, 177–178
  generation of OSM, 173–174
  implications for the SDGs, 174–177
  MDGs' implementation, 172
  open-source geospatial data, 178, 179
  spatial data, 172, 173
  sustainable development, 172
  unemployment, 172
  University of Malawi YouthMappers, 178
YouthMappers in Nepal
  adaptability during COVID-19 global pandemic, 155
  collaborations, 156
  GESAN, 154, 157
  map literacy, 157
  SDGs, 158
  student-centered activities, 154–155

YouthMappers-led case studies, 303
YouthMappers movement
  community/participatory mapping, 368
  connectivity, 369
  contribution to SDGs, 366
  data informing decisions, 368
  Digital Twins, 368
  future, 369
  GeoAI, 367
  geospatial data, 369
  inclusivity, 368
  inspiring and engaging local
        community, 367
  OSM community, 367
  privilege, 369
  resource allocations, 368
  strengthen youth links to UN SDGs, 366
YouthMappers Research Fellowship program,
        218, 256
YouthMappers techniques
  nexus methodology, 113
Youth skills development, 354

Z
Zambia, 234
Zoom platform, 21
Z-score, 53

Printed in the United States
by Baker & Taylor Publisher Services